AF615108

Ecophysiology of VA Mycorrhizal Plants

Editor

Gene R. Safir, Ph.D.
Professor
Department of Botany and Plant Pathology
Michigan State University
East Lansing, Michigan

CRC Press, Inc.
Boca Raton, Florida

Library of Congress Cataloging-in-Publication Data

Ecophysiology of VA mycorrhizal plants.

Includes bibliographies and index.
1. Vesicular-arbuscular mycorrhiza. 2. Mycorrhizal plants--Ecology. 3. Mycorrhizal plants--Physiology.
I. Safir, Gene R., 1944—
QK918.E26 1987 589.2′0452482 86-32713
ISBN 0-8493-6492-2

Direct all inquiries to CRC Press, Inc., 2000 Corporate Blvd., N.W., Boca Raton, Florida, 33431.

International Standard Book Number 0-8493-6492-2

Library of Congress Card Number 86-32713
Printed in the United States

THE EDITOR

Gene R. Safir, Ph.D., Professor in the Department of Botany and Plant Pathology, Michigan State University, East Lansing, Michigan, was born January 7, 1944, in New York City. He received his B.S. degree in biology from Bates College, Lewiston, Maine in 1965, and his M.S. (1968) and Ph.D. (1970) degrees in Plant Pathology from the University of Illinois, Urbana. He was a Research Associate in the Department of Botany at the University of Illinois from 1970 to 1971 and also in his current department at Michigan State University from 1971 to 1972 in which he became an Assistant Professor in 1972. He is married to Colleen O'Keefe Safir and has one son.

Dr. Safir was first to show that VA mycorrhizal root colonization can protect a plant from a root pathogen when working with the pink root of onion. While working for his Ph.D. degree, with Dr. James W. Gerdemann and John S. Boyer, he demonstrated that VA mycorrhizal plants are different in their water absorption characteristics from nonmycorrhizal plants. Later he also demonstrated that VA mycorrhizal plants are more drought tolerant and recover faster from water deficits than nonmycorrhizal plants, largely because of improved plant nutrient uptake.

Dr. Safir has also studied the water relations of several pathogen-host interactions. In addition, he was one of the first plant pathologists to study remote sensing of plant diseases.

Dr. Safir received the 1984 Ciba-Geigy Award from the American Phytopathological Society for outstanding research contributions to agriculture, primarily for his work with VA mycorrhizal associations. He has been an invited speaker at several universities as well as at many national and international conferences. He also has served as chairman of several committees of the American Phytopathological Society including the "Committee on Mycorrhizae".

While at Michigan State University, Dr. Safir's teaching responsibilities have included Environmental Plant Physiology and Epidemiology of Plant Disease.

CONTRIBUTORS

Glynn D. Bowen, D.Sc.
Chief Research Scientist
Division of Soils
CSIRO
Glen Osmond, Australia

David Harris, Ph.D.
Research Associate
Department of Crop and Soil Science
Michigan State University
East Lansing, Michigan

David S. Hayman, Ph.D.
Principal Scientific Officer
Department of Soil Microbiology
Rothamsted Experimental Station
Harpenden, England

David P. Janos, Ph.D.
Associate Professor of Biology
Department of Biology
University of Miami
Coral Gables, Florida

R. Michael Miller, Ph.D.
Soil Ecologist
Environmental Research Division
Argonne National Laboratory
Argonne, Illinois

Charles E. Nelsen, Ph.D.
Research Scientist
Plant Genetics, Incorporated
Davis, California

Stanley Nemec, Ph.D.
Research Plant Pathologist
Horticultural Research Laboratory
United States Department of Agriculture
Orlando, Florida

Eldor A. Paul, Ph.D.
Professor and Chairperson
Department of Crop and Soil Science
Michigan State University
East Lansing, Michigan

Gene R. Safir, Ph.D.
Professor
Department of Botany and Plant Pathology
Michigan State University
East Lansing, Michigan

David P. Stribley, Ph.D.
Senior Scientific Officer
Department of Soil Microbiology
Rothamsted Experimental Station
Harpenden, England

James M. Trappe, Ph.D.
Principal Mycologist
United States Department of Agriculture
Forest Service and
Professor of Forest Science and Botany-Plant Pathology
Oregon State University
Corvallis, Oregon

TABLE OF CONTENTS

Chapter 1

VA MYCORRHIZAE: AN ECOPHYSIOLOGICAL APPROACH

Gene R. Safir

TABLE OF CONTENTS

I. INTRODUCTION

Mycorrhizae are associations between plant roots and fungi and both potentially benefit from the association. This book will contain a thorough discussion of vesicular-arbuscular mycorrhizae from an ecological standpoint. Vesicular-arbuscular mycorrhizae are characterized by having fungal vesicles and arbuscules (haustoria-like structures) as well as both inter- and intracellular growth inside the roots of colonized plants. Also present is an extensive network of fungal hyphae in the soil. This network is believed to be responsible for increasing nutrient uptake by mycorrhizal plants (see Chapters 3 and 4) and it has been reported to comprise as much as 80 cm of fungal hyphae per centimeter of root tissue colonized.[1] The fungi that form this type of mycorrhizal association have been placed in several genera, such as *Glomus, Gigaspora, Acaulospora,* and *Sclerocystis* in the family Endogonaceae.[2,3] VA mycorrhizal associations are thought to be almost ubiquitous, having extremely wide host ranges, and are present on most existing species of vascular plants including agricultural crops grown throughout the world. There have been many demonstrations of dramatically improved plant growth, both in the laboratory and the field, which have been attributed to the presence of these mycorrhizal associations. These demonstrations have stimulated a great deal of recent scientific study and commercial interest in using VA mycorrhizal fungi as "biofertilizers". It should be pointed out, however, that some greenhouse demonstrations of VA mycorrhizal plant growth stimulations have not been reproducible under field conditions. In addition, it is not yet been possible to grow vesicular-arbuscular mycorrhizal fungi in axenic culture, and this greatly limits our ability to study these fungi and to more fully understand the role they play in the growth and development of plants in natural and agroecosystems. Soil inoculations with VA mycorrhizal fungi have been shown to promote the growth of infected plants in pot experiments and in the field, primarily when soil nutrients (particularly phosphorus) are in short supply and when indigenous populations of these fungi are small or limited by the use of pesticides (see Chapter 9). For example the use of these fungi, when growing containerized seedlings or cuttings in sterile media or potting mixes in nurseries, has been suggested. In addition, these fungi are currently being used to some extent in orchards, such as citrus orchards, following applications of biocides that restrict the indigenous VA mycorrhizal fungal populations (Chapter 10).

It is of immense interest that VA mycorrhizal root colonization has been associated with severe plant growth reductions in some instances.[4,5] For example, the VA mycorrhizal fungus *Glomus macrocarpum* has been suggested as the cause of stunt disease of burley tobacco, which greatly reduces tobacco quality and yield.[5] The existence of such phenomena should caution one against the indiscriminate use of VA mycorrhizal fungi in commercial situations without extensive field testing. It also emphasizes the need for studying a given mycorrhizal fungus and its plant host not only in isolation, but also in the ecosystem in which they are to function (see Chapters 4 and 9).

It is intended that this book should provide a thought-provoking discussion of what is known and what needs to be known about the role VA mycorrhizae play in natural and agroecosystems. It should also compliment two recent books which discuss VA mycorrhizae in different contexts.[6,7] This book contains an analysis of over 3000 publications concerning the mycorrhizal status of plant species from a phylogenetic, evolutionary, and ecological standpoint (Chapter 2). The biology, physiology, and biochemistry of the various stages of VA mycorrhizal root colonization are discussed and are related to the potential for quantification or modeling of VA mycorrhizal infection (Chapter 3). The mineral nutrition, water relationships (including drought tolerance), and carbon balance of VA mycorrhizae are discussed in depth in Chapters 4, 5, and 6. The last four chapters (7 to 10) of this book deal with the role of VA mycorrhizae in humid tropical ecosystems, grass and shrublands, field crop systems, and horticultural systems, respectively.

The authors of the various chapters have not only reviewed the pertinent published information relating to their subjects, but have speculated freely and often controversially.

REFERENCES

1. Sanders, F. E. and Tinker, P. B., Phosphate flow into mycorrhizal roots, *Pest. Sci.*, 4, 385, 1973.
2. Gerdemann, J. W. and Trappe, J. M., The Endogonoceae of the Pacific northwest, *Mycol. Mem.*, 5, 1, 1974.
3. Hall, I. R. and Fish, B. J., A key to the Endogonaceae, *Trans. Br. Mycol. Soc.*, 73, 261, 1979.
4. Buwalda, J. G. and Goh, K. M., Host fungus competition for carbon as a cause of growth depressions in vesicular-arbuscular mycorrhizal ryegrass, *Soil Boil. Biochem.*, 14, 103, 1982.
5. Modjo, H. S., Relationship between *Glomus macrocarpum*, an Endomycorrhizal Fungus, and Burley Tobacco Stunt Disease in Kentucky, Ph.D. dissertation, University of Kentucky, Lexington, 1983.
6. Harley, J. L. and Smith, S. E., *Mycorrhizal Symbiosis*, Academic Press, London, 1983.
7. Powell, C. Ll. and Bagyaraj, D. J., Eds., *VA Mycorrhiza*, CRC Press, Boca Raton, Fla., 1984.

Chapter 2

PHYLOGENETIC AND ECOLOGIC ASPECTS OF MYCOTROPHY IN THE ANGIOSPERMS FROM AN EVOLUTIONARY STANDPOINT

James M. Trappe

TABLE OF CONTENTS

I. INTRODUCTION

It has been estimated that "about 95 percent of the world's present species of vascular plants belong to families that are characteristically mycorrhizal...."[36] That quotation has been permuted and misapplied on occasion to a less careful, quite different, unsupported statement that "about 95 percent of the world's plant species are mycorrhizal." The distinction is meaningful: although the evidence for mycotrophy in families and orders of plants has been compiled in a broad sense by workers such as Kelley,[15] Gerdemann,[8] and several others, the data on individual species scattered through the more than 6000 papers and books published about mycorrhizae have never been systematically examined.

Important compilations of the mycorrhizal status of species have appeared over the decades since 1885, when the term "mycorrhiza" was coined. First among the major contributers to knowledge on the distribution mycotrophy in plant communities were Schlicht[28,29] and Stahl,[31] who examined many plants from many kinds of habitats. Surveys of mycotrophy continued over the first half of the 20th century, with particularly important contributions by many workers such as Fries in Sweden, Yachevsky in Russia, Klecka and Vukolov in Czechoslovakia, Gallaud in France, Peyronel in Italy, Asai in Japan, Janse in Java, and McDougall and his students and Thomas in America. After World War II, this tradition continued through the 1950s and up to the present with such workers as Dominik and his students in Poland, Boullard in France, Selivanov, Kryuger, and their colleagues in the Soviet Union, Read and his students and colleagues in the U.K. and Austrian alps, Khan and Saif in Pakistan, Maeda in Japan, Malajczuk and Warcup in Australia, and Alden, Bethlenfalvey, Molina, Pendleton, Rothwell and Williams in the U.S., to name a few.

Experimental work on mycorrhizae gained appropriate ascendency in the postwar period, partly at the expense of the field surveys and ecological studies that are still badly needed to expand and sharpen our knowledge of the place of mycotrophy in our planet's ecosystems. In any event, no one has hitherto brought together the extensive but scattered information already available on the mycorrhizal status of plant species to test the contention that "...mycorrhizae have evolved as the norm of terrestrial plant nutrition, not the exception"[36] or that "...mycorrhizae... are nearly universal in terrestrial plants...."[18] Indeed, it was barely possible to undertake such a task until the advent of computerized library searches, rapid photocopy devices, and microcomputers that enable registering, organizing, and manipulating large amounts of data. My personal interests lie especially in understanding the evolutionary ecology of mycorrhizae as a basis for learning how to simultaneously use and protect the ecosystems of the earth. For that purpose, I am fortunate that electronic instruments and a century of field research by my predecessors permit a start at interpreting how mycotrophy really exists in the world and at testing hypotheses on where it came from and where it is headed.

The objectives of my studies, as reported below in preliminary fashion, were fourfold: (1) to determine (insofar as available reports permit) the relationships of taxonomic and ecologic groupings of plants to their mycotrophic status; (2) to interpret these data in terms of evolutionary trends; (3) to interpret these data in terms of some broad ecological habits or habitats; and (4) to identify some areas of research needed to better understand the evolution of mycotrophy and its implications to phylogenetic, ecologic, and resource management theory.

II. METHODS

Data on mycorrhizal status of plant species have been extracted from about 3000

Table 1
NUMBERS AND PERCENTAGES OF SPECIES OF SUBCLASSES AND CLASSES OF ANGIOSPERMAE EXAMINED FOR MYCORRHIZAE AND PERCENTAGE OF EXAMINED SPECIES BY TYPE OF MYCORRHIZAE[a]

Taxon	Total species	Species examined	Percent examined	Percent with mycorrhiza types[b] Z only	AB only	Z + AB	F	N only
Division Angiospermae	223,400	6,507	3	50	15	5	12	18
Class Dicotyledonae	173,500	5,020	3	50	14	6	13	17
Subclass								
Magnoliidae	12,000	270	2	66	3	4	10	17
Hamamelidae	3,400	265	8	27	44	11	12	6
Caryophyllidae	11,000	317	3	14	4	2	21	59
Dilleniidae	25,000	792	3	33	29	7	11	20
Rosidae	62,100	1,838	3	56	16	5	11	12
Asteridae	60,000	1,538	3	63	2	5	15	15
Class Monocotyledonae	49,900	1,487	3	49	18	2	10	21
Subclass								
Alismatidae	500	26	5	4	0	0	8	88
Arecidae	5,600	61	1	56	3	3	8	30
Commelinidae	15,000	826	6	55	1	2	14	28
Zingiberidae	3,800	28	1	71	4	0	14	11
Liliidae	25,000	546	2	37	48	2	6	7

[a] Classification system and estimated numbers of species according to Cronquist.[5]

[b] Z = zygomycotous (vesicular-arbuscular); AB = asco- + basidomycotous (ecto, ericoid, orchidoid, other); F = facultative (can function with or without mycorrhizal fungi); N = nonmycorrhizal (autotrophic).

published papers, theses, and books. Each report was recorded in a computer file by plant species whether it was mycorrhizal or not; if mycorrhizal, what type of colonization (e.g., zygomycotous, asco- or basidiomycotous, ecto-, ericoid, orchidoid, "dark-septate"); whether the plant was annual or perennial; and what, if any, broad ecological traits it might have (e.g., occurring in the tropics or arctic-alpine, halophyte, hydrophyte, xerophyte, epiphyte, achlorophyte, parasite, weed, etc.). Terms such as "tropics", "halophyte", etc. are used to include both obligate and facultative species. Thus, for example, *Oryza sativa* L. is grown in both temperate and tropical habitats and may be in either aquatic or terrestrial culture or a combination of the two. It is included in the data base for each of those categories. A species such as *Elodea canadensis* Michx. which occurs only in water in temperate habitats is included only in the data base of those categories. A species was regarded as facultatively mycotrophic if it was reportedly able to function without mycorrhizae in some situations and if it formed mycorrhizae in others.

Special attention was devoted to verifying plant names and synonyms because many plant species have been designated by two to several different names in the literature; "Hortus Third"[2] was the basic reference for sorting out nomenclature, but numerous local and regional floras and taxonomic monographs were also regularly consulted.

The data were entered first on microcomputer disc files and then incorporated into the Synopta microcomputer program (written by Frank Evans, Corvallis, Ore.) for sorting and counting data by key words and manipulating that data by various utility programs. This effort, which has included about half the literature on mycorrhizae and all of the major works on occurrence of mycotrophy in plant communities, has been underway for 4 years. The major results to date are presented in Tables 1 to 5. The

Table 2
NUMBERS AND PERCENTAGES OF SPECIES OF ORDERS OF THE DICOTYLEDONAE EXAMINED FOR MYCORRHIZAE, AND PERCENTAGE OF EXAMINED SPECIES BY TYPE OF MYCORRHIZAE[a]

				Percent with mycorrhiza types[b]				
Taxon	Total species	Species examined	Percent examined	Z only	AB only	Z + AB	F	N only
			Subclass Magnoliidae					
Order								
Magnoliales	3,000	29	1	87	3	7	3	0
Laurales	2,500	36	1	72	6	6	11	6
Piperales	2,000	11	1	73	27	0	0	0
Aristolachiales	600	12	2	66	0	17	0	17
Illiciales	90	3	3	100	0	0	0	0
Nymphaeales	60	9	15	0	0	0	0	100
Ranunculales	3,150	149	5	70	1	6	12	11
Papaverales	600	21	4	5	5	5	24	61
			Subclass Hamamelidae					
Order								
Trochodendrales	2	1	50	100	0	0	0	0
Hamamelidalaes	110	15	12	72	7	7	14	0
Daphniphyllales	10	1	10	100	0	0	0	0
Didymelales	2	0	0	—	—	—	—	—
Eucommiales	1	0	0	—	—	—	—	—
Urticales	2,200	84	4	51	0	10	19	20
Leitneriales	1	0	0	—	—	—	—	—
Juglandales	64	15	23	47	20	20	13	0
Myricales	50	6	12	33	0	17	50	0
Fagales	910	135	15	2	83	11	4	0
Casuarinales	50	8	16	38	25	12	25	0
			Subclass Caryophyllidae					
Order								
Caryophyllales	9,700	234	2	13	4	1	20	62
Polygonales	900	77	9	17	7	2	25	49
Plumbaginales	400	6	2	17	0	33	0	50
			Subclass Dilleniidae					
Order								
Dilleniales	400	9	2	89	11	0	0	0
Theales	3,500	52	1	54	33	2	5	6
Malvales	3,260	91	3	69	13	3	8	7
Lecythidales	400	14	4	57	0	0	0	43
Nepenthales	200	6	3	16	0	0	16	68
Violales	5,000	107	2	57	8	13	7	15
Salicales	350	90	26	7	55	27	10	1
Capparales	4,200	142	3	7	1	1	16	75
Batales	20	0	0	—	—	—	—	—
Ericales	4,000	171	4	2	79	4	13	2
Diapensiales	20	3	15	0	100	0	0	0
Ebenales	1,750	45	3	84	0	2	2	11
Primulales	1,900	62	3	55	2	10	17	16

Table 2 (continued)
NUMBERS AND PERCENTAGES OF SPECIES OF ORDERS OF THE DICOTYLEDONAE EXAMINED FOR MYCORRHIZAE, AND PERCENTAGE OF EXAMINED SPECIES BY TYPE OF MYCORRHIZAE[a]

				Percent with mycorrhiza types[b]				
Taxon	Total species	Species examined	Percent examined	Z only	AB only	Z + AB	F	N only
			Subclass Rosidae					
Order								
Rosales	6,600	380	6	55	7	6	16	16
Fabales	17,000	528	3	64	12	5	9	10
Proteales	1,000	21	2	33	5	5	10	47
Podostemales	200	2	1	0	0	0	0	100
Haloragales	200	3	2	33	0	0	0	67
Myrtales	9,000	305	3	22	56	6	7	9
Rhizophorales	100	3	3	67	0	0	0	33
Cornales	150	21	14	81	5	0	9	5
Santalales	2,000	7	<1	43	0	0	0	57
Rafflesiales	60	1	2	0	0	0	0	100
Celastrales	2,000	36	2	77	0	6	6	11
Euphorbiales	7,540	96	1	71	6	5	4	14
Rhamnales	1,700	63	4	65	13	6	8	8
Linales	550	11	2	37	9	0	27	27
Polygalales	2,300	28	1	71	4	0	11	14
Sapindales	5,400	167	3	74	2	3	15	6
Geraniales	2,600	45	2	53	0	2	27	18
Apiales	3,700	121	3	63	3	4	18	12
			Subclass Asteridae					
Order								
Gentianales	5,500	111	2	73	5	3	11	8
Solanales	5,000	140	3	52	2	1	14	31
Lamiales	7,800	206	3	62	2	4	18	14
Callitrichales	50	2	4	50	0	0	50	0
Plantaginales	250	15	6	73	0	0	20	7
Scrophulariales	11,000	232	2	51	5	5	14	25
Campanulales	2,500	56	2	57	11	5	16	11
Rubiales	6,500	80	1	60	3	9	16	12
Dipsacales	1,000	68	7	67	0	3	19	11
Calycerales	60	0	0	—	—	—	—	—
Asterales	20,340	628	3	66	1	7	15	11

[a] Classification system and estimated numbers of species according to Cronquist.[5]
[b] Z = zygomycotous (vesicular-arbuscular); AB = asco- + basidomycotous (ecto, ericoid, orchidoid, other); F = facultative (can function with or without mycorrhizal fungi); N = nonmycorrhizal (autotrophic).

project will be continued indefinitely to incorporate data from past literature not yet examined and new data as they appear. As the data base continuously expands, so will the means of testing hypotheses.

The taxonomic-phylogenetic system for flowering plants proposed by Cronquist[5] has been followed in this work. Other systems will be examined in respect to mycotrophic-phylogenetic relationships at a later time.

Statistical analyses of the data are not presently appropriate. The available data are not randomly gathered; indeed, rather strong biases towards agronomically useful

Table 3
NUMBERS AND PERCENTAGES OF SPECIES OF ORDERS OF THE MONOCOTYLEDONAE EXAMINED FOR MYCORRHIZAE, AND PERCENTAGE OF EXAMINED SPECIES BY TYPE OF MYCORRHIZAE[a]

Taxon	Total species	Species examined	Percent examined	Percent with mycorrhiza types[b]				
				Z only	AB only	Z + AB	F	N only
		Subclass Alismatidae						
Order								
Alismatales	100	8	8	12	0	0	12	76
Hydrocharitales	100	7	7	0	0	0	0	100
Najadales	230	9	4	0	0	0	11	89
Triuridales	70	2	3	0	0	0	0	100
		Subclass Arecidae						
Order								
Arecales	2,900	22	1	64	9	0	9	18
Cyclanthales	200	0	0	—	—	—	—	—
Pandanales	700	2	<1	50	0	50	0	0
Arales	1,800	37	2	51	0	11	0	38
		Subclass Commelinidae						
Order								
Commelinales	1,000	15	2	80	0	0	13	7
Eriocaulales	1,200	1	<1	0	0	0	0	100
Restionales	470	0	0	—	—	—	—	—
Juncales	300	42	1	10	5	2	10	73
Cyperales	12,000	761	6	24	1	2	14	59
Hydatellales	10	0	0	—	—	—	—	—
Typhales	20	7	35	0	0	0	14	86
		Subclass Zingiberidae						
Order								
Bromeliales	2,000	3	<1	0	33	0	33	34
Zingiberales	1,800	25	1	80	0	0	12	8
		Subclass Lilidae						
Order								
Liliales	9,000	275	3	71	2	4	10	13
Orchidales	16,000	271	2	1	94	1	3	1

[a] Classification system and estimated numbers of species according to Cronquist.[5]
[b] Z = zygomycotous (vesicular-arbuscular); AB = asco- + basidomycotous (ecto, ericoid, orchidoid, other); F = facultative (can function with or without mycorrhizal fungi); N = nonmycorrhizal (autotrophic).

plants prevail. Thus, a species such as *Zea mays* L. has been examined for mycorrhizae and mycorrhizal responses by dozens of workers in well over 100 habitats or experimental treatments. In contrast, only one specimen of *Elmera racemosa* (Wats.) Rydb., an alpine member of the Saxifragaceae of the Cascade and Olympic Ranges of the Pacific northwestern U.S., has been examined, and that specimen happened to be nonmycorrhizal. Can it be mycorrhizal under other circumstances, i.e., is it a facultative

Table 4
NUMBERS OF SPECIES OF VARIOUS, BROAD, HABIT AND HABITAT GROUPS EXAMINED FOR MYCORRHIZAE, AND PERCENTAGE OF EXAMINED SPECIES TYPE OF MYCORRHIZA[a]

Habit or habitat group	Species examined	Percent with mycorrhiza types[b]				
		Z only	AB only	Z + AB	F	N only
Angiospermae						
Dicotyledonae	5020	50	14	6	13	17
Monocotyledonae	1487	49	18	2	10	21
Perennials						
Dicotyledonae	4199	53	15	8	8	16
Monocotyledonae	1336	43	20	8	10	19
Annuals						
Dicotyledonae	821	42	0	7	21	30
Monocotyledonae	151	64	0	2	21	13
Achlorophytes						
Dicotyledonae	19	16	69	15	0	0
Monocotyledonae	47	6	88	0	4	2
Tropics						
Dicotyledonae	1035	65	12	3	8	12
Monocotyledonae	316	44	28	2	8	18
Arctic-Alpine						
Dicotyledonae	541	37	10	6	22	25
Monocotyledonae	133	37	6	7	22	28
Geophytes						
Dicotyledonae	4644	55	13	6	10	16
Monocotyledonae	1228	50	21	12	7	10
Hydrophytes						
Dicotyledonae	450	20	7	6	35	32
Monocotyledonae	259	17	6	2	22	53
Xerophytes						
Dicotyledonae	377	55	3	2	16	24
Monocotyledonae	78	72	0	5	16	7
Halophytes						
Dicotyledonae	201	46	0	3	24	27
Monocotyledonae	92	33	0	1	27	39
Epiphytes						
Dicotyledonae	4	0	100	0	0	0
Monocotyledonae	49	0	98	0	2	0

[a] Classification system and estimated numbers of species according to Cronquist.[5]

[b] Z = zygomycotous (vesicular-arbuscular); AB = asco- + basidomycotous (ecto, ericoid, orchidoid, other); F = facultative (can function with or without mycorrhizal fungi); N = nonmycorrhizal (autotrophic).

mycorrhiza former? Only further sampling will provide an answer. In other words, the data available as summarized in Tables 1 to 5 represent a first approximation: a map with large areas of *terra incognita* beckoning the explorer onward.

III. THE DATA BASE FOR ANGIOSPERMAE

More than 6500 species of Angiospermae have been examined for mycorrhizae, about 3% of the estimated number of species in the world (Table 1). A large proportion

Table 5
NUMBERS OF WEED SPECIES EXAMINED FOR MYCORRHIZAE, AND PERCENTAGE OF EXAMINED SPECIES BY TYPE OF MYCORRHIZA[a]

Plant group	Species examined	Percent with mycorrhiza types[b]		
		M	F	N
All Angiospermae				
Dicotyledonae	5020	70	13	17
Monocotyledonae	1487	69	10	21
Perennial weeds				
Dicotyledonae	156	46	35	19
Monocotyledonae	42	55	28	17
Annual weeds				
Dicotyledonae	222	35	33	32
Monocotyledonae	56	59	25	16
Worst weeds[a]				
World-wide	18	33	56	11
Locally	66	38	38	24

[a] Weeds as defined by Bailey and Bailey[2] and various floras; "worst weeds" as designated by Holm et al.[11]
[b] M = mycorrhizal; F = facultative (can function with or without mycorrhizal fungi); N = nonmycorrhizal.

of them have been studied only from a single specimen, and it is demonstrable that, especially for annuals, the more a species is studied, the more likely it will be found to be a facultative mycotroph in at least some kinds of habitats. As it happens, both Dicotyledonae and Monocotyledonae have been examined in about the same proportion, i.e., 3% of the total number of species (Table 1).

In the "species examined" and "percent examined" columns of Tables 1 to 3, it is clear that different groups have been disproportionately sampled. Tables 2 and 3 present all data gathered, but it is inappropriate to draw conclusions about groups that have been only scantily sampled. As an arbitrary cutoff point, I have regarded either ten species or 10% of the total number of species as the minimum sampling for phylogenetic or ecologic interpretations of any given grouping.

For reasons unknown but probably pure chance, some sizable orders have been sampled only barely or not at all. In the Monocotyledonae, the orders Cyclanthales, Pandanales, Eriocaulales, Restionales, and Bromeliales total nearly 4600 species; only six species have been examined for mycorrhizal colonization!

IV. EVOLUTION OF MYCOTROPHY AND AUTOTROPHY

Pirozynski and Malloch[23] proposed a hypothesis that is attractive to those who view mycotrophy as the primary nutritional habit of most terrestrial plant systems, a view that is soundly supported by reams of solid evidence but is still resisted by some ecologists and physiologists (the resisters being either poorly read or afraid that tidy theories will be disrupted by the intrusion of facts in the form of fungi). Pirozynski and Malloch "suggest that the colonization of land and indeed the very evolution of plants (and, indirectly, of animals and 'higher' fungi) was possible only through the establishment (unique or repeated) of symbiotic association of a semi-aquatic ancestral alga and an aquatic fungus — an oomycete. In other words, terrestrial plants are the product of this ancient and continuing partnership.'' The arguments for this hypothesis and its implications are detailed by Malloch et al.[18] and Pirozynski[22] and need not be repeated here. We will return to the matter of the ''oomycete'' mycobiont shortly. Meanwhile, let us turn to the fossil record as the solid base for tracing the course of evolution of root-fungus associations.

The fossil record, alas, is scanty at best. The classic studies of Kidston and Lang[41] showed zygomycotous-like fungal colonizations of lycopsid and rhyniophyte rhizomes; these are among the most ancient of vascular plants, and their rhizomes are certainly the oldest known from the fossil record.[34] The ''spores'' or, to judge from the structures themselves, the vesicles in these rhizomes can be equated with vesicles formed by *Glomus* spp. in contemporary plants, as nicely illustrated by Pirozynski and Malloch.[23] The best judgment for now is that these early mycorrhizae were formed with imperfect Zygomycotina, such as are the present-day *Glomus* spp.[38] The ancestry of these fungi is open to debate, but unfortunately the Oomycotina connection postulated by Pirozynski and Malloch has been erroneously interpreted in some textbooks as meaning that the present-day fungi are Oomycotina rather than Zygomycotina.

The zygomycotous mycorrhiza typically occurs on contemporary taxa of the most archaic order of the Dicotyledonae (the Magnoliales) which are recorded as having infrequent asco- or basidiomycotous colonizations, but have yet to be found in an exclusively autotrophic state (Table 2). Baylis[3] proposed that the magnolioid root, with its minimal development of root hairs and strong development of zygomycotous mycorrhizae, represents the primitive condition of vascular plants; St. John's[32] test of this hypothesis with tropical trees supports it. Three hypotheses naturally follow: (1) asco- and basidiomycotous (AB) mycorrhizae are a more advanced symbiosis than the zygomycotous; this is supported by the fossil record, the relatively greater appearance of the AB type on the more advanced host taxa, and the general understanding that the fungi involved are more advanced than the Zygomycotina;[4,18,22] (2) the facultatively mycotrophic and autotrophic capabilities, as represented in the ''graminoid'' type of root,[3] with its well-developed root hairs and frequent absence of fungal colonization, are the most advanced characters of all. From these two hypotheses follows (3) vascular terrestrial plants evolved through a trophic association with primitive fungi and are progressing in their evolution through other, more advanced types of fungal associations to an ultimate independence of fungi.

These hypotheses can now be examined in light of the phylogenetic relationships of contemporary plant groups in reference to the occurrence of the different types of mycorrhizae. Caution must be exercised, however, in attempting to support or refute concepts of evolution by what occurs here and now. As Walker[37] cogently states it, ''Although a primitive character is usually also ancestral, a primitive taxon is not. Due to mosaic evolution, primitive taxa are rarely ancestral. A primitive taxon is simply one that retains a large number of primitive characters relative to some other taxon. Such retention of primitive characters is frequently used to infer a comparatively early evolutionary origin, but this may not always be the case.... The fossil record unquestionably provides the most decisive evidence on primitive characters.'' Primitive characters can have striking adaptive significance, as attested by the retention of zygomycotous mycorrhizae from the early fossil record to their continued predominance today. Still, consideration of phylogenetic relationships of mycotrophy among the plant taxa of today may be instructive in terms of evolutionary trends with particular reference to plant ecology.

V. PHYLOGENETIC RELATIONSHIPS OF MYCOTROPHY

Classes Dicotyledonae and Monocotyledonae strikingly resemble each other in proportions of species that are mycotrophic and of the different, broad types of mycotrophy (Table 1). The major difference lies in the nature of the AB types of mycorrhizae. In the Dicotyledonae, 20% of the known species are registered as either AB only or AB with the option of also forming zygomycotous (Z, more usually in the past referred to as vesicular-arbuscular or VA mycorrhizae). In this case, the AB refers primarily to ecto-, ectendo-, ericoid, and "dark-septate" mycorrhizae. In the Monocotyledonae, 20% also have either AB or AB + Z mycorrhizae. Here, however, the AB component is with the Orchidales as well as the "dark-septate" group of colonizations in the other orders. In both classes, about half the species examined are recorded as Z-mycorrhizal, and close to a fifth have been found to be nonmycorrhizal. Within the classes of Angiospermae it appears that mycotrophy and autotrophy have evolved in parallel patterns. This phenomenon suggests that at the class level the evolution of plant nutritional modes has been more strongly influenced by environmental and competitional selection pressures than by phylogenetic relationships.

More striking differences appear between subclasses (Table 1) or orders (Tables 2 and 3). The Caryophyllidae, which contains many ruderals, and the Alismatidae, preponderantly of hydrophytes, both have a particularly high proportion of nonmycorrhizal or facultatively mycorrhizal species. Here, Gerdemann's[8] observation that the Caryophyllales (Centrospermae) are notably autotrophic is reinforced by the data: 82% of the species examined has been reported as nonmycorrhizal or facultatively mycorrhizal (Table 2). At this taxonomic level, ecological traits begin to converge with phylogeny to reveal patterns of mycotrophy.

In all cases recorded, however, exceptions to the rules appear as more and more specimens and species are carefully examined. This is especially true when the colonizations originally described by Peyronel[21] as the "*Rhizoctonia*" type and later by Haselwandter and Read[10] as the "dark-septate" type are taken into account. This type of colonization has often been dismissed as weakly pathogenic or parasitic, but now it seems likely to be mutualistic and to produce a positive host response such as the better understood types of mycorrhizal colonizations.[10] In any event, considering Walker's[37] admonition about "mosaic evolution", it seems predictable that modern plant groups would not conform to clean and clear-cut patterns but instead would show great diversity of adaptive mechanisms. To the degree that phylogeny coincides with such adaptations, however, we can explore retention or discard of primitive characters such as zygomycotous mycorrhizae. Aside from the fossil record, we cannot determine how these traits evolved over time. We can, however, learn the present-day traits that have evolved in concert with phylogeny and thereby better predict how plants, in particular taxonomic groupings, will react to given environmental circumstances.

Phylogenetic dendrograms are illustrated in Figures 1 to 4 for selected groupings of plants, according to the system of Cronquist.[5] These particular groupings were selected as having sufficient data on mycorrhizae to permit first-approximation inferences. In each case, the mycorrhizal status of the taxon is indicated by percentages of species with zygomycotous mycorrhizae (Z) in circles, asco- and basidiomycotous (AB) in triangles, and nonmycorrhizal (N) in rectangles. The percentage totals exceed 100, because species that can be either Z or AB, e.g., members of the Salicaceae, are included in both categories, and facultatively mycotrophic-autotrophic species such as some *Salix* spp. may be in all three.

The Magnoliidae are regarded by Cronquist[5] as a primitive and possibly the ancestral subclass of the Dicotyledonae. The magnolioid root described by Bayliss[3] was perhaps derived from the Devonian lycopsid and rhyniophyte rhizomes and persists as a feature common to nearly all terrestrial Dicotyledonae today, most strikingly and appropriately in the genus *Magnolia* itself. In Figure 1, the high Z-mycorrhiza percentage of the Magnoliidae is not surpassed by any of the more advanced subclasses; indeed, it is

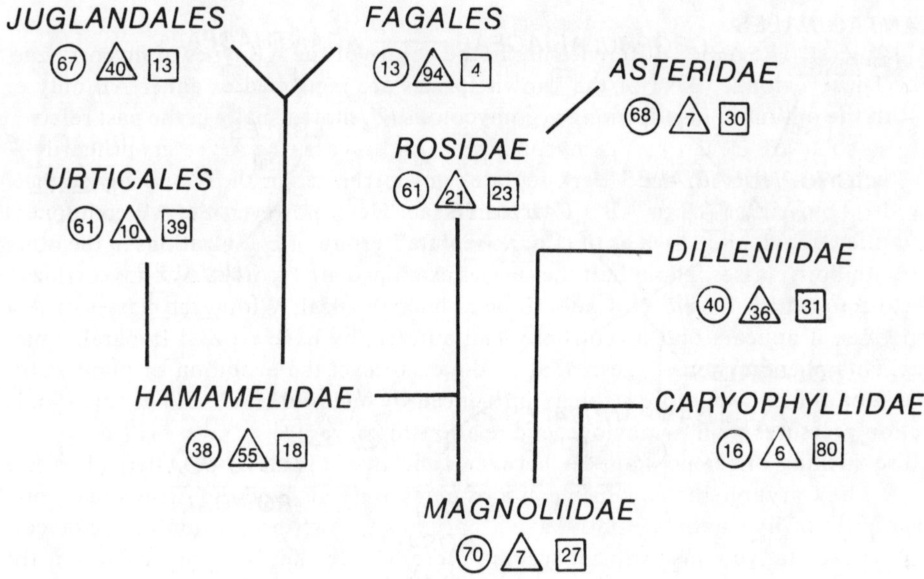

FIGURE 1. Phylogenetic dendrogram (after Cronquist[5]) for the subclasses of Dicotyledonae, with percentages of species with zygomycotous mycorrhizae (circles), asco- and basidiomycotous mycorrhizae (triangles), or no mycorrhizae (squares); many species may have more than one of these categories, so percentages total more than 100.

strikingly less in the Caryophyllidae, Dilleniidae, and Hamamelidae. The more advanced AB-mycorrhizae strikingly increase their presence over that in present day Magnoliidae in the Dilleniidae, Hamamelidae, and Rosidae. A high Z-mycorrhiza incidence seems to have been retained in the line to the Asteridae through the Rosidae.

Because the Asteridae are otherwise regarded by Cronquist[5] as the most advanced of the Dicotyledonae, it is interesting to note the strong retention of the primitive Z-mycorrhiza character. The Asteridae seem to have not only retained the primitive mycorrhiza type but have also developed no greater incidence of the AB type than the Magnoliidae, even though the Asteridae are derived through the Rosidae, which shows increased incidence of AB-mycorrhizae (Figure 1). If, however, we examine the specifics of the pathway from the Magnoliidae through the Rosidae to the Asteridae, i.e., through the order Rosales,[5] we see no suggestion of atavism of the Asteridae to a more primitive habit (Figure 2). Rather, we see that the Z-mycorrhizal character has been retained all the while in two major evolutionary lines, as has a low incidence of AB-mycorrhizae. Autotrophy has also been maintained at a relatively constant level within the evolutionary line leading from the Rosales to the Asterales and Dipsacales. In the other line, to the Scrophulariales and thence in divergent branches to the Plantaginales on the one hand and families of the Scrophulariales on the other, we see maintenance of both Z-mycorrhizae and autotrophy. The autotrophy, however, markedly increases in the families Scrophulariaceae and Acanthaceae as opposed to the other branch leading to the Plantaginales. The striking increase of AB-mycorrhizae in the Rosidae over that of the Magnoliidae is explained by other lines of evolutionary advance from the archaic order Rosales. The overall means for the Rosiidae in Table 1 and Figure 1 include strongly AB-mycorrhizal groups such as the Myrtales.

Seeming anomalies in evolution of orders of the Hamamelidae (Figure 1) can be similarly explained. The Urticales and Juglandales appear to have retained a substantial Z-mycorrhizal tendency, but the former followed a line towards autotrophy,

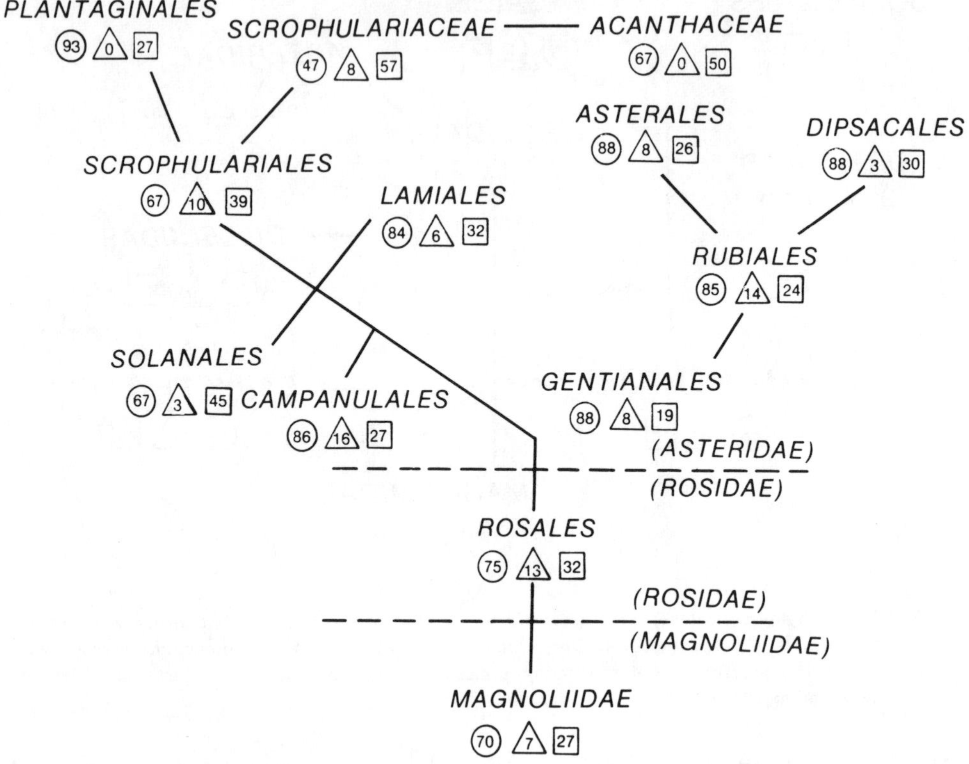

FIGURE 2. Phylogenetic dendrogram (after Cronguist[5]) for selected families of the Caryophyllales as derived from the Magnoliales, with percentages of species with zygomycotous mycorrhizae (circles), asco- and basidiomycotous mycorrhizae (triangles), or no mycorrhizae (squares); many species may have more than one of these categories, so percentages total more than 100.

whereas the Juglandales trended along a separate phylogenetic line towards AB-mycorrhiza formation. The Fagales branch from the same line of evolution as the Juglandales, but in that process they have lost much of their Z-mycorrhizal habit, which is expressed only as occasional Z-colonizations in predominantly ectomycorrhizal hosts such as *Alnus* spp.

Similar kinds of interpretations are possible with other evolutionary lines, such as those originating in the Magnoliales and leading through the Dilleniidae, on the one hand, and the Caryophyllidae on the other (Figures 3 and 4). In the case of the Dilleniidae, the increase in AB-mycorrhizae and autotrophy is strikingly expressed in pathways through the Theales and Violales, with some branches leading to increased AB-mycorrhiza formation and others to autotrophy (Figure 3). Similar advances are represented in the different end groups of the Caryophyllales, although the trend to autotrophy reaches its zenith here. Only one of the five families shown in Figure 4 (the Nyctaginaceae) shows strong development of AB-mycorrhizal habit; it is a tropical and subtropical family which represents one of the strongly ectomycorrhizal groups in those regions.

The other families of the Caryophyllales (Caryophyllidae, Figure 4) contain a disproportionate share of ruderals, as do the families of Capparales (Dilleniidae, Figure 3). Excepting the Nyctaginaceae of the Caryophyllales, both orders are consistently strongly facultatively mycotrophic or autotrophic.

Mycorrhiza-phylogeny interrelationships in the Monocotyledonae cannot be exam-

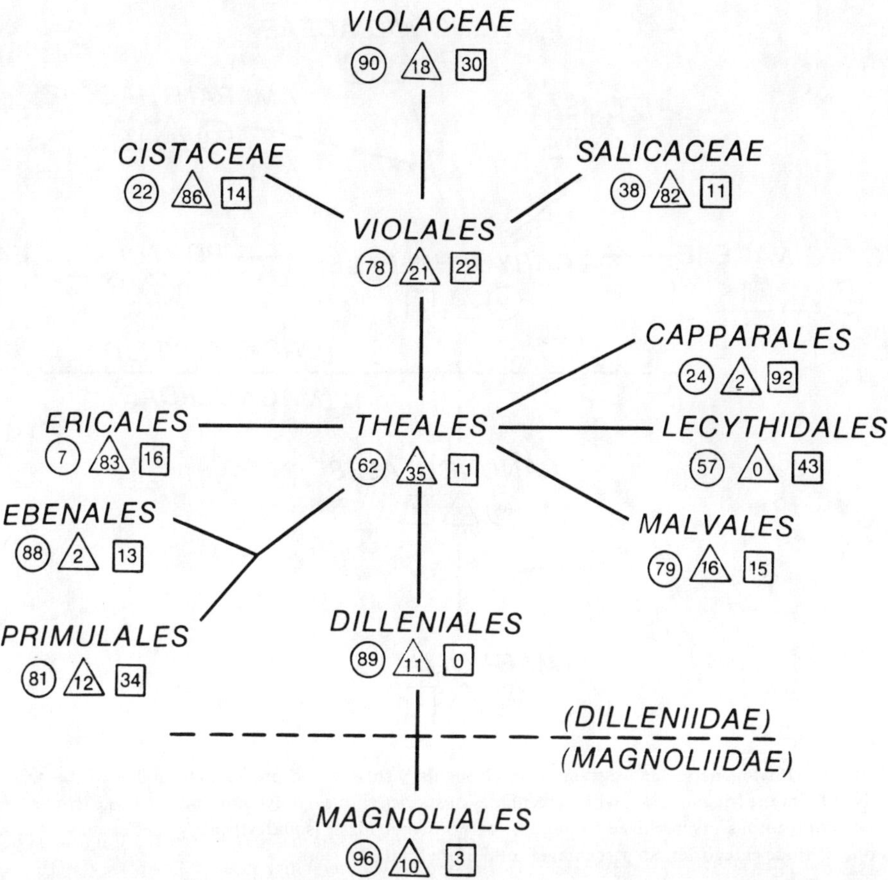

FIGURE 3. Phylogenetic dendrogram (after Cronquist[5]) for selected orders and families of the Dilleniidae as derived from the Magnoliales, with percentages of species with zygomycotous mycorrhizae (circles), asco- and basidiomycotous mycorrhizae (triangles), or no mycorrhizae (squares); many species may have more than one of these categories, so percentages total more than 100.

ined the same as were the Dicotyledonae by the system of Cronquist,[5] because he proposes that the subclasses have each evolved along independent lines off an ancestral main line rather than one deriving from another. The Alismatidae are strongly autotrophic (Table 3), but then that subclass runs heavily to hydrophytes. The Cyperales contains the two large and widely distributed families, Cyperaceae and Poaceae. Both show great autotrophic capability, although the Poaceae are also strongly Z-mycorrhizal and one genus in the Cyperaceae, *Kobresia,* is ectomycorrhizal. Both can be regarded as "closely related lines diverging at an angle from a common source ... too similar to be dissociated; they must stand side by side in the system."[5]

In the Liliidae, on the other hand, a well-developed evolutionary advance of the Orchidales from the Liliales is supported by morphological and ecological characters. The case is cogently put by Cronquist:[5] "The Orchidales differ from the Liliales essentially in their strongly mycotrophic habit, and in their very numerous, tiny seeds with a minute, mostly undifferentiated embryo and no endosperm. The reduction of the embryo is at least in part a consequence of mycotrophy .... The combination of mycotrophy and numerous, tiny seeds offers certain evolutionary opportunities as well as imposing some limitations. The plants are physiologically dependent on their fungal symbionts, sometimes even for food, sometimes only for other factors as yet not fully

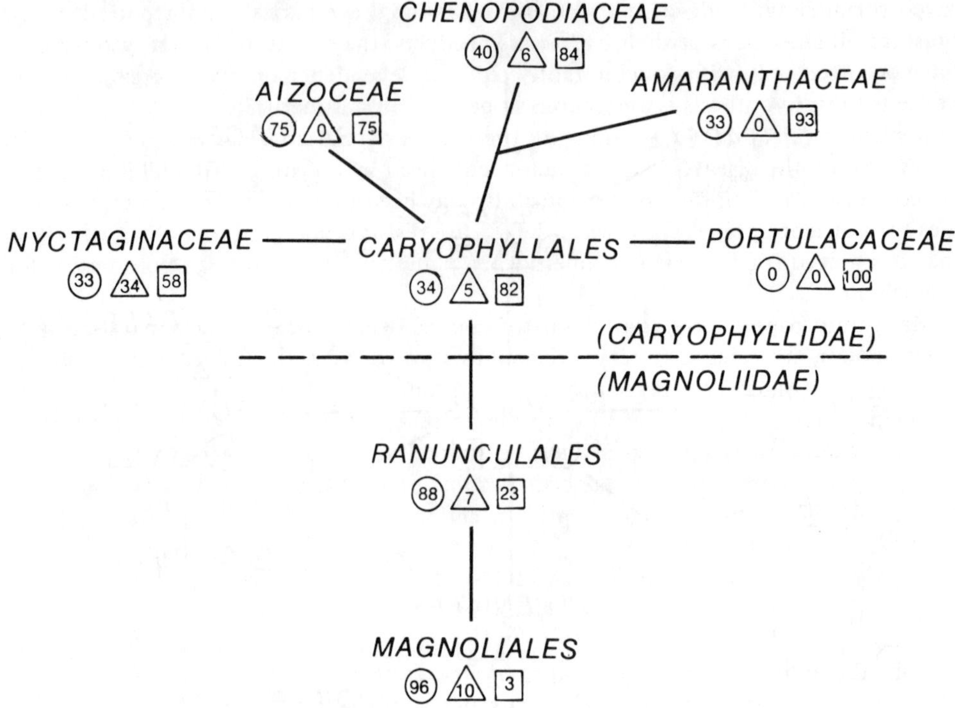

FIGURE 4. Phylogenetic dendrogram (after Cronguist[5]) for selected orders and families of the Asteridae as derived from the Magnoliidae, with percentages of species with zygomycotous mycorrhizae (circles), asco- and basidiomycotous mycorrhizae (triangles), or no mycorrhizae (squares); many species may have more than one of these categories, so percentages total more than 100.

understood, but in any case they can grow only where the fungal symbiont finds suitable conditions .... The Orchidales are evidently derived from the Liliales. All the characters in which the Orchidales differ from the Liliales represent evolutionary advances...." In Table 3, we see that the more primitive Liliales are strongly Z-mycorrhizal, with some autotrophic and facultative capabilities but little of the AB-mycorrhiza type. The more advanced Orchidales, in contrast, are primarily AB-mycorrhizal, having practically no Z-mycorrhizae or autotrophic capability.

Phytochemical properties of plants offer another way to explore the relationship of mycorrhizal habit to phylogeny and of mycorrhizal type to other primitive or advanced characters. The mycorrhizal attributes of plant groups, however, do not for the most part seem to relate to distribution of particular phytochemical characters such as those reviewed by Dahlgren et al.[6] One interesting exception is the glucosinolate-myrosinase system, which produces the mustard oil glucosides and is found in its most advanced state in the strongly autotrophic order Capparales.[6,27] The hydrolysis products of glucosinolates are virtually all "... physiologically potent compounds, indeed, destructive in sufficient quantity, for microorganisms, plants, and animals .... The ecological hypothesis of chemical defense by glucosinolates has been argued cogently and frequently ... while insect herbivores have been the focus of much of this literature, fungal pathogens may prove to be more important objects of glucosinolate-mediated plant defenses...."[27] The Capparales retain little of the primitive Z-mycorrhizal habit; over 90% of the individuals examined for mycorrhizal status are reported to be either nonmycorrhizal only or facultatively mycorrhizal (Table 2). One might infer that the evolution of a glucosinolate defense mechanism was accompanied by reduced dependence on mycorrhizal fungi. Indeed, if the mycorrhizal fungi are partially antagonized by the mustard oil glucosides or their hydrolysis products, the evolution of that system could have occurred only with concomitantly reduced dependence on mycotrophy. The evidence for this hypothesis is ambiguous at best.[35] If it has merit, however, it could have important implications in programs to breed disease resistance into crops such as *Zea mays.* Should the disease-resisting factors enhanced by breeding also inhibit mycorrhizal colonization, then the breeder must also pay heed to breeding in alternative strategies for nutrient uptake. Otherwise, the disease resistance might bring along with it an increasing need for fertilizers in an era when plant breeders ought to be promoting the opposite.

Many autotrophic terrestrial plants, of course, occur outside the Capparales; if the glucosinolate-myrosinase system is related to autotrophy in that order, it is clearly not the cause of it in other orders such as the Caryophyllales. Most families in this order are characterized phytochemically by the presence of betalains instead of anthocyanins.[6] Betalains are reported to have antifungal properties,[16] although how consistently they or related compounds occur in roots of the Caryophyllales and whether or not they affect mycorrhizal colonization is entirely unknown.

VI. MYCOTROPHY IN RELATION TO ECOLOGICAL HABITS AND HABITATS

Since the extremely important paper of Reeves et al.,[25] the role of mycorrhizae in plant competition and succession has been thoughtfully reviewed in several sources.[9,13,14,19,33,39] The ecological significance of autotrophy vs. mycotrophy has received increasing attention in a number of recent studies. My present purpose is not to re-review the topic, but rather to consider how mycorrhizal status of plants relates to some major ecological habits and habitats from the data available. As in the case of evolutionary relationships, one must be careful about drawing inferences on cause and effect from the data, which at this point only suggest relationships or lack thereof. It is fair, however, to erect hypotheses about the meaning of those relationships to plant ecology.

A. Annual vs. Perennial Habit

A number of ecological traits relates to whether plants are annual or perennial (biennials for our purposes are included in the data base for annuals of species that can complete their life cycles either as annuals or biennials, otherwise the biennials are included in the data base of perennials — the numbers of biennials are relatively small and have little impact on final conclusions). Annuals include many of the weedy opportunists that quickly invade disturbed ground and thrive in the absence of strong competition. The annual habit can be a strategy to avoid stress from seasonal heat, cold, drought, or flooding. Many annuals are characterized by prolific production of small seeds that often have adaptations for long-distance dispersal. Perennials, in contrast, must have strategies for competition as well as for coping with environmental stress in the niches they occupy, if they are to survive over extended periods.

Of the plants sampled for mycorrhizal status, about 15% are annuals and 85% perennials; the percentage of annuals is 16 for the Dicotyledonae and 10 for the Monocotyledonae, hardly a striking difference (Table 4). The mycotrophic status of the two classes is quite similar for the perennials. For the annuals, the Z-mycorrhizal habit is rather lower and the incidence of autotrophy rather higher in the Dicotyledonae than in the Monocotyledonae. The striking difference in the data, however, is between the annuals and perennials when the classes are combined. No AB-mycorrhizae are reported for the annuals, and those which are present in the Z + AB category are nearly all the dark-septate (*"Rhizoctonia"*) type of colonization. The high incidence of AB-mycorrhizae in the perennials is mostly due to the ecto-, ectendo-, ericoid, and orchidoid mycorrhiza types, all of which are advanced as compared to the more primitive Z-mycorrhizae.

The exclusive occurrence of AB-mycorrhizae in perennials coincides with the ability of AB-mycorrhizal hosts, especially ecto- and ericoid species, to pioneer in and dominate many kinds of stressful, disturbed, and low nutrient habitats.[12,18,33] AB-mycorrhizal fungi generally have more effective mechanisms for long-distance dispersal than do the Z-mycorrhizal fungi. Partly as a result of that, the AB-mycorrhizal hosts can be pioneers in disturbed sites along with autotrophic or facultatively mycotrophic species. This pioneering or weedy capability occurs with a number of advanced characters, including AB-mycotrophy or autotrophy, and the plants involved are to a large degree relatively recent in the fossil record. We can reasonably conclude, therefore, that pioneering capability in the Angiospermae has evolved concomitantly with an evolutionary advance away from Z-mycotrophy and, indeed, necessarily so in view of the spore dispersal limitations of the Z-mycorrhizal fungi.

In the case of autotrophs vs. AB-mycotrophs, the former give way to the latter in long-term succession, probably because the AB-mycotrophs are generally more effective in extraction of nutrients from low-fertility soils.[12,18,33] The association of free-living, nitrogen-fixing bacteria with ectomycorrhizal fungi[17] and with the ectomycorrhizae themselves,[40] and of nitrogen fixation in the ectomycorrhizosphere[26,40] may represent a complex co-evolution of host, fungus, and bacterium that enables the AB-mycorrhizal perennial to pioneer in low-nitrogen habitats and to persist as a community dominant.

B. Weeds

The extensive data available on weeds in agriculture present opportunity to explore the relationship of annual vs. perennial and mycotrophy vs. autotrophy in reference to initial succession on disturbed land. This could be the topic for a major paper in its own right, because mycotrophy and autotrophy are critically important to the weedy habit but have been virtually ignored in weed science. For present purposes, however, we will just briefly examine the mycorrhizal status of "the world's worst weeds" as described by Holm et al.[11] Table 5 presents the mycotrophic status of all angiosperms for comparison with all perennial and annual weeds in the data base as well as the "the world's worst weeds". In this case, the categories of comparison have been simplified to mycorrhizal, facultatively mycorrhizal, and nonmycorrhizal, because only Z-mycorrhizae are common on weeds.

Percentages of annuals of all species represented in the data base for the categories presented in Table 5 are Angiospermae, 15%, weeds, 58%, and world's worst weeds, 53%. Insofar as these data may be taken to represent the Angiospermae as a whole, then the weedy habit is associated with an increase in the annual habit. A large share of the perennial weeds are rhizomatous grasses that, once established, spread asexually and aggressively.

The mycorrhizal habit is rather reduced in all three weed categories as compared to the Angiospermae as a whole, rather more so for the Dicotyledonae than the Monocotyledonae; and for the Dicotyledonae, mycorrhizae occur less frequently in annual weeds than in perennials. The lowest incidence of mycorrhizae occurs in the worst weeds. Turning to the nonmycorrhizal category, we discover no striking differences between any of the weeds and the Angiospermae as a whole, with the possible exception of the annual Dicotyledonae. This leads us to examine the facultative column; it is here that the truly strong differences are found. All categories of weeds show substantially higher facultative mycotrophy than the Angiospermae as a whole, with the most striking of all being the "world's worst weeds".

Many of the weeds have been examined for mycorrhizal colonization in only a few habitats and for only a few specimens. Experience with plants that have been extensively studied (such as *Triticum aestivum* L. or *Zea mays*) suggests that the more a usually mycorrhizal species is examined, the more likely it will be found to occasionally lack mycorrhizae. Similarly, the more a usually nonmycorrhizal species such as *Chenopodium album* L. or *Rumex crispus* L. is studied, the more likely it will be found to occasionally form mycorrhizae. We can anticipate that the proportion of facultative mycotrophs in all categories will increase with expansion of the data base. Because mycotrophy seems to be more an obligate character in woody perennials than in annuals or herbaceous perennials, however, we may predict that the proportion of facultative mycotrophs will increase disproportionately more in the weed data base than in that of the Angiospermae as a whole. It thus appears that facultative mycotrophism, though not requisite for success of weeds, is associated with a sizable group of weeds, particularly the successful "world's worst". This conclusion entails an ecological logic, in that facultative mycotrophy offers the advantages of nutrient uptake in soils devoid of mycorrhizal fungi and the potential for utilizing mycotrophy in competition with other plants where the fungi are available. It would be interesting to learn if the facultative mycotrophs tend to be the worst weeds in agricultural land and if the autotrophs tend to be the more successful colonizers of degraded land such as mine spoils or landslides.

C. Achlorophytes

The achlorophyllous habit represents a highly advanced co-evolution of certain members of the Angiospermae with woody perennial hosts connected by hyphae of Asco- and Basidiomycotina or of direct parasitism of certain Angiospermae on roots of other host plants.[7,9] In the first case, AB-mycotrophy is the usual situation (Table 4), in the Dicotyledonae with Monotropaceae and a few other families and in the Monocotyledonae with the Orchidales. The little-studied direct parasites, on the other hand, tend to be either nonmycotrophic or facultatively mycotrophic. In some cases, such as *Orobanche* spp., direct parasites may be Z-mycorrhizal on occasion, although the documentation for that is not entirely convincing.

D. Tropics and Arctic-Alpine

The data base for Dicotyledonae that occur in the tropics shows an increased incidence of Z-mycorrhizae over that of all Dicotyledonae, but no substantial difference between the tropical vs. all Monocotyledonae (Table 4). The incidence of AB-mycorrhizae of Monocotyledonae in the tropics, on the other hand, is markedly higher than that for the Monocotyledonae as a whole, due largely to the enormous diversity of orchids in the tropics. In other respects, the tropical plants seem not to differ markedly from the Angiospermae as a whole even for ectomycorrhizae. On a species basis, the proportion with ectomycorrhizae is not appreciably reduced in the tropics, although it may be low in terms of ecosystem populations.

The picture that emerges for arctic-alpine species differs somewhat from that of the tropics. Mycorrhiza incidence tends to give way to increased facultative and autotrophic habit for both the Dicotyledonae and Monocotyledonae in comparison to the Angiospermae over all habitats (Table 4). I am inclined to interpret this difference as reflecting special adaptations evolved to cope with raw, nearly sterile substrates with minimal organic matter and short growing seasons, because my unpublished data on alpine mycotrophy in the Cascade Range of Oregon and Washington show autotrophy to be the norm for alpine plants in the most extreme habitats at the highest elevations and in snow-patch communities. Mycotrophy, in contrast, is abundant above timberline in well-established heaths and meadows. These observations in the Cascades are the same as reported for the Alps[10,24] and the Tatras.[20]

E. Geophytes, Hydrophytes, Xerophytes, and Halophytes

In comparing the data base for these groupings by major ecological habit, the most apparent differences are between the hydrophytes and all the others (Table 4). The incidence of mycotrophy on hydrophytes is low compared to geophytes, although higher than once thought. Autotrophy or facultative mycotrophy are the most common hydrophytic modes. Xerophytes tend to have a high incidence of Z-mycorrhizae, but also a fairly high incidence of autotrophy or facultative mycotrophy. The large proportion of Caryophyllales numbered among the xerophytes accounts in large part for the autotrophic and facultatively mycotrophic component of the Dicotyledonae, as is the case with annual grasses for the xerophytic Monocotyledonae. The relatively high proportions of autotrophic and facultatively mycotrophic species in the halophytes reflect particularly the marine or alkaline-aquatic plants. The evidence so far available does not suggest that reasonable salinity in terrestrial habitats prevents Z-mycotrophy, although some members of the Endogonaceae may be more salt tolerant than others. Especially in the Monocotyledonae the halophytic sedges, rushes, and marine or aquatic grasses tend to be auto- or facultatively mycotrophic.

AB-colonizations are very low in the hydrophytes, xerophytes, and halophytes as compared to the geophytes. Even the AB-mycotrophs present (mostly members of the Salicaceae in the case of the hydrophytes) will probably shift from AB status to Z + AB or F status as they are more extensively examined. The lack of AB-endophytes in hydrophytic situations probably relates to their strongly aerobic physiology, as opposed to some members of the Endogonaceae which can form Z-colonizations even in permanently ponded soils.[1,30] The AB-mycorrhizae found with xerophytic Angiospermae are primarily those formed with certain Fagaceae or by desert truffles with associated annual Cistaceae or desert perennials.[42] It would seem that from the data base, the Ascomycotina and Basidiomycotina are just not well-adapted to high-salt substrates which, of course, often involve very wet or very dry soils.

VII. CONCLUSIONS

The data presented in the tables and the discussion of their meaning in this paper must be viewed as a first approximation. At least the inferences drawn are based on data, imperfect though they may be. Such data, taken by many different researchers for many different purposes in many different habitats, are not necessarily a representative sample of any of the categories considered. More questions are raised than are answered, but this is a prime reason for the exercise.

Two major hypotheses were proposed in the section on evolution of mycotrophy and autotrophy. The first, that zygomycotous mycorrhizal associations are primitive and asco- and basidiomycotous are advanced, is supported by the fossil record and for the most part by present-day mycotrophic patterns. The exceptions of advanced plants retaining a predominantly primitive type of mycotrophy might be used to negate the hypothesis. Their considerable representation of facultatively mycotrophic or autotrophic species, on the other hand, could be said to support it. More and better data could resolve the matter to greater satisfaction. The second major hypothesis, that facultative mycotrophy and autotrophy are the most advanced trophic types, is rather well born out by the data presented in Figures 1 to 4.

The third hypothesis, that evolution is progressing from trophic associations with primitive fungi through more advanced fungal associations to ultimate independence of fungi, is not supported for most groups. In the Caryophyllales (Figures 1 and 4) most families show advance from Z-mycotrophy to autotrophy with no suggestion of an intervening AB-mycotrophic stage. Only the Nyctaginaceae as now existing suggests the possibility of the full sequence as hypothesized, but only 12 species in that family have been examined. In the Hamamelidae, in contrast, the Fagales and Juglandales show little autotrophy now, so the "end point" of the moment appears to be AB-mycotrophy. We will need to take another look in a million years or so to see if their AB-mycotrophy shows signs of giving way to autotrophy.

Once we discover from the data what we do not know, we can erect additional specific hypotheses for testing. A number of these are implicit in the discussion of the data, and many more can be formulated. A few of the many possible questions yet to be answered are cited below by way of example. What are the major gaps in our knowledge of the occurrence of mycotrophy in plants? What data on mycorrhizae need to be gathered on weeds to fill that glaring gap in weed theory, and how can such information be applied to benefit weed control measures? Are the hydrolysis products of the glucosinolates common to the autotrophic Capparales, or are the betalains that occur in the Caryophyllales responsible for lack of mycorrhizal colonization? If so, is autotrophism also the consequence of evolved chemical defense systems where it occurs elsewhere in the plant kingdom, and what are the implications of that to breeding plants for disease resistance? What do evolutionary patterns of mycotrophy tell us of plants likely to succeed in establishing vegetation on disturbed sites, and how can that information be put to better use? Is facultative mycotrophy always an ecological advantage, or does the compromise extract an ecological price? How do evolutionary patterns of mycotrophy explain patterns of plant succession? Lastly, how can field or forest management be improved by obtaining even clearer understanding of those evolutionary patterns?

Data collection on occurrence of mycorrhizae on individual plant species over many kinds of habitats or on the many plant species, in particular habitats or successional stages, needs to be continued and, for best effect, with a quickening pace. We now are at the stage where such data should be gathered by studies designed to answer specific questions to test specific hypotheses. In so doing, as we answer the questions and test the hypotheses, we will simultaneously add to the larger data base now summarized in Tables 1 to 5. That in turn will lead to sharpened inferences about a wide variety of evolutionary and ecological phenomena.

REFERENCES

1. Bagyaraj, D. J., Manjunath, A., and Patil, R. B., Occurrence of vesicular-arbuscular mycorrhizas in some tropical aquatic plants, *Trans. Br. Mycol. Soc.*, 72, 164, 1979.
2. Bailey, L. H. and Bailey, E. Z., *Hortus Third*, Macmillan, New York, 1976.
3. Baylis, G. T. S., The magnolioid mycorrhiza and mycotrophy in root systems derived from it, *in Endomycorrhizas*, Sanders, F. E., Mosse, B., and Tinker, P. B., Eds., Academic Press, London, 1975, 373.
4. Berch, S. M., Miller, O. K., Jr., and Thiers, H. D., Evolution of mycorrhizae, in Proc. *6th North American Conf. on Mycorrhizae*, Molina, R., Ed., Oregon State University Forest Research Laboratory, Corvallis, 1985, 189.
5. Cronquist, A., *An Integrated System of Classification of Flowering Plants*, Columbia University Press, New York, 1981.

6. Dahlgren, R. M. T., Rosendahl-Jensen, S., and Nielsen, B. J., A revised classification of the angiosperms with comments on correlation between chemical and other characters, in *Phytochemistry and Angiosperm Phylogeny,* Young, D. A. and Seigler, D. S., Eds., Praeger Publishers, New York, 1981, 149.
7. Furman, T. E. and Trappe, J. M., Phylogeny and ecology of mycotrophic achlorophyllous angiosperms, *Q. Rev. Biol.,* 46, 219, 1971.
8. Gerdemann, J. W., Vesicular-arbuscular mycorrhiza and plant growth, *Ann. Rev. Phytopathol.,* 6, 397, 1968.
9. Harley, J. L. and Smith, S. E., *Mycorrhizal Symbiosis,* Academic Press, London, 1983.
10. Haselwandter, K. and Read, D. J., Fungal associations of roots of dominant and sub-dominant plants in high-alpine vegetation systems with special reference to mycorrhiza, *Oecology,* 45, 57, 1980.
11. Holm, L. R., Plucknett, D. L., Pancho, J. V., and Herberger, J. P., *The World's Worst Weeds,* University of Hawaii Press, Honolulu, 1977.
12. Janos, D. P.,. Mycorrhizae influence tropical succession, *Biotropica,* 12, (Suppl. 2), 56, 1980.
13. Janos, D. P., Tropical mycorrhizas, nutrient cycles and plant growth, in *Tropical Rain Forest: Ecology and Management,* Sutton, S. L., Whitmore, T. C., and Chadwick, A. C., Eds., Blackwell Scientific, Oxford, 1983, 327.
14. Janos, D. P., Mycorrhizal fungi: agents or symptoms of tropical community composition?, in *Proc. 6th North American Conf. on Mycorhizae,* Molina, R., Ed., Oregon State University Forest Research Laboratory, Corvallis, 1985, 98.
15. Kelley, A. P., *Mycotrophy in Plants,* Chronica Botanica, Waltham, Mass., 1950.
16. Kimler, L. M., Betanin, the red beet pigment, as an antifungal agent, *Bot. Soc. Am. Abstr.,* p. 36, 1975.
17. Li, C. Y. and Castellano, M., Nitrogen-fixing bacteria isolated from within sporocarps of three ectomycorrhizal fungi, in *Proc. 6th North American Conf. on Mycorrhizae,* Molina, R., Ed., Oregon State University Forest Research Laboratory, Corvallis, 1985, 264.
18. Malloch, D. W., Pirozynski, K. A., and Raven, P. H., Ecological and evolutionary significance of mycorrhizal symbioses in vascular plants (a review), *Proc. Natl. Acad. Sci. U.S.A.,* 77, 2112, 1980.
19. Molina, R., Ed., *Proc. 6th North American Conf. on Mycorrhizae,* Oregon State University Forest Research Laboratory, Corvallis, 1985.
20. Nespiak, A., Mycotrophy of the alpine vegetation of the Tatra Mountains (transl.), U.S. Off. Tech. Service; *Acta Soc. Bot. Pol.* 22, 97, 1953.
21. Peyronel, B., Prime ricerche sulla micorize endotrofiche e sulla microflora radicola normale della fanerogame, *Rev. Biol.,* 6, 17, 1924.
22. Pirozynski, K. A., Interactions between fungi and plants through the ages, *Can. J. Bot.,* 59, 1824, 1981.
23. Pirozynski, K. A. and Malloch, D. W., The origin of land plants: a matter of mycotrophism, *BioSystems,* 6, 153, 1975.
24. Read, D. J. and Haselwandter, K., Observations on the mycorrhizal status of some alpine plant communities, *New Phytol.,* 88, 341, 1981.
25. Reeves, F. B., Wagner, D., Moorman, T., and Kiel, J., The role of endomycorrhizae in revegetation practices in the semi-arid West. I. A comparison of incidence of mycorrhizae in severely disturbed vs. natural environments, *Am. J. Bot.,* 66, 6, 1979.
26. Richards, B. N. and Voigt, G. K., Role of mycorrhiza in nitrogen fixation, *Nature (London),* 201, 310, 1964.
27. Rodman, J. E., Divergence, convergence, and parallelism in phytochemical characters: the glucosinolate-myrosinase system, in *Phytochemistry and Angiosperm Phylogeny,* Young, D. A. and Siegler, D. S., Eds., Praeger Publishers, New York, 1981, 43.
28. Schlicht, A., Ueber neue Fälle von Symbiose der Pflanzenwurzeln mit Pilzen, *Ber. Dtsch. Bot. Ges.,* 6, 169, 1888.
29. Schlicht, Z., Beiträge zur Kenntnis der Verbreitung und der Bedeutung der Mykorrhizen, *Landwirtsch. Jahrb.,* 18, 477, 1889.
30. Söndergaard, M. and Laegaard, S., Vesicular-arbuscular mycorrhiza in some aquatic vascular plants, *Nature (London),* 268, 232, 1977.
31. Stahl, E., Der Sinn der Mykorrhizenbildung, *Jahrb. Wiss. Bot.,* 34, 539, 1900.
32. St. John, T. V., Root size, root hairs and mycorrhizal infection: a re-examination of Baylis's hypothesis with tropical trees, *New Phytol.,* 94, 483, 1980.
33. St. John, T. V. and Coleman, D. C., The role of mycorrhizae in plant ecology, *Can. J. Bot.,* 61, 1005, 1983.
34. Taylor, T. N., *Paleobotany — an Introduction to Fossil Plant Biology,* McGraw-Hill, New York, 1981.
35. Tester, M. A., Smith, S. E., and Smith, F. A., The phenomenon of "non-mycorrhizal" plants, *Can. J. Bot.,* in press.

36. Trappe, J. M., Selection of fungi for ectomycorrhizal inoculation in nurseries, *Ann. Rev. Phytopathol.*, 15, 203, 1977.
37. Walker, J. W., Comparative pollen morphology and phylogeny of the Ranalean complex, in *Origin and Early Evolution of Angiosperms*, Beck, C. B., Ed., Columbia University Press, New York, 1976, 241.
38. Weijman, A. C. M. and Meuzelaar, H. L. C., Biochemical contributions to the taxonomic status of the Endogonaceae, *Can. J. Bot.*, 57, 284, 1979.
39. Williams, S. E. and Allen, M. F., Eds.,VA Mycorrhizae and Reclamation of Arid and Semi-Arid Lands, Wyo. Agric. Exp. Stn. Sci. Rep. SA1261, Laramie, 1984.
40. Li, C. Y., unpublished data.
41. Kidston, R. and Lang, W. H., On the old red sandstone plants showing structure from the Rhynie chert bed, Aberdeenshire. Part 5, *Trans. R. Soc. Edinburgh*, 52, 855, 1921.
42. Trappe, J. M., Mycorrhizae and productivity of arid and semiarid rangelands, in *Advances in Food-Producing Systems for Arid and Semiarid Lands*, Manassah, J. T. and Briskey, E. J., Eds., Academic Press, New York, 1981, 581.

Chapter 3

THE BIOLOGY AND PHYSIOLOGY OF INFECTION AND ITS DEVELOPMENT

Glynn D. Bowen

TABLE OF CONTENTS

I. INTRODUCTION

Given a soil-plant combination responsive to the VA mycorrhizal symbiosis (VAM), the speed and size of the response is governed largely by the rapidity of the initial infection and the extent of the infection in the root.[1] The five major developmental components are

1. Encounters between the fungus and the root in soil (''preinfection'')
2. Entry into the root (the ''primary'' infection)
3. The development and longevity of arbuscules (the major physiological interface between the plant and fungus partners)
4. Spread of the infection both within the root and rhizosphere, the latter leading to secondary infections
5. The growth of the fungus into soil, absorption of nutrients, and transfer to the plant

The first stage is affected largely by soil factors which affect propagule germination and growth, by rooting intensity of the plant, and possibly by root exudates. The plant, which is the major provider of growth substrates for the symbiosis, dominates the other stages, but soil conditions may also directly affect stage 5.[2,3] Organic matter in soil *may* contribute something to the fungus nutrition, as proposed for other mycorrhizal types,[4] but the evidence of this for VA endophytes is negative so far.[5]

Figure 1 indicates the result of all these processes, indicating plant response *(Medicago truncatula)* as a function of numbers of propagules in the soil. This was done under ideal soil moisture and temperature conditions for infection. Under less conducive conditions to infection and spread within the root, it is probable that the curve would have been displaced and skewed to the right (curves a′ a″), eventually becoming nonexistent.

This chapter addresses the ways in which factors affecting the development of the infection can be defined. It also addresses some of the physiological (and ecophysiological) aspects of the development of the symbiosis. The physiology of uptake and transfer of nutrients from fungus to host and vice versa is not examined. I have structured the discussion under the five components above; in general, I deal first with biology and dynamics and then with what is known of the physiology.

II. PREINFECTION

A. The Analysis of Preinfection and Primary Infection

Environmental and plant factors can have marked effects on infection. For example, Reid and Bowen[6] found that reducing soil moisture from −0.19 to −0.91 MPa reduced the numbers of primary infections on *Medicago truncatula* from 4.1 to 0.5 cm^{-1}, and Smith and Bowen[7] recorded a 90% decrease in primary infections at 8 days when plants were grown at 12°C soil temperature rather than 20°C soil temperature. Such data do not distinguish between effects of environmental factors on the growth of the fungus and effects via plant susceptibility to infection. They summate the effects of several stages: the germination of the propagule (spore or infected root or hyphae in soil), its growth and contact with the root, and finally the entry into the root. Each or any of these may be a rate-limiting step in the numbers of infection formed. Below I discuss the ways in which effects on these various stages can be distinguished.

Note that the basic measurement in the analysis of *primary* infection should be the *number* of infections. The infections may be simple with a single infection point, or compound with more than one closely placed point of entry.[8,10] If one assumes each

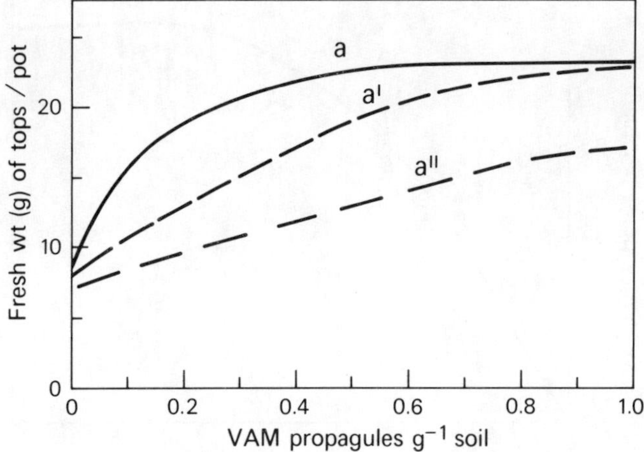

FIGURE 1. Growth response of *Medicago truncatula* at different inoculum levels in soil. Curve a was derived experimentally.[26] Curves a′ and a″ are suggested possible response curves with conditions less suitable to propagule germination and infection.

infection develops subsequently, counts of "infection units" — a short length of root from a well-defined entry point from which mycelium has spread[11] — are an easier measurement. Most of the mycorrhizal literature records not the numbers of infection, but the percentage of root infected. This is quite a different parameter, embracing spread of the infection from primary infections and secondary infections arising from growth of the fungus in the rhizosphere. Similarly, studies on the epidemiology of root disease[12] are of limited applicability here as they deal mainly with relations between inoculum density and numbers of infected plants or disease severity in a crop.

1. Dose-Response Curves

The definition of curves of inoculum density-primary infection numbers is probably as basic to analyses of infection as generation times are to population biology. Figure 2 from one such study[12] was obtained by diluting soil with sand (possibly with some concomitant change in physical properties) in order to vary propagule density. The soil had 1.8 to 4.0 propagules g^{-1} soil (Most Probable Number assay) and shows a saturation of infection (a plateau) under the conditions of that experiment at 50% soil dilution. This curvilinear-linear relation between the numbers of infection at low numbers of propagules followed by a plateau at higher propagule numbers has also been observed in other VAM studies[14] and, for the time being at least, should be regarded as the reference curve. Soils with a low population of organisms may not achieve the saturation level, and in those cases only linear relationships would be obtained on dilution.

In practice, I favor diluting soil with the same soil which has been sterilized (provided this does not release excessive nutrients or toxins[19,20]), as this gives fewer physical changes than diluting with sand. In addition, by keeping extra replicates for a longer period, an assessment of the dose-plant growth response can be obtained for that soil and hence, the level of population needed to be maintained by management practices in order to maximize the mycorrhizal response.

The dose-response curves will be affected by the time of observation: propagules in soil may be at quite different distances from the root and take longer to reach it (see Figure 5 in Sanders and Sheikh[15]); spores of the same species in a soil may be of different ages and may germinate at different rates;[16] in the one species propagules of

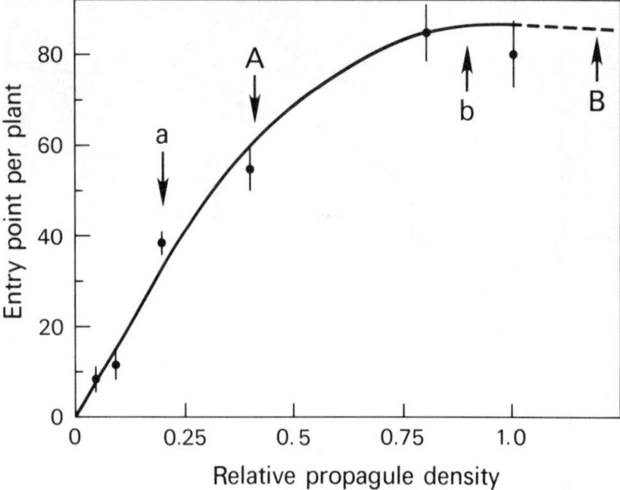

FIGURE 2. The effects of inoculum density on numbers of primary infections. a and b identify the key components of the curve; A and B are suggested soil dilutions to use to screen treatments for environmental/plant effects on a and b.[13]

different types may germinate at different times, e.g., although infected roots of *Acaulospora laevis* may infect readily, spores of the fungus may not germinate for several weeks or months.[17] Because of such factors and the possibility that lateral roots may sometimes be more readily infected than others,[18] analysis of environmental effects on dose-infection curves should be performed on specified parts of a root and at a particular reference time which is short enough to preclude interference by spread and secondary infections, e.g., 10 to 12 days with rapidly infecting fungi. Performed in this way, dose-infection curves can be an important approach in epidemiological studies and in examining the influence of various soil, climate, and plant factors. Similarly, dose-infection curves performed at different times after some soil treatment (e.g., storage of soil, moistening soil, etc.) can yield important information on matters such as survival of propagules and rates of germination of propagules in soil.

The important parameters of the response curve are the initial slope (a in Figure 2) and the level of primary infection at saturation (b in Figure 2). Two approaches can be employed in investigation of the impact of an experimentally imposed condition on a and b. The first is to define the whole curve by the use of several levels of inoculum (few replications are needed in such studies) following which the curves can be mathematically described using computer methods for function minimization similar to those used to define fertilizer responses.[21] The second approach is to use a reference curve, derived under perceived "optimum" condition for mycorrhiza formation, and then to select two (or three) dilutions giving populations most adequate for examining significant departure from the reference curve. For example, in the soil of Figure 2 (and assuming an imposed treatment had no effect on the susceptibility of the host to infection) a decrease of propagule germination of some 50% would be needed from an imposed treatment before an effect on numbers of infections would be noticed and, therefore, the undiluted soil would not be a sensitive level for examining effects on a. A more appropriate dilution for this would be at A. For soils with high populations of VAM fungi, either naturally occurring or with a high population of fungus added, I suggest two appropriate levels for experimentation would be those indicated by A and B. A would be the most sensitive position at which to impose several treatments to see

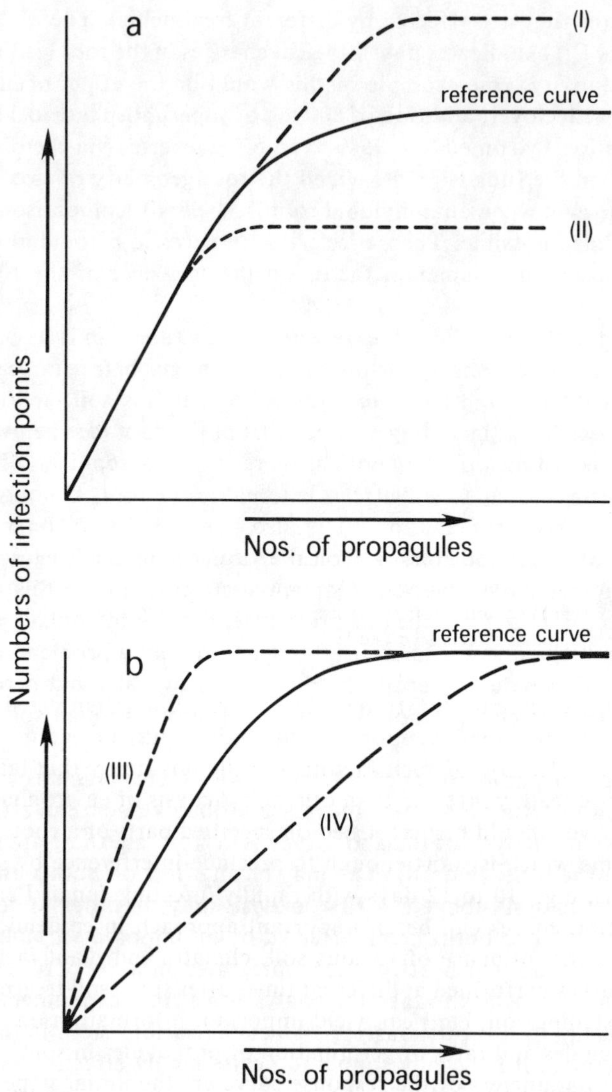

FIGURE 3. Hypothesized simple variations from the "standard" or "reference" curve. (a) Variation in phase a; (b) variation in phase b — see text. Combinations may also occur.

if significant vertical displacement of a occurs in either direction; to use high dilutions leading to low numbers of infections would result in unacceptably high statistical variability. Experiments at B (undiluted soil) would possibly indicate difference in saturation levels (due to a plant physiological factor) in this soil. The first approach (definition of the whole curve) is the sounder of the two approaches, but for particular purposes investigation at one level (e.g., B) imposing many different treatments (with several replications) may have logistic advantages.

Can any biological deductions be made from variation in the values of a and b caused by imposed treatments? In Figure 3a and b, I have drawn the four basic variations from the "standard" curve (other variations would be a combination of a and b, Figure 3). I propose that the number of infections at the saturation level, b, is a measure of plant receptivity (or susceptibility to infection) and is determined by plant physiological factors (as they are affected by different treatments). Therefore, variation in b (Figure 3a, lines 1, 11) indicates physiological changes in the root leading to increased or decreased receptivity. One example of this would be the effect of increasing phosphate; in studies with clover, addition of 500 mg of superphosphate to 3 kg soil reduced numbers of infection by some 80 to 95%,[22] an effect exerted via the plant, not via an effect of the soil on the fungus[23,24] Provided the root geometry of two plant species is identical (as is the case when an individual root is assessed), comparisons of b between plant species are also possible. Theoretically, differences in b could also be due to an effect of an imposed environmental factor on the *virulence* of the fungus, but this seems unlikely.

I propose that the slope of the curve (or line) a and changes in it (Figure 3b) indicate the effect of the soil conditions on the phases of the fungus before infection, i.e., spore germination, growth through soil, etc. Note also that this soil environment is also affected by root exudates (including volatiles) from the root (see below). As the data are obtained by recording *infection* points, there is a philosophical difficulty in allocating differences in a solely to soil effects external to the root, although this seems to be the most appropriate explanation. It is also possible that differences in a could reflect an influence of the soil condition on the virulence of the fungus, but this seems to be a special, improbable situation. One apparent difficulty in this analysis is that several infections can arise from the one propagule, but would not basically change the conclusions from the analysis. Theoretically, it would pose a problem if, for example, an imposed condition reduced germ tube production by half, and if remaining germ tubes then produced twice the amount of hyphal growth through soil to reach roots further away.

The large reductions in numbers of infections with factors such as low soil temperature and low soil moisture indicated above are probably due to an effect on a rather than b, but unless an analysis of the above type is performed, there are no grounds for supposing one or the other (indeed, there may be an effect on both a and b).

I stress that the analysis above is a first analysis only, in order to define the group of components (a or b or both) being affected by an imposed variable. The analysis gives no information on which of the steps integrated in a are affected, and for this finer resolution it is necessary to take the analysis to a finer experimental level. I have discussed approaches to this previously;[25,26] once the various steps in the overall process have been defined, it is usually simple to devise highly effective experimental approaches to examining them individually (aware as always in a reductionist approach that the whole may be greater than the sum of the component parts). For example, one can germinate and recover spores from soil both in the presence and absence of roots; to separate effects on growth of the hyphae through soil from effects on germination one can place germinated spores at set distances from a root; to distinguish effects on infection processes, pregerminated spores can be applied to the root surface in soil.

It is necessary to sound a caution in ecological studies as opposed to basic physiological studies. Wherever possible, ecological studies such as those above should be performed in a natural "synthetic" soil, for it has been shown that in other associations in phenomena as diverse as effects of temperature and antibiosis, studies in laboratory media often have little or no relevance to what is observed in studies in soil.[26] However, it is quite valid to use agar and solution culture methods in examining the detailed *physiology* of a process, especially since these usually require well-defined, constant conditions.

B. Propagule Germination

Hyphae in soil and roots, and spores are the major infecting propagules; these may have very different properties with regard to germination, longevity, and tolerance to adverse conditions. Spores are considered to be the long-term survival propagules;[27] in one study hyphae of *A. laevis* and *Glomus caledonius* in root fragments could not survive drying of the soil, but hyphae of *G. fasciculatum* and *Gigaspora* sp. could.[28] In some cases, the infectivity of VAM roots can be quite short lived, e.g., root segments infected with *A. laevis* have low infectivity as soon as spore formation occurs.[29]

Most research has been on the germination of spores, and little is known about factors stimulating growth of hyphae from infected roots; perhaps they are similar to those enhancing hyphal growth through soil (section C).

1. Methods of Study

Two main substrates have been used in studies of spore germination — agar and soil, with studies on agar being more numerous. The difficulties of extrapolating from agar to soil in ecological research are indicated in various studies: four collections of *Glomus mosseae* spores germinated in soil, while only one source germinated on agar;[30] the presence of germination inhibitors in some commercial agars;[31] and a marked interaction between agar medium pH, composition, and germination of spores of *G. margarita*.[32] Indeed, nutrition appears to be critical in germination of spores on agar — in one study,[33] germination both in agar alone and with dialyzable extracts of air-dried, autoclaved, or fumigated soil (all of which release nutrients) was poor, but germination was excellent with the addition of dialyzed soil extracts of field soils. Additions of nonsterile soil to media for germination of spores has been used since 1962,[8] but this is not always necessary.[34] Attempts to stimulate germination of spores in agar by the addition of a specific organic compound have been inconclusive.[32] Concentrations of ions (e.g., phosphate) observed to inhibit spore germination on agar are frequently far in excess of those in soils[34] and in such cases have little ecological relevance.

Nevertheless, some germination studies on agar would be expected to have some ecological relevance, e.g., germination as a function of aging of spores,[33] soil temperatures,[35] pH,[32,36] and O_2 and CO_2 concentration.[37] Some caution is needed though: pH/nutrition interactions occur in agar[32] and there may well be difficulty in translating from O_2/CO_2 in agar to concentrations occurring in microniches in soil. There is no doubt about the high relevance of studies in laboratory media to eventually being able to grow and mass produce VAM fungal inocula or to examining the basic *physiology* of germination. For *ecological* studies, simple, effective methods for experiments in soil have been devised and one must question whether one *needs* to perform ecologically oriented studies in laboratory media.

Spore germination in soil with or without plants is conveniently studied by placing the spores between polycarbonate or cellulosic membrane filters[34,38,39] or within small stainless steel mesh or nylon containers filled with soil and mesh apertures smaller than the test spores (e.g., 70 μm) from which the spores can be recovered.[40,44]

2. Dormancy

Tommerup[41] pointed out the considerable variability in germination between (and within) populations of spores of the one VAM fungus species reported by various authors, and suggested part of this may be accounted for by the need for a "dormant" (or aging) period of spores of some species during which they would not germinate. Following this dormancy period, they became "quiescent", i.e., they could germinate when conditions were appropriate. Spores of *Glomus* sp. passed through dormancy to the quiescent state after 1 week in dried soil and 6 weeks in moist soil (−0.15 MPa).[41] For *Gigaspora calospora* these periods were 6 and 12 weeks, respectively, and for *A. laevis* the dormancy period was 6 months under all conditions. Transition from dormant to quiescent states occurred over a matter of days, and populations of quiescent

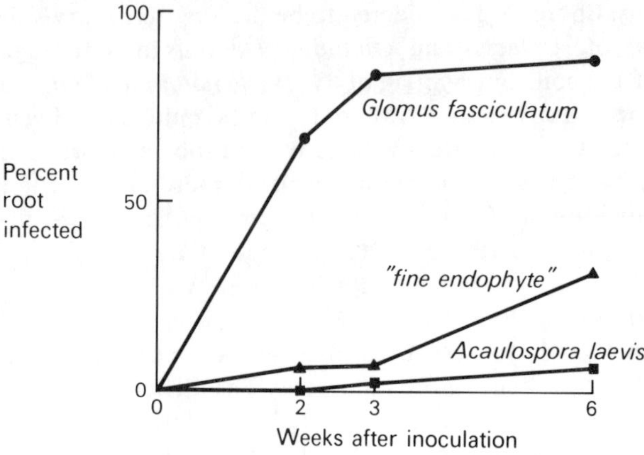

FIGURE 4. Development of infection by three endophytes on *Trifolium subterraneum*. (From Abbott, L. K. and Robson, A. D., *Aust. J. Agric. Res.*, 33, 1049, 1982. With permission.)

spores exhibited almost uniform 90 to 100% germination. Germination times for quiescent *A. laevis* spores were some few days slower than for the other species. Neither a range of soil temperatures (10 to 22°C) nor the presence of host plant roots had an effect on the dormancy time. There are several reports of spore germination improving with storage.[41]

Nothing is known of the physiological changes occurring over the dormancy period. It may be difficult sometimes to distinguish between essential physiological changes occurring during dormancy from the diffusion of naturally occurring germination inhibitors from the spore over time. These occur in some biotrophic fungi[42] and can also be associated with quiescent spores of VAM fungi.[53] However, it could not apply in the cases above, for dormancy periods were much shorter in dry soils than in moist soils.

As the major mycorrhizal responses (at least in annuals) are related to speed of infection, the dormancy phenomenon has particular importance to the mycorrhizal response, especially with fungi such as *A. laevis,* in which a large part of the population in soil could have a protracted period before germination. This no doubt is important in the contrasting progress of infection by *Glomus fasciculatum* and *Acaulospora* sp. indicated in Figure 4. Tommerup[41] suggested short dormancy periods would prevent spores from germinating immediately after formation around the root; such germination would thwart their roles as infection agents for later crops. Tommerup also suggested long dormancy periods may protect spores against early false breaks in a season; selection pressure may have operated to select such a population of VAM fungi.

3. Soil Temperature

Daniels and Trappe[44] considered soil temperature, soil moisture, and soil pH (to a lesser degree) dominated germination of *G. epigaeus* spores; this probably holds for most of the VAM fungi. However, there is some variation between species in spore germination (Table 1). On so few data it is difficult to be dogmatic and, indeed, there could be temperature ecotypes within species, but in these published studies generally the *Glomus* sp. and *Acaulospora* had a minimum for germination between 10 and 18°C and maxima of approximately 30°C but with sharp optima usually at 20 to 25°C. As a group, species of *Gigaspora* appear to have a higher temperature range and higher optima. The absence of germination, or its delay, with many of the fungi in the lower

Table 1
REPORTED TEMPERATURES FOR GERMINATION OF SPORES OF VAM FUNGI

Species	Medium	Min	Max	Optimum	Ref.
Glomus epigaeus	Soil	10—15	30	18—25	44
G. mosseae	Agar	15	34	20—25	35
G. caledonium	Soil	5[a]	25	20—25	38
Acaulospora laevis	Soil	10[b]	25	20	38
Gigaspora calospora	Soil	10	>30	20—>30	38
G. gigantea	Sand	>15	35	20—30	45
G. corraloidea	Agar	15	>34	25—>34	35
G. heterogama	Agar	25	>34	>34	35

[a] Some germination after 7 weeks.
[b] Germination reduced and prolonged.

end of the range (some 10 to 15°C) could be of great consequence for plant response under field conditions, and especially as root growth frequently occurs at lower temperatures than these. It would be particularly interesting to examine the temperature characteristics for spore germination of VAM fungi obtained from alpine soils, on the one hand, and tropical soils on the other.

4. *Soil Moisture, O_2, CO_2*

Large differences in the effects of soil moisture (or water potential) on germination of spores have been reported between workers using different methods and different VAM fungi: a continual decline of germination of *G. gigantea* in polyethylene glycol from 80% at −0 MPa to 15% at −10 MPa;[115] little effect on *Glomus epigaeus* in soil more moist than −0.03 MPa (60% germination), but a marked decline beyond that (e.g., 20% at −1.8 MPa);[44] and high germination of *G. caledonium, Gigaspora calospora,* and *A. laevis* spores in soil from −0 to −1.4 MPa, but delayed germination (some days) at −2.2 MPa.[41] On this data, some spore germination can probably occur at moistures in soils at which roots do not normally grow[47,151] (see below, also).

Delayed germination of spores in dry soils is due to the protracted time for their hydration before germination processes commence; as soils become drier and water recedes into smaller and smaller pores there is lowered contact between the spores and water films in soil. At −0.1 MPa only soil pores smaller than 3 μm will be filled with water, and at approximately field capacity this is 0.30 μm.[46] Indeed, much of the hydration of the spore may arise via the vapor phase of water in soil[41] and water potential may need to be below −3 to −4 MPa to completely inhibit germination. Although readily available soil moisture no doubt plays an important role in the rapid germination of spores, the germination of spores at low soil moistures, albeit slow, indicates water per se may not be the most important regulator of germination in the field.

The indications above of high germination of spores at 0 MPa suggest the poor infection of roots under waterlogged conditions[6] may be due largely to effects on the infection process. However, in waterlogged conditions in soil it is necessary to draw a clear distinction between zero water potential and associated phenomena of anerobiosis and increases of carbon dioxide. The effects of the many changes which occur on waterlogging[48] (including elevated ethylene levels and possible production of toxins from organic matter) have not been fully investigated for their impact on root physiology, let alone the mycorrhizal state. On agar, 5% carbon dioxide had little effect on germination of *Glomus mosseae* spores, but inhibition occurred variously below 0.1 to 3% oxygen.[37] It was suggested that oxygen was necessary for mitochondria formation in quiescent spores (and subsequent germination) and that this explained why inhibition of germination by low oxygen was much less if spores were first incubated in air.

5. pH

The relatively few studies on effects of pH on spore germination of VAM fungi have mostly been on agar and are subject to the caution that the composition of the medium has a marked interaction with pH in this regard.[32] Good germination usually occurs between pH 6 and 7, but there can be large differences between fungi in their germination at pH 5 (and below) and pH 8 (and above). Optima for germination of *Gigaspora corraloidae, G. heterogama,* and *Glomus mosseae* on agar have been recorded at pH 5, 6, and 7, respectively.[36]

In soil, more than 40% germination of *G. epigaeus* spores was found over the pH range 4.8 to 8.0 (optimum 7.0).[44] There is an obvious need for more studies in soil, and if marked pH effects occur, this could be an important factor in the speed of infection by different fungi. Spores of some VAM fungi do not occur below pH 5.3, but others are less affected.[1] The greater plant response to some VAM fungi in acid soils than in alkaline soils and vice versa[49] (which is related partly to the extent of infection of different pHs) may be partly due to effects of pH on soil phases of the fungi, but this requires further study.

6. Soil Inorganic Nutrition

Although a depressive effect of inorganic nutrients on spore germination in agar can be demonstrated, the relevance to soil conditions is questionable. Usually they occur at levels far beyond those expected in soils, e.g., phosphate inhibition.[34] Further differences arise in relating concentration of some substances (e.g., heavy metals) in agar to their effective concentrations in soil because of adsorption to clay and organic matter in soil. In agar, spore inhibition has been demonstrated with Mn^{2+}, Cu^{2+}, and Zn^{2+}, and there are some differences between the source of spores[30,50] in this. Heavy metal-tolerant strains of *G. mosseae* have been obtained from heavy metal-polluted areas.[51] The stage of development, i.e., spore germination, hyphal growth, infection, etc. at which nontolerant strains succumb has not been examined. Similarly, extremely high concentrations of VAM spores can be found in highly saline soils.[52] It is not known if these are specifically adapted to high salinity; in such cases, osmotic factors may also operate.

7. Fungistasis

Tommerup[53] defined fungistasis as a form of exogenous dormancy associated with biological properties of a soil resulting in suppression of germination (or suppression of hyphal growth) when the physical and chemical properties of the soil are otherwise conducive to germination or growth.

Spores of some VAM fungi will germinate in some natural soils but not others.[41,54] Tommerup[53] found quiescent spores of *G. caledonium, Gigaspora calospora,* and *A. laevis* were completely suppressed in *some* soils under crops or pasture, and that soil from under a virgin forest became suppressive after one crop of subterranean clover. In contrast to the usual findings that plant roots do not enhance germination of VAM spores, the growth of seedlings in these suppressive soils relieved fungistasis: less than 10% of spores of the *Glomus caledonium* and *Gigaspora calospora* and less than 20% of spores of *A. laevis* germinated when placed in undisturbed, unplanted weeded cores, but four to six times as many germinated in planted, unweeded cores. This study provided one of the first convincing demonstrations of the importance of fungistasis in the epidemiology of the VAM infection.

To what can we ascribe fungistasis, both at the biological level and the biochemical level? We have the following information:[53] fungistasis occurred under pasture but not under adjacent fallow; the pasture effect decreased with depth; cold water (20°C)-dialyzed extracts of suppressive soils inhibited germination of spores on agar, but only if the spore surface microflora had been eliminated by surface sterilization; the suppressiveness of the soil dialyzate was eliminated by heating at 60°C for 5 min; fungistatis was overcome by the presence of seedling roots (which lead to a high microbial population), by steaming at 60°C for 30 min, or by fumigation, neither of which eliminates the microflora, but changes its composition radically (e.g., large numbers of pseudomonads soon reestablish[58]); suppression was not eliminated by drying and wetting soils (which would be expected to release nutrients from the soil biomass). Of the two major explanations of fungistasis, deprivation of nutrients (particularly energy sources)[56] and the occurrence of inhibitory compounds,[57,58] the former seems unlikely in this case because the spores themselves have an adequate endogenous supply of energy for germination — they can germinate on water agar. The evidence suggests the production of low-molecular, heat-labile compounds from older roots of particular species or from decomposition products of these, although (as Tommerup indicated) leachate from decomposing above-ground litter should not be overlooked. It is apparent that these hypothesized compounds can also be broken down by soil organisms (including those associated with the spore) and which are enhanced by root exudates coming from seedlings. A role of soil organisms in spore germination has been indicated previously.[44]

The imposition of fungistasis by roots of annual crops, but not perennial tree species (as in this case) and its elimination by seedlings would serve an important inoculum conservation function.

8. *Biochemistry of Spore Germination*

The biochemistry of spore germination is outside the scope of this chapter. Recent pertinent literature is given in the references.[50,59,60,69]

C. The Growth of Hyphae from Propagules

Following the production of the germ tube, growth through soil occurs and, if the fungus is in luck (or is it entirely luck?), it encounters the root.

1. *Methods of Study*

The growth of VAM fungi from spores and root fragments through soil has received little experimental study. One approach has been to examine the hyphal extension rate from germinating spores between membrane filters.[38,39] Although a soil environment of a type is thus provided, growth is along a plane surface and the physical tortuosity, the physical and chemical heterogeneity of soil, and the environmental microgradients which occur across pores are not present. However, the method is relatively simple, and good comparative data on factors such as temperature, water potential, and other environmental/plant factors are obtained. A variation of this is to sow spores on a glass slide over which a root is grown and to examine growth across the slide.[16] This also is subject to the artifacts of growth along the surface of a slide, but can give answers to specific questions such as the occurrence of directed growth to roots. The artifacts caused by the use of inert surfaces can be avoided by placing (germinated) spores accurately in soil and measuring hyphal growth from them into soil. One way to do this is to place spores between stainless steel mesh with apertures large enough for the mycelium to pass through, germinate them, and then bury the assembly in the soil. A complete description of growth distribution through soil would demand a soil-embedding sectioning technique coupled with morphological, immunological, or other tracer methods[61,147,148] to identify the organism. However, in many cases all that is required is information on net linear growth through soil to a root. An appropriate method, therefore, is to place germinated propagules at one or several preset distances from roots growing along a defined plane (e.g., behind stainless steel mesh) and, by progressive harvests, to record time for arrival at the root or commencement of infection. In this way, the growth of *Glomus caledonium* hyphae through a soil has been calculated at 0.7 mm day^{-1}.[15] This was only a quarter of that found by Tommerup[38] using the membrane filter method, the same fungus species (different strain), and a different soil.

2. Effects of Moisture and Temperature

The data of Tommerup[39] show a greater effect of soil water potential on hyphal growth of *A. laevis* and, to a lesser extent, *G. caledonium* than on spore germination. Some hyphal extension occurred at −2.18 MPa with both species. In studies on temperature[38] there was some 50% decline of hyphal elongation of *Gigaspora calospora* between 20 and 15°C, but a major effect on *Glomus caledonium* hyphal elongation only between 10 and 5°C, with the particular collections of spores used. As with spore germination, it would be interesting to examine possible ecotypic variation within fungal species.

3. Biological Interactions

The data of Tommerup[31] show the presence of plant roots consistently increased hyphal elongation with one fungus species, had no effect with another, and had quite inconsistent effects with a third fungus. There has been little study of the impact of other soil organisms on growth of hyphae through soil, and comparisons of hyphal growth through sterilized soil and nonsterile soil are long overdue. It is difficult to establish VAM fungi in some soils unless the soil is first sterilized.[40] This could be due to lability of fungistatic compounds, but it may also be due to biological control of some VAM fungi, in particular soils. The phenomena of soils which are suppressive to certain pathogens and of microbial control of soil pathogens are well established.[62] Such control could occur by competition for substrates, by antibiosis, or by direct attack by organisms such as protozoa,[63] and could occur via attack on the spore or on hyphal growth through soil or in the rhizosphere. Marked reduction of growth of ectomycorrhizal fungi in the rhizosphere by nonspecific mycophagous amoebae has been demonstrated[64] and there is little reason to expect this will not also occur with VAM fungi, the spores of which have been shown to be attacked by amebae.[63] The importance of this interaction could be much greater in the two-dimensional space of the rhizosphere than in soil, but this needs quantitative experimental study. A mixed soil flora has been shown to halve VAM hyphal growth from roots into organic matter.[3] Parasitism of VAM fungal spores by some other fungi also occurs.[65,66,156]

4. Sources of Nutrients for Growth through Soil

Spore reserves, microbial substrates in soil organic matter, and root exudates are all possible sources of organic nutrients for hyphal growth from the spore in soil. Hyphae from germinating spores can use glycine, cystine, and lysine (but not some other amides/amino acids) in laboratory media;[67] but in soil, competition for these from other organisms would be great. Furthermore, free amino acids are likely to be very low in soil, except perhaps in organic matter layers. Nonvolatile root exudates in solution probably diffuse for only very small distances from the root (except in very wet soils), and are soon depleted by a developing rhizoplane/rhizosphere microflora; under moist soil conditions the microbial substrate 0.3 mm from the root surface has been calculated to be possibly only 1% of that at the root surface.[68] Therefore, the available substrate in soil (either from organic matter or root exudates) is small and the spores (or the infected root in the case of root fragments) are probably the main energy sources for growth of hyphae through soil. Spores have very high concentrations of lipids which could serve as an energy source.[69]

Hyphae from spores of *G. caledonium* can grow through soil for at least 8 mm and produce mycorrhizas.[15]

5. *Chemotropism*

Generally, chemotropism has not been regarded as important in the growth of mycorrhizal propagules to the root, but this needs more research. In laboratory studies, germ tubes of *Gigaspora gigantea* are attracted to roots of bean and corn by volatiles produced by the roots.[70] These volatiles may act over much larger distances in soil than solutes because of their more rapid diffusion in soil. In other studies,[16] by placing spores and roots on glass slides in soil, it was found initially that there was no preferred directional growth of mycorrhizal fungi to roots, but attraction occurred as hyphae radiating from the spore approached a root. The data show directional growth to onion occurred for 1.6 mm from spores of two species of VAM fungi, but for 3 to 4 mm with another species. The artifacts of the experiment could have exaggerated these distances, but if this occurs naturally, it could be of considerable importance. As the half distance between axial parts of neighboring roots is often only 1.5 mm in an established grass sward,[71] and as lateral roots of clovers and grasses often emerge only a few millimeters apart, directional growth over 1 to 3 mm to roots may occur more frequently than is generally supposed.

6. *Models of Growth to the Root*

On the basis that the propagule provides the main source of energy for growth of the hyphae, it has been suggested[25] that the distance over which propagules could grow to the root would be proportional to the size (volume) of the propagule. That is, very small propagules (such as small fungal spores and bacteria) move over only small distances to the root, and infection occurs mostly by interception of the spore by the growing root. Given optimum condition, therefore, for fungal germination and growth to the root, one might expect a relation between the numbers of infections and numbers of propagules such as given in Figure 5. These curves are drawn by an analogy with the relations between uptake by plant roots of highly mobile ions or poorly mobile ions through soil.[71] In most cases, VAM fungi will fall into categories (a) and (b). Support for this thesis comes from plant pathology,[73-75] e.g., the minimum numbers of *microslerotia* of *Verticillium dahliae* required to give 100% infection of tomato were 100 g^{-1} of soil, but 5×10^5 *conidia* g^{-1} of soil was needed for the same level of infection. As soil conditions are less optimal for fungal germination and/or growth to the root, the curves would move downward, i.e., (a) will approach (b) and eventually (c). Note that by the same reasoning, the same types of curves would be obtained when the *y* axis is the percentage of a fixed number of propagules causing an infection, and the *x* axis is the plant rooting intensity. A knowledge of these types of relations is essential to analyzing and modeling infection and plant response with respect to rooting intensity of different plant species in various soils. These relationships should also be the jumping-off point for models of spread of infection in roots.

III. THE INFECTION PROCESS

Earlier I have indicated[76] the need to distinguish between infection processes and those processes which relate more to spread and longevity of the infection — an im-

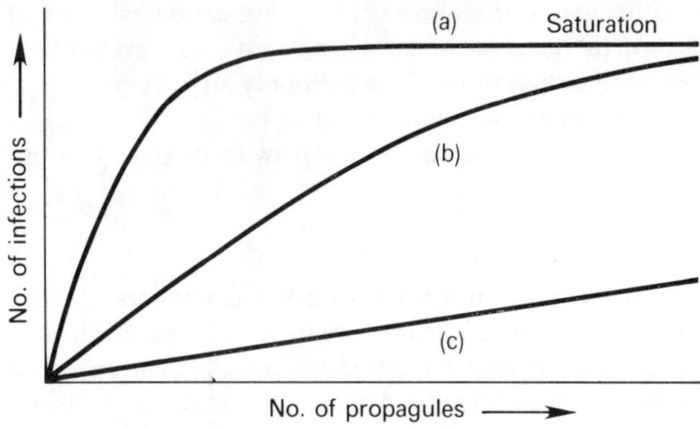

FIGURE 5. Hypothesized relations between numbers of infections and number of propagules for (a) propagules growing through soil for long distance, (b) for moderate distances, and (c) for very short distances. (From Bowen, G. D., *Contemporary Microbial Ecology*, Ellwood, D. C., Hedger, J. N., Latham, M. J., Lynch, J. M., and Slater, J. H., Eds., Academic Press, London, 1980, 283. With permission.)

portant basic distinction which is frequently not appreciated. The infection process itself may be considered as two-phase, viz. (1) the attachment to, and the entry of the root and (2) the infection of cortical cells to give arbuscules (and vesicles). This latter phase has received much study at the electron-microscope level, including histochemical study,[77-80] but the initial penetration of the root has received little study. Very little is known of the physiology and biochemistry of penetration; much of what follows is therefore conjecture.

A. Recognition

The term "recognition" has been used in two quite different ways. To some people it refers to *specific* situations such as the narrow host specificity developed between rhizobia and legumes, or perhaps the rather more coarse specificity between a number of basidiomycetes and ascomycetes and the relatively small group of plants forming ectomycorrhizas; that is, particular host specificity, recognition, or receptiveness factors are involved. This connotation of recognition may have rather less meaning for the relatively nonspecific infections between VA endophytes and the enormous range of plants forming VAM. Another definition of recognition is that of Heslop-Harrison,[81] that "a cell that reacts in a special way in consequence of association with another must do so because it 'acquires' information from that other, information that must be conveyed through chemical and/or physical signals." This is probably the most generally acceptable definition. With VAM we are concerned with questions such as why is there such lack of specificity in VAM? and why is the development of arbuscules frequently restricted to inner cortical cells? We must also inquire into the physiological reasons for the lack or sparsity of infection with certain plant groups such as the Chenopodeaceae, Brassicaceae, Caryophyllaceae, Cyperaceae, and Commelinaceae, and the depression (or elimination) of infection under some environmental conditions such as high phosphate.[22]

Keen[82] listed the requirements of successful pathogens and concluded that in mutualism with higher plants, microorganisms have acquired all necessary properties of both specialized and nonspecialized pathogens.[41] For pathogens these included:

1. Pectic enzymes produced for maceration of host tissue

2. Toxins for killing host cells
3. Substances destroying permeability barriers
4. Degradation of preformed toxins or phytoalexins
5. Production of surface molecules which prevent detection of underlying polymers which would otherwise elicit a hypersensitive reaction
6. Production of substances which suppress expression of host hypersensitive reactions
7. Spore adhesion to the host surface
8. Formation of haustoria and other specialized structures

With the exception of (2) and to some extent (3), analogous phenomena can occur with VAM.[80] These factors are discussed later.

Keen[82] commented that in mutualism, plant and microorganisms have evolved such that specific recognition does *not* occur to evoke host defense. It will be seen below that this is too simplistic a view. For example, the frequent formation of an appressorium[8,10] is a type of recognition. Other reactions also occur (see below); Anderson[83] has discussed several aspects of recognition and compatibility with respect to mycorrhizas in general.

B. Entry

It is now considered that increases in root exudation associated with phosphorus deficiency are not related to the infection process.[76,84] However, are there properties of root exudates which affect the infection process? It is probable, but unproven, that pectic enzymes are produced to assist penetration of the root, but we do not know if VAM fungi have such constitutive enzymes of if they are induced by root exudates or by cell wall contact. There appear to be no enzyme studies so far on external hyphae to act as controls for settling this question.

Are there other "messages" between the fungus in the rhizosphere and the root? A protein in root exudates of soybean can stimulate early nodule formation in rhizobia.[85] Could similar phenomena occur with VAM infection? If specific growth substances needed by the VA endophytes (in addition to energy sources) are synthesized by the plant, does it do this in response to a signal from the fungus? It is possible, even probable, that the fungus also increases the permeability of the host plasmalemma; cytokinin-type activity has been associated with VAM infection,[11,86] and other growth-regulating compounds may also occur. If the fungus increased the permeability of the plasmalemma locally and hence the supply of substrate for its own growth, this would be extremely important in the highly competitive environment of the rhizosphere. Such permeability changes could be achieved by hormone production, but also a component of hyphal cell walls of *Phytophthora infestans* has been shown to affect the membrane of potato tuber cells,[87] a factor which probably affects permeability.

It is usually considered that if the spore germinates but does not encounter a root, it retracts cytoplasm from the hyphae and resumes the quiescent state.[88] However, hyphae from germination spores of *Glomus caledonium* and *Acaulospora laevis* can retain their infectivity for up to 4 months,[19] but decreasing infectivity of hyphae after 4 to 6 weeks correlated with loss of ability to form prepenetration and penetration structures on the root. This could indicate some transient factor associated with hyphae which enhances infection. Germinated spores lost their infectivity more rapidly in unsterilized soils than in steamed soils.

Why is there such a wide host range for VAM fungi? It is possible that the infecting hyphae do not absorb or can degrade constitutive toxins or elicited phytoalexins formed by the plant cell as a defense mechanism. However, the wide range of such toxins and phytoalexins in the plant kingdom would require a very versatile array of enzymes in VAM fungi to degrade them all. Anderson[83] has indicated the frequent production of peroxidase activity of roots as a defense against pathogens and suggested mycorrhizal fungi should possess catalase and superoxide dismutase activities. It is likely that the fungi elicit neither phytoalexins (but this has not been tested experimentally) nor hypersensitive reactions to infection. This could occur by the absence of elicitors from VAM fungi (which seems unlikely considering the wide range of compounds which can be elicitors[89]), the occurrence of surface molecules on the hyphae which mask underlying polymers leading to a hypersensitive response, or to the production of suppressors of hypersensitive responses. As hypothesized by Sequeira[90] for *Agrobacterium tumefaciens,* the wide host range for VAM would be consistent with nonspecific recognition phenomena involving polymers common to many plant species. This is consistent with the early evolution of VAM symbioses in relation to land plants.[91]

The lack (or sparsity) of infection of Chenopodiaceae, Brassicaceae, Caryophyllaceae, and a few other families could be due to any of numerous reasons, ranging from a physical barrier of the cell wall, to an absence of essential nutrients, to production of toxins by the plant. An absence of essential nutrients is not likely to be the major reason, for VA endophytes can grow in the rhizosphere of nonhost plants.[44,92-94] However, in studies by Tommerup,[94] spore germination and hyphal growth from spores of *G. caledonium* placed with *Brassica napus* were much slower than when placed with *Trifolium subterraneum;* this may be due to a partial lack of nutrients from *B. napus* or production of mild toxins. Inhibition or depression of mycorrhiza formation on susceptible plants by accompanying or preceding plantings of Brassicaceae has been recorded on some occasion, but not others,[93,95-97] lending further weight to a toxin theory. The occurrence of a wide range of S compounds in Brassicaceae and of betalins (related to phenols and anthocyanins) in Chenopodiaceae, both of which have fungistatic activity, might be important.

Infection of some species in these two families has been recorded, but usually infection is light and often confined to the older roots;[76] often arbuscules do not form. *G. caledonium* formed fewer appressoria on *B. napus* than on *T. subterraneum;* it formed penetration pegs, but fewer than a third of the infections produced arbuscules.[97] The production of some typical infection of *Chenopodium quinona* following simazine application[98] indicates an interesting tool for further study of infection.

It may be that noninfection of some species is due to a necessary fungus enzyme not being elicited by the plant, that enzymes are inactivated or that they are largely inactive toward the plant cell wall, but the weight of evidence is presently pointing to a toxin theory rather than one based on physical failure to penetrate cell walls.

The decrease in numbers of infections caused by factors such as high phosphate operating via the plants' physiology needs explanation. Is this due to enhancement of plant resistance, to an effect of nutrition on plant factors affecting fungus physiology (e.g., induction of pectic enzymes), or simply to a failure to produce substrates needed for the growth of the fungus? This may be related to the effect of phosphate on spread of the fungus in the root (see later).

C. Development of the Primary Infection

After hyphae penetrate the root (sometimes through a root hair, but commonly between epidermal cells) they spread intercellularly along the cortex and also become intracellular on reaching the second layer of cortical cells. The biology and physiology of the infection following root penetration have received much study. Considerable interaction between the fungus and the higher plant occurs.

Most, if not all, studies have been on a few species of *Glomus,* only one of four genera producing VAM. Most of the ultrastructural studies are on the penetration of cortical cells and the subsequent formation of the arbuscule, the highly branched organ involved in the exchange of nutrients and metabolites with the plant. Vesicles, in which lipid storage is prominent and glycogen sometimes occurs, may develop both within and between plant cells and appear to be incipient chlamydospores. Detailed, excellent reviews of the ultrastructure of arbuscule development are given elsewhere.[77-79]

In arbuscule development, penetration starts as a constricted hyphal peg causing the cortical cell wall to stretch and invaginate.[99] After penetration, the hypha resumes its normal size and dichotomizes frequently to produce the arbuscule. This is multinucleate, is always enclosed in the fungal cell wall (sometimes very thin), and has a relatively short life span, usually from 4 to 15 days.[42,100] A stretching of the host cell wall and alteration of the middle lamella during penetration[101,102] suggest a fungus enzyme action in penetration of the cell wall rather than a mechanical rupture of the cell wall. During arbuscule development the fungus plasmalemma invaginates to produce multivesiculate paramural bodies or lomasomes, which are particularly abundant in the arbuscule branch. These are rich in polysaccharides and have been suggested as important in exchanges of metabolites between fungus and plant.[103] The arbuscule has many organelles and has inclusions of glycogen, lipid, and polyphosphate.[77,100,102] In the root, the fungus cell wall changes from a layered, stratified wall to a monostratified wall. It thins considerably and polymerization of *N*-acetyl-glucosamine units, which form the chitin of the fungus cell wall, is blocked.[104]

What about the plant's reaction to the infection? Many, but not all, of the reactions to biotrophic pathogens occur. Papilla formation[105,106] by the plant does not occur, but the trunk of the arbuscule is surrounded by a "collar", continuous with the plant cell wall and containing glycoproteins which are the same as those of the cell wall.[78,103] As the arbuscule develops, this collar becomes thinner and is replaced by a matrix of plant cell origin consisting largely of disorganized polysaccharide fibrils[100] that surround the entire arbuscule. The amount of this interfacial matrix diminishes with development of the arbuscule. Histochemical tests show that the matrix also contains some protein and possibly glycoprotein.[107] Host cell wall deposition is not rapid enough to contain the fungus, which apparently disorients the ability of the plant to organize the fibrillar network into a cell wall. When growth of the arbuscule ceases, this fibrillar layer encases it.

Although phenolic-type deposition in the plant cell does not occur,[99] there are several detectable changes in the host cytoplasm. The highly invaginated host plasmalemma is elaborated into paramural bodies which can contain cytoplasm and are cut off and occur near the fungus. Many changes reflect a greatly increased metabolic activity in the cell: host cytoplasm is increased by up to 20-fold;[100] the nucleus becomes highly polyploid and the nucleolus is greatly enlarged; Golgi activity increases markedly (no doubt involved in the secretion of the polysaccharide matrix), and the endoplasmic reticulum is also greatly increased. VAM have been found to respire at a rate six times that of roots lacking the VA endophyte.[108] Plastid development in the cell is blocked at the protoplastid phase or plastids turn into chromoplasts, which suggests a modification of the carbohydrate metabolism of the root cell. Although starch is usually absent from invaded cells, minute amyloplasts have been observed and, therefore, the fungus may not prevent starch accumulation totally.[78] In sharp contrast to plant pathogenic infections, no observable cytochemical or morphological difference in the host plasmalemma occur after infection,[107] a fundamental difference which may be crucial to symbioses and the maintenance of a two-directional transfer of nutrients and metabolites between fungus and root.

The many and varied changes in the root cell with arbuscule formation are interesting mainly in their amount, rather than their nature. This and the reversion of the host

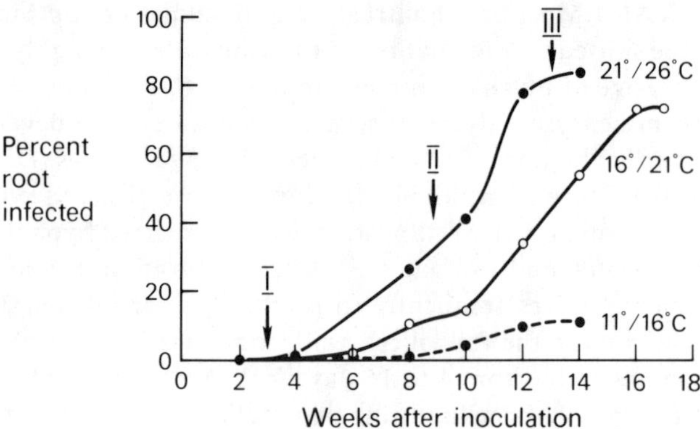

FIGURE 6. Development of VAM infection in *Allium cepa* by *Gigaspora calospora*. Phases of infection are indicated for night/day temperatures of 21°C/26°C. (From Bowen, G. D., *Current Perspectives in Microbial Ecology,* Klug, M. J. and Reddy, C. A., Eds., American Society of Microbiology, Washington, D.C. With permission.)

cell to "normality" after arbuscule degeneration suggest that the fungus may affect the production of growth regulators. The production of cytokinin and growth regulator compounds by VA mycorrhizal fungi is indirectly evidenced by the enlarged nucleolus of the plant cell and a possible effect on stomatal physiology.[86]

D. Saturation

Figure 2 indicated a saturation level of infections on a root. It is pertinent to ask what controls this physiologically. So far, there are few experimental answers to this (note we are discussing *infection* numbers, not the length of root infected). Are only certain cells on the root susceptible to infection? In organ tissue culture and on buried slides in soil, hyphae of the VAM fungus often by-pass particular cells or parts of the root.[109] Is this a function of root cell age? Calculations by Walker and Smith[18] indicated lateral roots are more "susceptible" than are main axes; why is this? Do the same saturation levels of main areas and laterals occur under ideal conditions? (In their studies the main axes had ceased growth but lateral roots were growing actively.) Does the number of infection points on laterals or main axes relate to plant assimilate available for successful development of the infection? Do fungi differ in their saturation level? Existing infections on the root restrict the growth of other infections (see later), but do they actually induce local resistance to later infections? These are emerging questions. In many cases, answers to these questions can be by simple experimental approaches. It is now an appropriate time for such studies, defining the biology, ontogeny, and physiology of infection with greater resolution.

IV. THE GROWTH OF THE INFECTION

Three cardinal phases in the development of the infection have been long recognized and are shown in Figure 6: a lag phase (I), a period of rapid development largely due to spread of the infection (II), and a plateau phase (III) which is the maximum percentage of the root which becomes mycorrhizal. Phase I is related to the primary infection (discussed above); phase II (spread) and phase III (maximum infection level) are addressed particularly in this section. Note, we are now discussing the *length* of root infected and this arises from spread from primary infections both internally and externally, the latter giving rise to secondary infections either on the same root or on adjacent roots.

A. Spread

The dynamics of spread have been approached in two ways: (1) modeling, or producing mathematical analyses of the spread, based on progressive measurement of infection and infection points on the growing root, and certain assumptions.[13,15,18,110-112] These have led to a number of hypotheses on the biology of the system, which some workers wrongly accept as proof but which need experimental testing. Not surprisingly, the modeling approach has highlighted the need for more detailed experimental study of the biology of individual infections; (2) detailed experimental study of particular questions, e.g., the rate of spread of individual infections, the mode of spread (internal and external to the root), the susceptibility of different parts of the root, and ages of root to infection, etc.[40,92] The former approach has tended to dominate to this stage.

Smith and Walker[13] modeled infection of *Trifolium subterraneum* in soil with inoculum dispersed through it, using the assumptions that any uninfected portion of the root was equally likely to be infected and that all infections grow at the same rate, regardless of age. They derived the rate of new infections cm^{-1} root and the rate of spread of existing infections day^{-1} and found that fertilizing with nitrate decreased the rate of production of new infections and the spread of existing infections by some 50 to 60%, but that added NaCl affected only the rate of infection. Note that this is a very different type of analysis from that in the first part of this chapter e.g., Figure 2; it does not separate the components a and b given there and their analysis does not distinguish between primary and secondary infections. The large effect of nitrate on rate of spread suggests root physiological factors which could well have affected both growth of the fungus in the root and in the rhizoplane (and, thus, secondary infection.) The nitrate effect on the amount of root infected is conditioned greatly by the phosphate status of the plant.[113] The indicated effect of nitrate on rate of spread demonstrates one of the values of this approach, and although it can be argued that this could be achieved equally well by the experimental approach of using point inoculation and following the spread, the Smith and Walker approach gives additional information on rate of production of new infections. Their data also suggested the young regions of the root were some ten times more susceptible to infection than the average for the whole root system. They clearly indicated the need to test this experimentally, but the need to test hypotheses experimentally seems to escape many workers, for within a year this suggestion was being quoted as a proven fact. Indeed, later studies[114] showed that at 16 days the older and younger halves of lateral roots are equally "infectible", but that the basal 2 cm of the main axes is not infected. Other studies with clover[109] suggest recently developed lateral roots to be preferred infection sites.

The data of Walker and Smith[18] show a marked decline in new infections cm^{-1} root for main roots after 14 days, but no such trend with lateral roots. This appears to correspond with cessation of main root growth at this time. The rapid increase in lateral root length at that time may well have masked a similar behavior in older parts of lateral roots, for the modeling approach is an averaging approach and too broad a sampling can miss important information. A marked reduction in the growth rate of existing infections in the main root after 14 days and at later times in lateral roots was indicated. A slowing of growth rate of older infections on laterals could have occurred, but large numbers of new infections on the rapidly expanding lateral root growth may have masked that. (The slowing of rate of spread for older infections would be counter to one of the basic initial assumptions of the model.) The reasons for the slowing of rate of spread are unclear, for we have little idea of what substrates (or energy sources and growth factors) "drive" growth in the root (see Section IV.C). The growth in the root of individual infections can vary with fungus,[92,115] to some extent with host[92,111] and nutrition,[113,115] but on the present (sparse) information, it appears to be limited to some 1 cm in each direction.[13,15] The symbiosis is obviously a "contained" infection, but we do not know what contains it — a plant defense or a shortage of energy substrate for fungus maintenance and growth.

Competition — Buwalda et al.[110] mathematically analyzed the spread of infection in leek *(Allium porrum* L.) from a point source of inoculum and later[112] from both a point source and inoculum dispersed in soil. The rate of increase in the total length of infected root was similar for both inoculum placement treatments, but it may be that this was due to a fortuitous choice of plant/fungus combination. Their calculations showed a somewhat slower rate of spread of infection in the root from a point source of inoculum than when there were several primary infection points spread along the root when inoculum was dispersed through soil. They deduced the maximum rate of spread of a front decreased roughly in proportion to the numbers of points, i.e., infections have a competitive interaction. This may be consistent with other data[18] on clover showing a greater mean size of (fewer) infections on the main root axis compared with those on the laterals which had more infection units cm^{-1}. Wilson[115] noted mutual inhibition of infection and such phenomena are, no doubt, important in the "saturation" phenomenon in primary infection. By contrast, Mosse and Hepper[109] found that once an infection was established in a root, it seemed to become more prone to further new infections. Walker and Smith[18] found at high concentrations of propagules in soil, new infections cm^{-1} declined after 13 to 16 days; data on total infections were not presented, but it is possible that the roots were reaching saturation infection so that it was difficult to impose new infections on the existing infections. Inhibition or competition for space (resources) between different fungi is an important applied aspect for study. Abbott and Robson[116] found that the fungus which infects most rapidly in a mixture had a competitive advantage in occupancy of the root.

B. Maximum Infection Level

The attainment of a maximum, sustained percentage of root infected occurs, (see Figure 6) and this varies with plant species, nutrition,[111,118] and the fungus.[22,115,118] At this stage, an equilibrium is reached between the growth rate of roots and that of the fungus, and any factor changing the relative growth rate of the root or of the fungus will change the equilibrium, and, therefore, the plateau level.

Buwalda et al.[110,111] identified this as n (the fraction of the root becoming mycorrhizal), a factor inserted into their equation, not generated. At low phosphorus levels, n for leek was 0.95 and 0.47 for wheat. Increasing the levels of soil phosphate to 30 mg P kg^{-1} soil had only a slight effect of this on leek, but reduced it to 0.07 in wheat. Studies on infection of soybean at four temperatures with six VAM fungi[118] and the data of Figure 6 suggest a large effect of soil temperatures and fungus species on n.

The arbitrary insertion of n into modeling equations has been criticized as unnecessary,[18] for it can be derived from calculations of rate of spread of the fungus in the root. Be this as it is, all models at present give good descriptions of the spread, not physiological explanations.

Fungi vary in their extent and mode of spread and in the final balance between fungus and root; e.g., in one study[115] *Glomus fasciculatum* spread within the root more extensively than *Gigaspora decipiens* and *Glomus tenue.* In some cases the length of the developed infection unit was three times that of the other two fungi (this is also influenced by the absolute numbers of infection points). *Gigaspora decipiens* and *Glomus tenue* were more dependent on external spread along the root and production of secondary infections. Similarly, *Gigaspora calospora* has been found to produce about twice the amount of external hyphae cm^{-1} that *Glomus fasciculatum* does under the same conditions.[117] If this is largely in the rhizosphere, one could expect an effect on the rate of external spread of the hyphae along the root and on secondary infection.

C. Some Physiological Considerations

I have indicated some of the dynamics of spread of VAM infection and variables affecting them. Complete physiological explanations must await more knowledge on what "drives" the symbiosis, both in terms of the spread of the fungus and the longevity of arbuscules (both may have the same basis).

A considerable amount has been written about the possible roles of root exudates in VAM symbioses, especially as low phosphorus and other factors which increase the percentage of root becoming mycorrhizal, often increase losses of organics from roots, which are collectively referred to as "root exudates".[84,119-122] This greater root exudation was initially thought to be involved in the infection process, but this is now not considered to be so.[84] I have discussed the role of exudates in primary infection in detail elsewhere.[76] Some of the difficulties in interpretation of root exudate studies are because what is usually collected as "exudates" comes from a variety of sources.[123] Although increases in losses of amino acids, organic acids, and sugars are recorded with various treatments increasing the percentage of root becoming mycorrhizal, the relative amounts of these differ from experiment to experiment; often a *rise* in infected root length occurs with the addition of suboptimal levels of phosphate for plant growth.[124,125] Often the correlation between exudate measured under varying conditions is much tighter with the percentage of root infected than with the numbers of infections.[22,76,125]

There is a temptation to correlate external spread with root exudates, but this also presents some difficulties: (1) much of the energy for spread of the fungus on the outside of the root may come largely from translocation from the arbuscule within the host. The only definitive way to examine the role of root exudates quantitatively would be to study rhizosphere growth with hosts which were not infected, and this itself may lead to other problems (see Section III.B); and (2) root exudates probably largely reflect the composition of the cytoplasm of the cortical cells, and any correlations between collected root exudates and percentage root infected are just as likely to be correlations between cytoplasmic composition of the cortical cells and fungal spread.

Because of our inability to grow VA endophytes using a range of substrates commonly used by fungi, we must assume for the time being that (as yet) unknown growth factors are required in addition to energy sources. We have, therefore, two considerations: this unknown factor(s), and commonly used substrates supplying energy. Because most research on root composition and root exudates focuses on easily assayed, common constituents of the cytoplasm, it is almost certain that only the second factor (commonly used substrates) is examined, not both factors. Such possible growth substances as glycerols, sterols, and lipids have also been largely ignored.

Focusing on commonly used fungus growth substrates, what do we know of their use by VA endophytes? Not very much, but studies of limited independent growth of hyphae[67,127] may give a method for examining this. It would be interesting to examine the effects of root exudate fractions and root extracts using these techniques. It is interesting that a large increase in glycine and lysine in root exudates of P-deficient sudan grass has been found[120] and that these (and cystine) have been shown to stimulate growth of *G. caledonius* hyphae considerably, compared with several other amino acids/amides.[67]

If fungal spread along the root and into soil is driven by energy (and growth factors) derived from the arbuscule, what determines the rate of growth and the restriction on growth of each infection unit, and what determines the longevity of arbuscules? The simplest theory would be to relate both of these to availability of substrate or growth factors from the host cell.

Until we know more about the substrate preferences of VA endophytes, it is unwise to be too dogmatic in conclusions drawn from analyses of extracts of roots and exudates. However, if we knew more about the various substrates used by VA endophytes we might be in a better position to understand spread of the infection and longevity of arbuscules and to relate these to root physiology and the effects of environment on cytoplasm composition along the root. Until then, it might be worthwhile to have some study of distribution of translocated assimilate (regardless of identity) along the root in relation to mycorrhizal growth and arbuscule longevity.[80] Any theory of growth and longevity of VAM infection must eventually accomodate observations such as the different final size of infection units with different fungi, differences between fungi in their infection of plants grown with added phosphate,[22,127] and that intensity of fungus development cm^{-1} can differ markedly with the inoculum level.[115,121]

Developmental variation must also be considered. The domination of the infection by arbuscules suggests these are formed when the fungus is in its rapid phase of extension; at least in many species of *Glomus* the vesicles form after arbuscule development.[129] Because vesicles are rich in lipids,[60,69,130] their formation has been considered to be governed by excess carbohydrate[131] no longer required for exponential growth or arbuscule formation. (This would argue against arbuscule senescence being associated with lowered supplies of carbohydrate.) The production of vesicles has been correlated with increasing light intensity,[132] and their absence with a low supply of root carbohydrate.[133] Abbott and Robson[128] suggested high inoculum levels, leading to high intensity of infection, and a high production of external hyphae led to delayed vesicle formation via reduction of available carbohydrates for their formation.

Bethlenfalvay and colleagues[134,135] found cessation of endophyte growth in the root to occur with the onset of the reproduction phase of the plant, and considered this to be due to a change in the source-sink with relationships in the plant, i.e., a direction of assimilates to reproduction rather than to growth of shoots and roots. The direction of assimilates to reproductive growth rather than to roots and shoot is consistent with existing plant physiological knowledge.

V. FUNGAL GROWTH INTO SOIL

Although the increases in nutrient uptake by VAM are due largely to the growth of extraradical hyphae — this leading to an increased nutrient inflow[136-139] — there has been relatively little study of this important phase of the symbiosis, due partly to the difficulty of its quantitative study. Two basic parameters need to be examined: the amount of extraradical hyphae and its disposition in soil, for this will influence the volume of soil used by the fungus. Restriction of growth of hyphae to the rhizosphere is likely to give much less increase in nutrient uptake than if the hyphae grow into soil appreciably.[117,140,144]

A. Methods of Study

Measurements of external hyphae have been made by:

1. Removing and weighing the hyphae attached to roots.[141] This method is tedious and prone to large errors due to the difficulty in recovering all of the hyphae, and difficulty in removing sand grains and organic matter from them.

2. Assaying chitin in soil,[135,142] adjoining mycorrhizal roots and calibrating this against the chitin of a known quantity of VAM fungus; chitin being a component of the fungal cell wall. The non-VA endophyte component of the soil (other fungi and soil microfauna) needs to be corrected for, and seiving of soil reduces this factor to some extent.[142] However, the method may have its greatest potential in experimental situations where the growth medium is simple, e.g., washed sand or synthetic soil so that these background factors are minimized.
3. Weighing of soil adhering tightly to hyphae[143,144] — an attractive simple method, best suited for comparison of relative amounts of hyphae under various treatments, and especially for soils of a sandy loam texture. In clay soils difficulties would be expected in the recovery of hyphae. Calibration of known lengths of hyphae (see below) against weights of adhering soil is needed for each soil for conversion to something approaching absolute values. An assumption implicit in this and other methods is that hyphae growing in the rhizoplane are also easily detached from the root for sampling, and, furthermore, that they have the same amount of soil attached to them as do hyphae in the soil. The method may be more sound for measuring the growth into soil rather than for soil and rhizoplane together.
4. Extracting hyphae from soil by dispersion in water, filtering hyphae onto a membrane filter, and measuring length by a line intersect method[117,124,145] or by weighing.[146] Sampling procedures and errors in this have been discussed by Abbott et al.[124] This direct method, or modifications of it, can be employed in any soil. Despite claims to the contrary by some authors, it is often difficult to recognize the VA endophyte from other soil fungi, especially after dispersion,[117] and it is essential to correct for this background. It would be useful to combine this technique with immunological techniques[61,147,148] which, although not perfect, would assist in the identification of VAM endophytes, and especially of inoculated endophytes for which highly specific serum had been prepared.

In considering nutrient uptake, length of hyphae is more important than weight. The weight/length conversion can differ greatly between fungi because of large differences in mean hyphal diameter.[124]

Methods for studying hyphal distribution in soil have been less well-developed than those for hyphal quantity. It should be possible to embed blocks of soil near the root and section for the occurrence of hyphae, or extract hyphae from portions of the block by using method (4) above. In either case, identification of the fungus as a VAM endophyte from the root is a problem, and the use of immunological or other techniques would be an advantage (see above). I suggest the position in soil of hyphae active in uptake at any time may be obtained quantitatively by placement of radioactive phosphate or other tracer elements in soil at known distances from the mycorrhizal root and examining appearance of the isotope in the root. A simple version of this[138] has demonstrated the translocation of phosphate to the root for 8 cm.

B. Factors Affecting Extraradical Growth of Hyphae

The amount of external hyphae is often correlated with the amount of infection,[141,146,149] but this is not always so,[117,144,150] for the stage of infection, fungus, plant physiological factors, and soil factors affect the growth of the fungus.

In studies on soybean,[134,135] the amount of external hyphae of *G. fasciculatum* was initially some ten times that of internal hyphae 1 to 2 weeks after infection, but approximately the same as the internal hyphae after a further 6 to 7 weeks. (One set of data is given in Figure 7.) After that the internal fungal biomass increased (in one

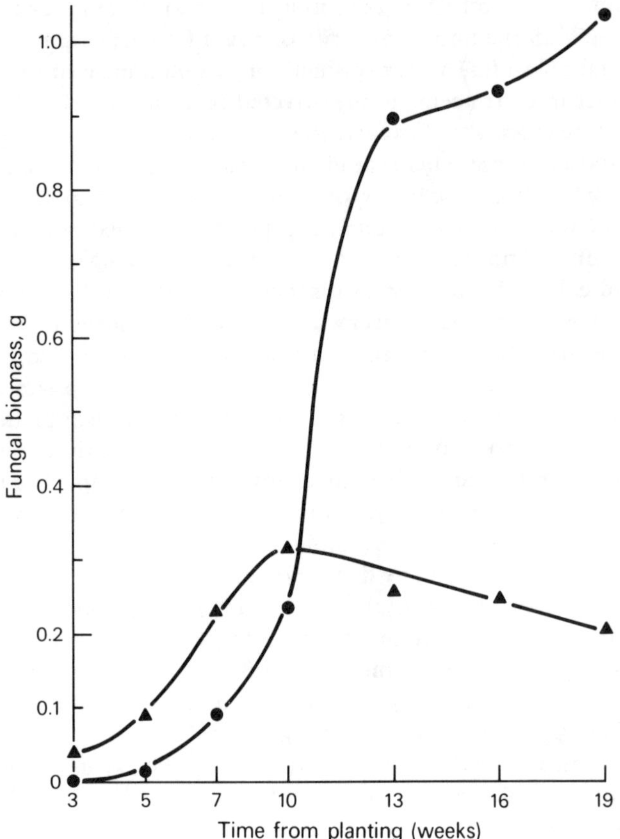

FIGURE 7. Biomass of extraradical and intraradical mycelia of *Glomus fasciculatum* in soybean; ▲ = extra radical and • = intraradical mycelium.[135]

instance some threefold over 3 weeks), while the amount of extramatrical hyphae declined by about one third during 4 to 9 weeks. At plant senescence the fungus/root dry weight ratio was 12.3%.[134] Not only do this data indicate that the time of sampling of extramatrical hyphae is important, but also that the dynamics of internal/external hyphae are continually changing. Other studies[107] suggest that the amount of external hyphae on a root may sometimes remain at approximately the same level over several weeks, although the amount of root infected is increasing. This and a decline in extramatrical hyphae with age of the root[134,135] suggest hyphae may be continually turning over and what is really being measured in extramatrical hyphae is the balance between production and death. This is a point of considerable importance for further examination; as with turnover of fine roots of perennial plants,[151] its implications are large for estimating the total energy cost of extramatrical hyphae and the volume of soil used by extraradical hyphae over a season.

It has now been established that fungi differ in the amount of external hyphae they produce.[117,144] Of eight *Glomus* species studied in neutral-alkaline soils,[144] four produced little or no external hyphae. In a study of four VA endophytes on *Trifolium subterraneum*,[117] *G. fasciculatum* produced only 20 to 40% of the external hyphae (cm^{-1} infected root) of *Gigaspora calospora, Acaulospora laevis,* and *Glomus tenue.* Nevertheless, *G. fasciculatum* is a highly effective fungus and this emphasizes the probable importance of distribution of external fungus — in the bulk of soil or in the rhizosphere. The lower external hyphal production by this fungus compared with *G. tenue* is consistent with the observation that spread of *G. fasciculatum* within the root was more important than for *G. tenue,* which formed smaller infection units but spread more in the rhizoplane, producing more secondary infections.[115]

Two basic factors probably affect extraradical growth, the "driving force", i.e., the availability of substrate in the root (and the effect of environment on this) and soil factors. Applied phosphorus (whether to soil onto the foliage) at relatively high levels decreases both the infection intensity and the production of external hyphae, but small amounts of phosphate increase these.[23,24,84,124] In studies on *T. subterraneum,*[124] external hyphal production by *G. fasciculatum* was some 50% greater with 0.5 and 1.0 g superphosphate per kilogram soil than with no applied phosphate or with 1.5 g. Percentage of root length infected was greatest at 0.5 g applied superphosphate. Similarly, in other studies[152] the ratio of extraradical/intraradical hyphae was also greatest at applied phosphate levels at about a quarter of that necessary for maximum growth of uninoculated soybean. It has been hypothesized that the external growth of hyphae is greatest at phosphate applications leading to high soluble sugars in the root,[124,153] and that in active stages of the fungus carbohydrates are used for external growth rather than for vesicle formation. Attractive though the carbohydrate theory is, it must be remembered that so far this is only correlative evidence, and it assumes carbohydrates are the sole driving force for the infection, which is probably not correct, e.g., amino acids and especially organic acids may also be important plus growth factors.

Regardless of the availability of substrates in the root, soil factors themselves may affect growth of hyphae into soil and some of the factors which affect growth of hyphae *to* the root would also operate in the phase. This has received little experimental study. It is probably that soil pH is a major factor in growth of hyphae into soil and that there are differences between VA endophytes in this.[140,153] Soil microorganisms and soil organics may also have an important effect; in one study[3] in the absence of organic matter, microorganisms had little effect on the penetration of soil by *G. fasciculatum;* the addition of sterilized organic matter increased in growth by some 250%, but the presence of other microorganisms reduced this by half. It is not known if this was due to antibiosis or reduction of available nutrients in the presence of other microorganisms. Similarly, soil physical factors have a marked effect on penetration of soil by ectomycorrhizal fungi; compaction (or associated changes in soil oxygen or carbon dioxide) reduced growth of these fungi into soil by some 80%,[2] and similar effects could well occur with VAM. Soil type can also have a marked effect on growth into soil by ectomycorrhizal fungi.[2] The microenvironment of the rhizosphere/rhizoplane is markedly different from the rest of the soil chemically and physically[154] (e.g., in pH, nutrient availability, and substrate availability) and all these may affect the distribution of extraradical hyphal dramatically.

Overall, there is a great need for more experimental study of soil factors affecting growth of fungi into soil and its uptake characteristics, i.e., the "functional" end of the symbiosis.

ACKNOWLEDGMENTS

It is with pleasure that I acknowledge the valued, creative discussion on many of the topics above with Lynette Abbott, Chris Gilligan, Alan Robson, Albert Rovira, and Inez Tommerup, and access to unpublished data so freely given.

REFERENCES

1. Abbott, L. K. and Robson, A. D., The role of vesicular arbuscular mycorrhizal fungi in agriculture and the selection of fungi for inoculation, *Aust. J. Agric. Res.*, 33, 389, 1982.
2. Skinner, M. F. and Bowen, G. D., The penetration of soil by mycelial strands of ectomycorrhizal fungi, *Soil Biol. Biochem.*, 6, 57, 1974.
3. St. John, T. V., Coleman, D. C., and Reid, C. P. P., Association of vesicular-arbuscular mycorrhizal hyphae with soil organic particles, *Ecology*, 64, 957, 1983.
4. Stribley, D. P. and Read, D. J., The biology of mycorrhiza in the Ericaceae. VII. The relationship between mycorrhizal infection and the capacity to utilize simple and complex organic nitrogen sources, *New Phytol.*, 86, 365, 1980.
5. Ames, R. N., Reid, C. P. P., Porter, L. K., and Cambardella, C., Hyphal uptake and transport of nitrogen from two ^{15}N-labelled sources by *Glomus mosseae*, a vesicular-arbuscular mycorrhizal fungus, *New Phytol.*, 95, 381, 1983.
6. Reid, C. P. P. and Bowen, G. D., Effects of soil moisture on v.a. mycorrhizal formation and root development in *Medicago*, in *The Soil-Root Interface*, Harley, J. L. and Russell, R. S., Eds., Academic Press, London, 1979, 211.
7. Smith, S. E. and Bowen, G. D., Soil temperature, mycorrhizal infection and nodulation of *Medicago truncatula* and *Trifolium subterraneum*, *Soil Biol. Biochem.*, 11, 469, 1979.
8. Mosse, B., The establishment of vesicular-arbuscular mycorrhiza under aseptic conditions, *J. Gen. Microbiol.*, 27, 509, 1962.
9. Gianinazzi-Pearson, V., Trouvelot, A., Morandi, D., and Marocke, R., Ecological variations in endomycorrhizas associated with wild raspberry populations in the Vosges region, *Acta Oecol.*, 1, 111, 1980.
10. Abbott, L. K., Comparative anatomy of vesicular-arbuscular mycorrhizas formed on subterranean clover, *Aust. J. Bot.*, 30, 485, 1982.
11. Cox, G. and Sanders, F. E., Ultrastructure of the host-fungus interface in a vesicular-arbuscular mycorrhiza, *New Phytol.*, 73, 901, 1974.
12. Gilligan, C. A., Modelling of soil borne pathogens, *Annu. Rev. Phytopathol.*, 21, 45, 1983.
13. Smith, S. E. and Walker, N. A., A quantitative study of mycorrhizal infection in *Trifolium:* separate determination of the rates of infection and of mycelial growth, *New Phytol.*, 89, 225, 1981.
14. Carling, D. E., Brown, M. F., and Brown, R. A., Colonization rates and growth responses of soybean plants infected by vesicular-arbuscular mycorrhizal fungi, *Can. J. Bot.*, 57, 1769, 1979.
15. Sanders, F. E. and Sheikh, N. A., The development of vesicular-arbuscular mycorrhizal infection in plant root systems, *Plant Soil*, 71, 223, 1983.
16. Powell, C. Ll., Development of mycorrhizal infections from *Endogone* spores and infected root segments, *Trans. Br. Mycol. Soc.*, 66, 439, 1976.
17. Tommerup, I. C., Spore dormancy in vesicular-arbuscular mycorrhizal fungi, *Trans. Br. Mycol. Soc.*, 81, 37, 1983.
18. Walker, N. A. and Smith, S. E., The quantitative study of mycorrhizal infection. II. The relation of rate of infection and speed of fungal growth to propagule density, the mean length of the infection unit and the limiting value of the fraction of the root infected, *New Phytol.*, 96, 55, 1984.
19. Rovira, A. D. and Bowen, G. D., The effects of micro-organisms upon plant growth. II. Detoxication of heat sterilized soils by fungi and bacteria, *Plant Soil*, 25, 129, 1966.
20. McLaren, A. D., Radiation as a technique in soil biology and biochemistry, *Soil Biol. Biochem.*, 1, 63, 1969.
21. Campbell, N. A. and Keay, J., Flexible techniques in describing mathematically a range of response curves of pasture species, in *Proc. 11th Int. Grassland Congr.*, 1970, 332.
22. Jasper, D. A., Robson, A. D., and Abbott, L. K., Phosphorus and the formation of vesicular-arbuscular mycorrhizas, *Soil Biol. Biochem.*, 11, 501, 1979.
23. Sanders, F. E., The effect of foliar-applied phosphate on the mycorrhizal infections of onion roots, in *Endomycorrhizas*, Sanders, F. E., Mosse, B., and Tinker, P. B., Eds., Academic Press, London, 1975, 261.
24. Menge, J. A., Steirle, D., Bagyaraj, D. J., Johnson, E. L. V., and Leonard, R. T., Phosphorus concentrations in plants responsible for inhibition of mycorrhizal infection, *New Phytol.*, 80, 575, 1978.
25. Bowen, G. D., Integrated and experimental approaches to the study of growth of organisms around roots, in *Soil-Borne Plant Pathogens*, Schippers, B. and Gams, W., Eds., Academic Press, London, 1979, 209.
26. Bowen, G. D., Misconceptions, concepts and approaches in rhizosphere biology, in *Contemporary Microbial Ecology*, Ellwood, D. C., Hedger, J. N., Latham, M. J., Lynch, J. M., and Slater, J. H., Eds., Academic Press, London, 1980, 283.

27. Bolan, N. S. and Abbott, L. K., Seasonal variation in infectivity of vesicular-arbuscular mycorrhizal fungi in relation to plant response to applied phosphorus, *Aust. J. Soil Res.*, 21, 207, 1983.
28. Tommerup, I. C. and Abbott, L. K., Prolonged survival and viability of va mycorrhizal hyphae after root death, *Soil Biol. Biochem.*, 13, 431, 1981.
29. Abbott, L. K. and Robson, A. D., Infectivity and effectiveness of five endomycorrhizal fungi; competition with indigenous fungi in field soils, *Aust. J. Agric. Res.*, 32, 621, 1981.
30. Daniels, B. A. and Duff, D. M., Variation in germination and spore morphology among four isolates of *Glomus mosseae, Mycologia*, 70, 1261, 1978.
31. Hepper, C. M. and Smith, G. A., Observations on the germination of *Endogone* spores, *Trans. Br. Mycol. Soc.*, 66, 189, 1976.
32. Siqueira, J. O., Hubbell, D. H., and Schenck, N. C., Spore germination and germ tube growth of a vesicular-arbuscular mycorrhizal fungus in vitro, *Mycologia*, 74, 952, 1982.
33. Daniels, B. A. and Graham, S. O., Effects of nutrition and soil extracts on germination of *Glomus mosseae* spores, *Mycologia*, 68, 108, 1976.
34. Hepper, C. M., Effect of phosphate on germination and growth of vesicular-arbuscular mycorrhizal fungi, *Trans. Br. Mycol. Soc.*, 80, 487, 1983.
35. Schenck, N. C., Graham, S. O., and Green, N. E., Temperature and light effect on contamination and spore germination of vesicular-arbuscular mycorrhizal fungi, *Mycologia*, 67, 1189, 1975.
36. Green, N. E., Graham, S. O., and Schenck, N. C., The influence of pH on the germination of vesicular-arbuscular mycorrhizal spores, *Mycologia*, 68, 929, 1976.
37. LeTacon, F., Skinner, F. A., and Mosse, B., Spore germination and hyphal growth of a vesicular-arbuscular mycorrhizal fungus, *Glomus mosseae* (Gerdemann and Trappe), under decreased oxygen and increased carbon dioxide concentrations, *Can. J. Microbiol.*, 29, 1280, 1983.
38. Tommerup, I. C., Temperature relations of spore germination and hyphal growth of vesicular-arbuscular mycorrhizal fungi, *Trans. Br. Mycol. Soc.*, 81, 381, 1983.
39. Tommerup, I. C., Effect of soil water potential on spore germination of vesicular-arbuscular mycorrhizal fungi, *Trans. Br. Mycol. Soc.*, 83, 193, 1984.
40. Bowen, G. D., unpublished data.
41. Tommerup, I. C., Spore dormancy in vesicular-arbuscular mycorrhizal fungi, *Trans. Br. Mycol. Soc.*, 81, 37, 1983.
42. Macko, V., Inhibitors and stimulants of spore germination and infection structure formation in fungi, in *The Fungal Spore: Morphogenetic Controls*, Turian, G. and Hohl, H. R., Eds., Academic Press, London, 1981, 565.
43. Rich, J. R. and Bird, G. W., Association of early-season vesicular-arbuscular mycorrhizae with increased growth and development of cotton, *Phytopathology*, 64, 1421, 1974.
44. Daniels, B. A. and Trappe, J. M., Factors affecting spore germination of the vesicular-arbuscular mycorrhizal fungus, *Glomus epigaeus, Mycologia*, 72, 457, 1980.
45. Koske, R. E., *Gigaspora gigantea:* observations on spore germination of a va-mycorrhizal fungus, *Mycologia*, 73, 289, 1981.
46. Leeper, G. W., *Introduction to Soil Science*, 4th ed., Melbourne University Press, Melbourne, 1964.
47. Hsiao, T. C., Plant responses to water stress, *Annu. Rev. Plant Physiol.*, 24, 519, 1973.
48. Cannell, R. Q. and Jackson, M. B., Alleviating aeration stresses, in *Modifying the Root Environment to Reduce Crop Stress*, Arkin, G. F. and Taylor, H. M., Eds., American Society of Agricultural Engineers, St. Joseph, Mich., 1981, 141.
49. Mosse, B., The influence of soil type and *Endogone* strain on the growth of mycorrhizal plants in phosphate deficient soils, *Rev. Ecol. Biol. Sol.*, 9, 529, 1972.
50. Hepper, C. M., Germination and growth of *Glomus caledonius* spores: the effects of inhibitors and nutrients, *Soil Biol. Biochem.*, 11, 269, 1979.
51. Gildon, A. and Tinker, P. B., A heavy metal-tolerant strain of a mycorrhiza fungus, *Trans. Br. Mycol. Soc.*, 77, 648, 1981.
52. Bowen, G. D., Mycorrhizal roles in tropical plants and ecosystems, in *Tropical Mycorrhiza Research*, Mikola, P., Ed., Oxford University Press, New York, 1980, 165.
53. Tommerup, I. C., Suppression and spore germination of VA mycorrhizal fungi in natural soil and pot cultures, *Proc. 6th N.A.C.O.M. Meeting, Bend, Oregon*, University of Oregon, Corvallis, 1985, 375.
54. Mosse, B., The regular germination of resting spores and some observations on growth requirements of an endogone sp. causing vesicular-arbuscular mycorrhiz, *Trans. Br. Mycol. Soc.*, 42, 273, 1959.
55. Mulder, D., *Soil Disinfestation*, Elsevier, Amsterdam, 1979.
56. Epstein, L. and Lockwood, J. L., Suppression of conidial germination of *Helminthosporium victoriae* in soil and in model fungistatic systems, *Phytopathology*, 74, 90, 1984.
57. Baylis, C. and Kouyeas, V., Contributions of chemical inhibitors to soil mycostasis, in *Soil-Borne Plant Pathogens*, Schippers, B. and Gams, W., Eds., Academic Press, London, 1979, 98.

58. Pavlica, D. A., Hora, T. S., Bradshaw, J. J., Skogerboe, R. K., and Baker, R., Volatiles from soil influencing activities of soil fungi, *Phytopathology*, 68, 758, 1978.
59. Beilby, J. P. and Kidby, D. K., The early synthesis of RNA, protein, and some associated metabolic events in germinating vesicular-arbuscular mycorrhizal fungal spores of *Glomus caledonius, Can. J. Microbiol.*, 28, 623, 1982.
60. Beilby, J. P., Effects of inhibitors on early protein, RNA, and lipid synthesis in germinating vesicular-arbuscular mycorrhizal fungal spores of *Glomus caledonium, Can. J. Microbiol.*, 29, 596, 1983.
61. Kough, J., Malajczuk, N., and Linderman, R. G., Use of the direct immunofluorescent technique to study the vesicular-arbuscular fungus *Glomus epigaeum* and other *Glomus* species, *New Phytol.*, 94, 57, 1983.
62. Cook, R. J. and Baker, K. F., *The Nature and Practice of Biological Control of Plant Pathogens,* American Phytopathological Society, St. Paul, Mich., 1983.
63. Old, K. M. and Wong, J. N. F., Perforation and lysis of fungal spores in natural soils, *Soil Biol. Biochem.*, 8, 285, 1976.
64. Chakraborty, C., Theodorou, C., and Bowen, G. D., Mycophagous amoebae reduction of root colonization by *Rhizopogon,* in *Proc. 6th N.A.C.O.M. Meeting, Bend, Oregon,* University of Oregon, Corvallis, 1985, 275.
65. Ross, J. P. and Ruttencutter, R., Population dynamics of two vesicular-arbuscular mycorrhizal fungi and the role of hyperparasitic fungi, *Phytopathology*, 67, 490, 1977.
66. Tzean, S. S., Chu, C. L., and Su, H. J., Spiroplasmalike organisms in a vesicular-arbuscular mycorrhizal fungus and its mycoparasite, *Phytopathology*, 73, 989, 1983.
67. Hepper, C. M. and Jakobsen, I., Hyphal growth from spores of the mycorrhizal fungus *Glomus caledonius:* effect of amino acids, *Soil Biol. Biochem.*, 15, 55, 1983.
68. Newman, E. I. and Watson, A., Microbial abundance in the rhizosphere. A computer model, *Plant Soil*, 48, 17, 1977.
69. Beilby, J. P. and Kidby, D. K., Biochemistry of ungerminated and germinated spores of the vesicular-arbuscular mycorrhizal fungus, *Glumus caledonius:* changes in neutral and polar lipids, *J. Lipid. Res.*, 21, 739, 1980.
70. Koske, R. E., Evidence for a volatile attractant from plant roots affecting germ tubes of a v.a. mycorrhizal fungus, *Trans. Br. Mycol. Soc.*, 79, 305, 1982.
71. Barley, K. P., The configuration of the root system in relation to nutrient uptake, *Adv. Agron.*, 22, 159, 1970.
72. Bowen, G. D. and Rovira, A. D., Microbial colonization of plant roots, *Annu. Rev. Phytopathol.*, 14, 121, 1976.
73. Baker, K. F. and Cook, R. J., *Biological Control of Plant Pathogens,* W. H. Freeman, San Francisco, 1974.
74. Green, R. J., Jr., Survival and inoculum potential of conidia and microslerotia of *Verticillium alboatrumo* in soil, *Phytopathology*, 59, 874, 1969.
75. Chuang, T. Y. and Ko, W. H., Propagule size: its value in the prediction of inoculum density and infection potential in soil, in *Soil-Borne Plant Pathogens,* Schippers, B. and Gams, W., Eds., Academic Press, London, 1979, 35.
76. Bowen, G. D., The development of vesicular-arbuscular mycorrhizas, in *Current Perspectives in Microbial Ecology,* Klug, M. J. and Reddy, C. A., Eds., American Society of Microbiology, Washington, D.C., 1984, 201.
77. Carling, D. E. and Brown, M. F., Anatomy and physiology of vesicular-arbuscular and non-mycorrhizal roots, *Phytopathology*, 72, 1108, 1982.
78. Scannerini, S. and Bonfante-Fasolo, P., Comparative ultrastructure analysis of mycorrhizl associations, *Can. J. Bot.*, 61, 917, 1983.
79. Gianinazzi-Pearson, V., Host-fungus specificity, recognition and compatibility in mycorrhizae, in *Genes Involved in Microbe Plant Interactions,* Verma, D. P. S. and Hohn, T., Eds., Springer-Verlag, Vienna, 1984, 225.
80. Bowen, G. D., Physiological factors in infection and spread of v.a.m., in *Proc. 6th N.A.C.O.M. Meeting, Bend, Oregon,* University of Oregon, Corvallis, 1985, 181.
81. Heslop-Harrison, J., *Cellular Recognition Systems in Plants. Institute of Biology, Studies in Biology 100,* Edward Arnold, London, 1978, 60.
82. Keen, N. T., Specific recognition in gene-for-gene host-parasite systems, in *Advances in Plant Pathology,* Vol. 1, Ingram, D. S. and Williams, P. H., Eds., Academic Press, London, 1982, 35.
83. Anderson, A. J., The problems of living with a plant root: factors involved in root colonization, in *Proc. 6th N.A.C.O.M. Meeting, Bend, Oregon,* University of Oregon, Corvallis, 1985, 175.
84. Schwab, S. M., Menge, J. A., and Leonard, R. T., Comparison of stages of vesicular-arbuscular mycorrhiza formation in sudangrass grown at two levels of phosphorus nutrition, *Am. J. Bot.*, 70, 1225, 1983.

85. Halverson, L. J. and Stacey, G., Host recognition in the *Rhizobium*-soybean symbiosis. Detection of a protein factor in soybean root exudate which is involved in the nodulation process, *Plant Physiol.*, 74, 84, 1984.
86. Allen, M. F. and Boosalis, M. G., Effects of two species of v.a. mycorrhizal fungi on drought tolerance of winter wheat, *New Phytol.*, 93, 67, 1983.
87. Katou, K., Tomiyama, K., and Okamoto, H., Effects of hyphal wall components of *Phytophthora infestans* on membrane potential of potato tuber cells, *Physiol. Plant Pathol.*, 21, 311, 1982.
88. Mosse, B., Vesicular-arbuscular mycorrhiza research for tropical agriculture in, *Res. Bull. Hawaii Inst. Trop. Agric. Hum. Resour.*, p. 194, 1981.
89. Tommerup, I. C., The persistence of infectivity by germinated spores of vesicular-arbuscular mycorrhizal fungi in soil, *Trans. Br. Mycol. Soc.*, 82, 275, 1984.
90. Sequeira, L., Defenses triggered by the invader: recognition and compatibility phenomena, in *Plant Disease,* Vol. 5, Horsfall, J. G. and Cowling, E. B., Eds., Academic Press, New York, 1980, 179.
91. Nicolson, T. H., Evolution of vesicular-arbuscular mycorrhizas, in *Endomycorrhizas,* Sanders, F. E., Mosse, B., and Tinker, P. B., Eds., Academic Press, London, 1975, 25.
92. Bevege, D. I. and Bowen, G. D., Endogone strain and host plant differences in development of vesicular-arbuscular mycorrhizas, in *Endomycorrhizas,* Sanders, F. E., Mosse, B., and Tinker, P. B., Eds., Academic Press, London, 1975, 77.
93. Ocampo, J. A., Martin, J., and Hayman, D. S., Influence of plant interactions on vesicular-arbuscular mycorrhizal infections. I. Host and non-host plants grown together, *New Phytol.*, 84, 27, 1980.
94. Tommerup, I. C., Development of infection by a vesicular-arbuscular mycorrhizal fungus in *Brassica napus* L. and *Trifolium subterraneum* L., *New Phytol.*, 98, 487, 1984.
95. Hayman, D. S., Johnson, A. M., and Ruddlesdin, I., The influence of phosphate and crop species on *Endogone* spores and vesicular-arbuscular mycorrhiza under field conditions, *Plant Soil*, 43, 489, 1975.
96. Ocampo, J. A., Martin, J., and Hayman, D. S., Influence of plant interactions on vesicular-arbuscular mycorrhizal infections. I. Host and non-host plants grown together, *New Phytol.*, 84, 27, 1980.
97. Ocampo, J. A., Effect of crop rotations involving host and non-host plants on vesicular-arbuscular mycorrhizal infection of host plants, *Plant Soil,* 56, 283, 1980.
98. Schwab, S. M., Johnson, E. L. V., and Menge, J. A., Influence of simazine on formation of vesicular-arbuscular mycorrhizae in *Chenopodium quinona* Willd, *Plant Soil,* 64, 283, 1982.
99. Holley, J. D. and Peterson, R. L., Development of a vesicular-arbuscular mycorrhiza in bean roots, *Can. J. Bot.*, 57, 1960, 1979.
100. Cox, G. and Tinker, P. B., Translocation and transfer of nutrients in vesicular-arbuscular mycorrhizas. I. The arbuscule and phosphorus transfer: a quantitative ultrastructural study, *New Phytol.*, 77, 371, 1976.
101. Kaspari, J., Eledtronenmikroscopische Untersuchung zur Feistruktur der Endotrophen Tabakmykorriza, *Arch. Mikrobiol.*, 92, 201, 1973.
102. Kinden, D. A. and Brown, M. F., Electron microscopy of vesicular-arbuscular mycorrhizae of yellow poplar. III. Host-endophyte interactions during arbuscular development, *Can. J. Microbiol.*, 21, 1930, 1975.
103. Dexheimer, J., Gianinazzi, S., and Gianinazzi-Pearson, V., Ultrastructural cytochemistry of the host-fungus interfaces in the endomycorrhizal association *Glomus mosseae/Allium cepa, Z. Pflanzenphysiol.*, 92, 191, 1979.
104. Bonfante-Fasolo, P. and Grippiolo, R., Ultra-structural and cytochemical changes in the wall of a vesicular-arbuscular mycorrhizal fungus during symbiosis, *Can. J. Bot.*, 60, 2303, 1982.
105. Aist, J. R., Papillae and related wound plugs of plant cells, *Annu. Rev. Phytopathol.*, 14, 145, 1976.
106. Dickinson, C. H. and Lucas, J. A., *Plant Pathology and Plant Pathogens,* 2nd ed., Blackwell Scientific, Oxford, 1982.
107. Bonfante-Fasolo, P., Dexheimer, J., Gianinazzi, V., Gianinazzi-Pearson, V., and Scannerini, S., Cytochemical modifications in the host-fungus interface during intracellular interactions in vesicular-arbuscular mycorrhizae, *Plant Sci. Lett.*, 22, 13, 1981.
108. Sanders, F. E., Martin, J. K., and Bowen, G. D., unpublished data.
109. Mosse, B. and Hepper, C., Vesicular-arbuscular mycorrhizal infections in root organ cultures, *Physiol. Plant Pathol.*, 5, 215, 1975.
110. Buwalda, J. G., Ross, G. J. S., Stribley, D. P., and Tinker, P. B., The development of endomycorrhizal root systems. III. The mathematical representation of the spread of vesicular-arbuscular mycorrhizal infection in root systems, *New Phytol.*, 91, 669, 1982.
111. Buwalda, J. G., Ross, G. J. S., Stribley, D. P., and Tinker, P. B., The development of endomycorrhizal root systems. IV. The mathematical analysis of effects of phosphorus on the spread of vesicular-arbuscular mycorrhizal infection in root systems, *New Phytol.*, 92, 391, 1982.

112. Buwalda, J. G., Stribley, D. P., and Tinker, P. B., The development of endomycorrhizal root systems. V. The detailed pattern of development of infection and the control of infection level by host in young leek plants, *New Phytol.*, 96, 411, 1984.
113. Hepper, C. M., The effect of nitrate and phosphate on the vesicular-arbuscular mycorrhizal infection of lettuce, *New Phytol.*, 92, 389, 1983.
114. Smith, S. E. and Walker, N. A., The spatial distribution of infection in *Trifolium subterraneum*, in *Proc. 6th N.A.C.O.M. Meeting, Bend, Oregon*, University of Oregon, Corvallis, 1985, 143.
115. Wilson, J. M., Comparative development of infection by three vesicular-arbuscular mycorrhizal fungi, *New Phytol.*, 97, 413, 1984.
116. Abbott, L. K. and Robson, A. D., Colonization of the root system of subterranean clover by three species of vesicular-arbuscular mycorrhizal fungi, *New Phytol.*, 96, 275, 1984.
117. Abbott, L. K. and Robson, A. D., Formation of external hyphae in soil by four species of vesicular arbuscular mycorrhizal fungi, *New Phytol.*, 99, 245, 1985.
118. Schenck, N. C. and Smith, G. S., Responses of six species of vesicular-arbuscular mycorrhizal fungi and their effects on soybean at four soil temperatures, *New Phytol.*, 92, 193, 1982.
119. Bowen, G. D., Nutrient status effects loss of amides and amino acids from pine roots, *Plant Soil*, 30, 139, 1969.
120. Schwab, S. M., Menge, J. A., and Leonard, R. T., Quantitative and qualitative effects of phosphorus on extracts and exudates of sudangrass roots in relation to vesicular-arbuscular mycorrhiza formation, *Plant Physiol.*, 73, 761, 1983.
121. Graham, J. H., Leonard, R. T., and Menge, J. A., Membrane mediated decrease in root exudation responsible for phosphorus inhibition of vesicular arbuscular mycorrhiza formation, *Plant Physiol.*, 68, 548, 1981.
122. Johnson, C. R., Menge, J. A., Schwab, S., and Ting, I. P., Interactions of photoperiod and vesicular-arbuscular mycorrhizae on growth and metabolism of sweet orange, *New Phytol.*, 90, 665, 1982.
123. Rovira, A. D., Foster, R. C., and Martin, J. K., Origin, nature and nomenclature of the organic materials in the rhizosphere, in *The Soil-Root Interface*, Harley, J. L. and Russell, R. S., Eds., Academic Press, London, 1979, 1.
124. Abbott, L. K., Robson, A. D., and deBoer, G., The effect of phosphorus on the formation of hyphae in soil by the vesicular-arbuscular mycorrhizal fungus, *Glomus fasiculatum*, *New Phytol.*, 97, 437, 1984.
125. Same, B. I., Robson, A. D., and Abbott, L. K., Phosphorus, soluble carbohydrates and endomycorrhizal infection, *Soil Biol. Biochem.*, 15, 593, 1983.
126. Hepper, C. M., Limited independent growth of a vesicular-arbuscular mycorrhizal fungus in vitro, *New Phytol.*, 93, 537, 1983.
127. Abbott, L. K. and Robson, A. D., Growth of subterranean clover in relation to the formation of endomycorrhizas by introduced and indigenous fungi in a field soil, *New Phytol.*, 81, 575, 1978.
128. Abbott, L. K. and Robson, A. D., The effect of root density, inoculum placement and infectivity of inoculum on the development of vesicular-arbuscular mycorrhizas, *New Phytol.*, 97, 285, 1984.
129. Douds, D. D. and Chaney, W. R., Correlation of fungal morphology and development to host growth in a green ash mycorrhiza, *New Phytol.*, 92, 519, 1982.
130. Cooper, K. M. and Lösel, D., Lipid physiology of vesicular-arbuscular mycorrhiza. I. Composition of lipids in roots of onion, clover and ryegrass infected with *Glomus mosseae*, *New Phytol.*, 80, 143, 1978.
131. Smith, S. E., Mycorrhizas of autotrophic higher plants, *Biol. Rev.*, 55, 475, 1980.
132. Bethlenfalvay, G. J. and Pacovsky, R. S., Light effects in mycorrhizal soybeans, *Plant Physiol.*, 73, 969, 1983.
133. Laws, H., unpublished data.
134. Bethlenfalvay, G. J., Brown, M. S., and Pacovsky, R. S., Relationships between host and endophyte development in mycorrhizal soybeans, *New Phytol.*, 90, 537, 1982.
135. Bethlenfalvay, G. J., Pacovsky, R. S., Brown, M. S., and Fuller, G., Mycotrophic growth and mutualistic development of host plant and fungl endophyte in an endomycorrhizal symbiosis, *Plant Soil*, 68, 43, 1982.
136. Tinker, P. B., Effects of vesicular-arbuscular mycorrhizas on higher plants, *Symp. Soc. Exp. Biol.*, 29, 325, 1975.
137. Hattingh, M. J., Gray, L. E., and Gerdemann, J. W., Uptake and translocation of ^{32}P-labelled phosphate to onion roots by endomycorrhizal fungi, *Soil Sci.*, 116, 383, 1973.
138. Smith, S. E., Inflow of phosphate into mycorrhizal and non-mycorrhizal plants of *Trifolium subterraneum* at different levels of soil phosphate, *New Phytol.*, 90, 293, 1982.
139. Rhodes, L. H. and Gerdemann, J. W., Hyphal translocation and uptake of sulphur by vesicular-arbuscular mycorrhizae of onion, *Soil Biol. Biochem.*, 10, 355, 1978.
140. Tinker, P. B., Soil chemistry of phosphorus and mycorrhizal effects on plant growth, in *Endomycorrhizas*, Sanders, F. E., Mosse, B., and Tinker, P. B., Eds., Academic Press, London, 1975, 353.

141. Sanders, F. E., Tinker, P. B., Black, R. L. B., and Palmerley, S. M., The development of endomycorrhizal root systems. I. Spread of infection and growth promoting effects with four species of vesicular-arbuscular endophyte, *New Phytol.*, 78, 257, 1977.
142. Pacovsky, R. S. and Bethlenfalvay, G. J., Measurement of the extraradical mycelium of a vesicular-arbuscular mycorrhizal fungus in soil by chitin determination, *Plant Soil*, 68, 143, 1982.
143. Sutton, J. C. and Sheppard, B. R., Aggregation of sand-dune soil by endomycorrhizal fungi, *Can. J. Bot.*, 54, 326, 1976.
144. Graham, J. H., Linderman, R. G., and Menge, J. A., Development of external hyphae by different isolates of mycorrhizal *Glomus* spp. in relation to root colonization and growth of Troyer citrange, *New Phytol.*, 91, 183, 1982.
145. Tisdall, J. M. and Oades, J. M., Stabilization of soil aggregates by the root systems of ryegrass, *Aust. J. Soil Res.*, 17, 429, 1979.
146. Kucey, R. M. M. and Paul, E. A., Biomass of mycorrhizal fungi associated with bean roots, *Soil Biol. Biochem.*, 14, 413, 1982.
147. Aldwell, F. E. B., Hall, I. R., and Smith, J. M. B., Enzyme-linked immunosorbent assay (ELISA) to identify endomycorrhizal fungi, *Soil Biol. Biochem.*, 15, 377, 1983.
148. Wilson, J. M., Trinick, M. J., and Parker, C. A., The identification of vesicular-arbuscular mycorrhizal fungi using immunofluorescence, *Soil Biol. Biochem.*, 15, 439, 1983.
149. Hepper, C. M., Techniques for studying the infection of plants by vesicular-arbuscular mycorrhizal fungi under axenic conditions, *New Phytol.*, 88, 641, 1981.
150. Bethlenfalvay, G. J., Pacovsky, R. S., and Brown, M. S., Parasitic and mutualistic associations between a mycorrhizal fungus and soybean: development of the endophyte, *Phytopathology*, 72, 894, 1982.
151. Bowen, G. D., Tree roots and the use of soil nutrients, in *Nutrition of Plantation Forests*, Bowen, G. D. and Nambiar, E. K. S., Eds., Academic Press, London, 1984, 147.
152. Bethlenfalvay, G. J., Bayne, H. G., and Pacovsky, R. S., Parasitic and mutualistic associations between a mycorrhizal fungus and soybean: the effect of phosphorus on host plant-endophyte interactions, *Physiol. Plant.*, 57, 543, 1983.
153. Abbott, L. K. and Robson, A. D., The effect of soil pH on the formation of va mycorrhizas by two species of *Glomus*, *Aust. J. Soil Res.*, 22, 253, 1985.
154. Rovira, A. D., Bowen, G. D., and Foster, R. C., The significance of rhizosphere microflora and mycorrhizas in plant nutrition, in *Inorganic Plant Nutrition. Encyclopedia of Plant Physiology; New Series*, Vol. 15A, Lauchli, A. and Bieleski, R. L., Eds., Springer-Verlag, Heidelberg, 1983, 61.
155. Skinner, M. F. and Bowen, G. D., The penetration of soil by mycelial strands of pine mycorrhizas, *Soil Biol. Biochem.*, 6, 57, 1974.
156. Daniels, B. A. and Menge, J. A., Hyperparasitization of vesicular-arbuscular mycorrhizal fungi, *Phytopathology*, 70, 584, 1980.

Chapter 4

MINERAL NUTRITION

David P. Stribley

TABLE OF CONTENTS

I. INTRODUCTION

This essay does not seek to summarize current knowledge of the role of vesicular-arbuscular (VA) mycorrhiza in mineral nutrition of plants because this has been done in several recent reviews. Rather, its purpose is to interpret this knowledge in an ecological context, drawing attention to recent studies that have particular relevance to ecology and which deserve further investigation.

The central question to be addressed is whether the ecologist needs to include mycorrhizae in his or her studies of the factors that influence plant performance in the field. This chapter will show how deficiencies in technique and problems of extrapolation currently allow us to answer this question only in a very limited way.

II. UPTAKE OF SOIL NUTRIENTS — A SUMMARY

This section describes the most important features of uptake of mineral nutrients by mycorrhizae. For a more comprehensive account the reader should refer to one of the several excellent reviews in recent literature.[1-8]

A. Phosphorus

Numerous pot experiments and some field experiments have demonstrated beyond all reasonable doubt that mycorrhizae are much more efficient at taking up soil P than are uninfected roots. Most observed changes in the physiology of plants as a result of mycorrhizal infection are a consequence of this increased uptake of P. This "null" hypothesis has not yet been seriously challenged, although recent work (to be discussed below) is beginning to show that there are important effects of mycorrhizae not wholly ascribable to enhanced P uptake.

The mechanism of the improved uptake of P is now fairly well agreed. Briefly, it is assumed that mycorrhizal and uninfected plants take up P from the same source, the soil solution. Because the concentration of P in soil solution is extremely low (typically less that 1 μM^9), mass flow is insufficient to adequately supply the plant with P: Itoh and Barber[10] estimated that mass flow contributed only 13% to P uptake by onion, and less than 3% of that of saltwort *(Salsola kali* L.). Diffusion is the rate-limiting step and major pathway for movement of P to the root: it is very slow in soil (diffusion coefficient approximately 10^{-9} cm^2/sec). Hyphae of mycorrhiza extend from the internal infection out into the surrounding soil and, like root hairs, effectively increase the absorbing area of the root. Phosphorus is taken up by this external mycelium and translocated, possibly as polyphosphate,[11,12] to the root. The narrow radius of hyphae implies that gradients in P concentration around them will be very flat. Little is known about competition for P uptake between neighboring hyphae, nor is the distance known to which mycelium extends from the root except in artificial conditions. It is of interest to note that the increase in the radius of the zone of depletion of P surrounding the roots caused by mycorrhizal infection appears to be about the same as that observed for root hairs.[13]

This "central dogma" of how mycorrhizae function has been supported by a series of experiments on ^{32}P-labeled soil which has shown no differences between mycorrhizal and nonmycorrhizal plants in the specific activity of absorbed P, implying uptake of phosphorus from the same soil fraction.[14] However, from an elegant study in which they used iron hydroxide to reduce the chemical availability of P, Bolan et al.[14] argue that experiments with ^{32}P are, in fact, incapable of revealing differences in the pattern of absorption of P from various soil fractions. The question of the phosphorus utilized by mycorrhizal plants, therefore, needs urgent reinvestigation, particularly on soils where much P is in forms other than Pi.[15]

B. Other Nutrients

It has been demonstrated that S[16,17] and N[18] are taken up by and translocated along mycorrhizal hyphae. It seems unlikely that mycorrhizae contribute directly to enhance uptake of these elements which are mainly conducted to plants by mass flow.[19] Increased uptake of these elements and also, in particular, K is probably a consequence of improved supply of P. There is, however, clear evidence that mycorrhizae efficiently take up zinc and copper from soil.[8,20] However, heavy metals can reduce mycorrhizal infection and may thus influence plant growth.[21] The role of mycorrhizae in uptake of heavy metals from contaminated soil is of considerable ecological interest. Bradley et al.[22] have provided clear evidence that ericoid mycorrhizae protect the host plant from the adverse effect of toxic metals such as copper; it has also been observed that ectomycorrhizae can protect the host against zinc.[23] Whether VA mycorrhizae function similarly is presently controversial; Tinker and Gildon[8] found little effect of mycorrhizal infection on heavy metal uptake, whereas Killham and Firestone[24] provided evidence that mycorrhizal plants took up more toxic metals when exposed to acid rain. This topic is worthy of further investigation.

III. SPECIAL ASPECTS OF MINERAL NUTRITION

A. Soil Water Status and Phosphorus Uptake

Infection by VA mycorrhizal fungi has been shown to affect several aspects of the water relations of the host, e.g., root hydraulic conductivity, leaf water potential, and leaf resistance.[25-32] In a critical review, Safir and Nelsen[32] argued that most of these phenomena are probably a result of improved P nutrition. However, this explanation may be incomplete, for Snellgrove et al.[33] reported that mycorrhizal plants of leek *(Allium porrum* L.) had a greater shoot fresh weight/dry weight ratio that did nonmycorrhizal plants that grew at the same rate because of addition of fertilizer P. Of great potential importance is the report by Nelsen and Safir[34] that mycorrhizal onion plants grew much better than did uninfected plants when both were subjected to droughting. These authors suggested that mycorrhizae, but not uninfected roots, can maintain supply of P when its rate of diffusion is markedly restricted by dry soil. Fitter[35] has recently turned this hypothesis into a mathematical model which suggests that in general the importance of mycorrhizae in P nutrition of plants will markedly increase as soil water content is reduced. These studies suggest that mycorrhizae may be critically important in establishment and survival of plants in the field under conditions of restricted water supply. Although it has been demonstrated that mycorrhizae improve establishment of transplanted horticultural crops,[36] their role in survival of seedlings in the field has not been tested and clearly merits investigation.

B. Rhizosphere pH

Buwalda[37] showed that VA mycorrhizal infection decreased the cation/anion ratio in shoot tissue of cereals and that this effect was not a simple consequence of improved P nutrition. He suggested that the relative increase in anion uptake resulted from a change in the internal mechanism of pH regulation of the plant caused by a restricted supply of organic acids, and that a result of this should be an increase in rhizosphere pH. Smith and co-workers[5,38] have also postulated that mycorrhizae could be involved in changes in soil pH, and an increase in pH in the rhizosphere of ectomycorrhizal Douglas fir has been demonstrated.[39] Changes in rhizosphere pH could have important implications for uptake of P, but the size of the change will critically depend on the buffer power of the soil.[19] There are interesting reports[40,41] that both ectomycorrhizae and endomycorrhizae allow plants to grow on calcareous soils that would not support uninfected plants, and it is possible that changes in rhizosphere pH may be involved.

It is of great importance to investigate the nutritional implications of altered rhizosphere pH, and whether VA mycorrhizae are widely involved in modifying the ecological amplitude of plant species.

IV. THE IMPORTANCE OF ROOT COLONIZATION

It is almost self-evident that if the physiological importance of mycorrhizae in ecology is to be assessed adequately, then reliable and detailed information is needed on the extent and variation of natural infections. But as St. John and Coleman[42] have pointed out, such data are "embarrassingly scarce." The few detailed studies that have been made have shown rather constant levels of infection, for example, in apple orchards on various soils in the U.S.[43,44] and in soils with a wide range of pH (in agricultural[45-48] and natural[49] systems) or P. One gains the impression that natural infections do not vary widely, but the data are too scanty to allow general conclusions to be drawn.

Pot experiments have clearly demonstrated that the rate of colonization of a developing root system is a very important determinant of the response extent of the plant to infection.[50] This will be particularly important for annual plants which may complete most of their life cycle before maximum levels of infection are attained. The colonization rate of a plant will be determined by the nature and number of propagules in the soil and the inherent characteristics of the fungal species involved. A clear distinction must be drawn between highly disturbed environments and those that are stable. In the former (exemplified by agricultural systems where annual crops are grown) inoculum will be scattered and mixed with soil, perhaps resulting in low inoculum densities and thus, slow colonization of plants. By contrast, seedlings in undisturbed vegetation may become rapidly infected when their roots contact existing mycorrhizae in the surface layers of soil. Read and Birch[114] have shown that seedlings transplanted into undisturbed vegetation become infected in 2 to 5 days.

There is little information on numbers of propagules in agricultural systems. It has been measured by the most-probable-number (MPN) technique,[51,52] in which samples of soil are sequentially diluted with a sterilized medium until only a fraction of the seedlings planted on them become infected. This fraction is then used to calculate the number of propagules. Agricultural soils in the U.K. that have been assessed in this way show a MPN of between 0.1 to 10 propagules per gram.[53] It is not known what determines this range of inoculum concentration, nor how stable these levels are. In several experiments on agricultural crops, increases in yield following artificial inoculation with VA mycorrhizal fungi have been observed,[37,54-63] indicating that natural levels of inoculum are often suboptimal. It must be noted that most reports are confounded by possible differences in efficacy between inoculant and natural fungi.

The MPN technique is clearly only applicable to highly disturbed soils, and even here it is possible that propagules are not evenly distributed, but are clumped: present techniques are inadequate to test this. In undisturbed systems, the inoculum level needs to be assessed either directly in the field, or perhaps by the use of soil cores[64,65] which cause minimum disturbance to the structure of the profile.

Another important variable in the field will be the numbers of species of indigenous mycorrhizal fungi present. Current evidence suggests that it is normal for several species to be present. Thus, Walker et al.[66] found up to 12 species in relatively undisturbed sites in central Iowa, and Wang et al.[47,48] described 8 species in agricultural soils of pH 5.5 upwards. It seems reasonable to assume that species revealed by the presence of spores are also present in plant roots, but the endophytes present in a root system cannot be accurately determined with present techniques. The only one which can be identified with certainty is "fine endophyte" *(Glomus tenuis)* which has internal hyphae of typically less than 3 μm diameter compared to the diameter of 5 μm for most other species of VA mycorrhizal fungi ("coarse endophytes"). Abbott and coworkers[67,68] claim to be able to identify species by their internal structures, but Miller et al.[43,44] found this was very difficult, especially at high levels of soil P. Immunofluorescence has been used to identify infecting fungi, but its specificity is low.[69,70] The lack of precise techniques for identification is a serious impediment to the study of mycorrhizae in ecology.

Even if the relative abundance of the individual species in the root cortex is known, there remains the question of possible inter- and intraspecies variation in physiological characteristics of the fungi. In detailed studies of a limited number of species in pot experiments, it has been shown that isolates may vary widely both in the rate at which they colonize the root system and in the amount of external mycelium they produce.[50,71] Unfortunately, most investigations on efficacy of isolates have been confounded with variations in inoculum density.[43,54,68,72-75] The only published investigation of fungal efficacy at known levels of inoculum is that of Haas and Krikun,[76] who showed that individual isolates of *G. macrocarpum* varied considerably in their effect on the host, but this variation was not clearly related to rate of colonization.

There is an almost complete lack of knowledge of the factors which determine the species constitution at a given site. Indeed, as Burnett[77] pointed out in regard to speciation in fungi, "virtually nothing is known about natural selection nor, indeed, about selection of any kind in fungi." The most systematic investigation of environment effects upon natural populations of VA mycorrhizal fungi is that of Wang,[47] who studied the effects of soil pH on sites that were initially homogeneous, but were subsequently maintained at four different levels of pH for more than 20 years by differential liming. On the most acid plots (pH 4.2), only *G. tenuis* was present in roots. At other levels of pH there were up to eight species in the soil, but the proportion of fine-to-coarse endophytes in the root cortex declined with increasing pH; at pH 7.8 fine endophyte was absent. There was little difference in tolerance to pH between isolates of *G. caledonicum* from sites of low and high pH, respectively, and the implication was that selection for tolerance of pH occurred at the level of the species. A sudden artificial change in the pH of a soil rapidly selected the species most tolerant to the new pH. This was true even for the severely acid plots where inoculum accidentally introduced by cultivation practices rapidly infected roots when the soil pH was increased by liming. Whether introduction of new genetic material from outside occurs in natural habitats is uncertain. Chlamydospores of VA mycorrhizal fungi are large and usually subterranean; they appear to be agents for survival rather than dissemination. Although it is known that inoculum may be transferred by earthworms, birds, and small mammals,[78] quantitative information is lacking.

A very pertinent question is which particular attributes of a given fungus are selected by a given soil property?[79] Clearly, as in the experiments of Wang, the absolute ability to infect a root in a particular soil is of major importance. Beyond this little is known, but it is tempting to speculate that the ability to infect and colonize a root system quickly must be important. The question of natural selection of VA mycorrhizal fungi is a fruitful one for further study.

V. PROBLEMS IN INTERPRETING CURRENT KNOWLEDGE

It is salutary to recognize that most of the corpus of knowledge on the effects of mycorrhizae on the mineral nutrition of plants comes from pot experiments where plants have been grown alone on sterilized or partly sterilized soil. Field experiments have been few and the results too variable[35] to allow general conclusions to be drawn. Yet extrapolation of results of pot experiments to an ecological context continues unabated. The dangers of this are worth emphasizing. Sterilized soil may differ greatly in chemistry to untreated soil. Nitrification is inhibited and mineralization of N is stimulated,[81-84] so that the supply of ammonium nitrogen to the plant is greatly increased, and P availability may be enhanced.[85a] A less well-appreciated effect is that formation of root hairs is often greatly stimulated.[83,84] These structures can, like mycorrhizae, make roots much more efficient at taking up P from the soil,[10,85] and may confound experiments designed to investigate effects of mycorrhizae.

Another potentially serious effect of sterilization is that it will eliminate fungivorous animals. Warnock et al.[86] have demonstrated that Collembola (Insecta) will eat external mycelium and thereby markedly reduce the efficacy of mycorrhizae. The importance of these fungivores in the field has not been investigated fully, but it is interesting that Finlay[87] observed an increase in the relative growth rate of plants per unit P content following application of an insecticide to field soil, suggesting that soil insects normally depress mycorrhizal activity.

Plants rarely grow alone in nature, yet many pot experiments have used one plant per pot. Root density may influence mycorrhizal activity because external mycelium in a dense root system may only be able to explore soil already depleted of P by neighboring roots. Evidence to support this is provided by the experiments of Bååth and Hayman,[88] where the mycorrhizal response of onion was greatly reduced at high plant density. Effects of root density may account for the apparent differences in response to mycorrhizal infection between different plant species. Thus, Buwalda et al.[56] found that wheat and barley species with dense root systems were much less responsive to mycorrhizal infection than they were to applied P. However, a sound theoretical basis for comparing the mycorrhizal responsiveness of various species has not been designed. Pairunan et al.[89] have pointed out that valid comparisons can only be made by the use of parameters that describe the whole P response curve; nowhere does this appear to have been done.

A major deficiency of many pot experiments is that they are done at soil temperatures which are quite unrepresentative of the field. For example, for about half the growth period of winter wheat in northern temperate latitudes, soil temperatures are in the range 0 to 10°C, yet soil temperatures in most pot experiments are at least twice this.

Finally, in most laboratory studies plants are well-watered and, as described above, the importance of mycorrhizae to the P nutrition of the host may increase rapidly with increasing water deficiency.

Apart from these specific problems associated with pot experimentation there are more general dangers in extrapolation. First, the greatly enhanced efficiency of P uptake by mycorrhizae has provoked many to draw the conclusion that *ipso facto* mycorrhizae must be important in the ecology of plants, particularly on P-deficient soils. It should however be noted that plants on nutrient-deficient soils generally show a strategy of conservation, that is, they have low rates of growth and low concentrations of nutrients in their tissues. There seems to be a general agreement that efficiency of uptake of soil nutrients is relatively unimportant in such situations.[90,91]

A second general error is to draw ecological inferences from species grown alone in pots. The folly of this was exposed in a cogent essay by Pigott,[92] who illustrated very clearly by several examples that "a change favourable to a plant in isolation may be disadvantageous or even disastrous in vegetation." One must conclude that experiments in comparative physiology of mycorrhizal plants cannot immediately be translated into an ecological context.

VI. MYCORRHIZAE AND THE PLANT COMMUNITY

The subjects covered in this section are those in the author's view that have particular relevance to plant ecology and warrant further investigation.

A. Transfer of Nutrients between Mycorrhizal Plants

It is well established that mycorrhizal infection can pass from root to root,[93] but this mode of spread appears to be very much slower than that along a continuous length of cortex.[94] Interplant infection implies the existence of hyphae that connect individual roots, and these have been observed.[95-97] By the use of radioactive tracers it has been demonstrated that carbon[98] and particularly phosphorus[97,99,100] can move from plant to plant along these hyphal bridges. Such movement is interesting because it implies a reversal in the normal direction of phosphorus movement (i.e., out of the donor plant) and it could have important consequences for the plant community since a young seedling rapidly infected from an existing root network would receive an immediate supply of phosphorus and, thus, an improved chance of establishment.[97] However, this topic is highly contentious. Ritz and Newman[100] have shown that ^{32}P could move in both directions between two plants, thus, movement observed in one direction only does not necessarily imply *net* movement. Further, it is important to determine whether significant quantities of phosphorus are moved. The experiments of Whittingham and Read[97] suggest that they are, since a growth response was observed in a recipient plant connected to a donor by VA mycorrhizal hyphae, when the latter plant was given additional P.

B. Mineral Nutrition and Plant Interaction

Competition for nutrients between roots is an important aspect of plant interaction in the field. It is now clear that mycorrhizal infection can modify root competition.[101-103] Thus, Fitter[80] showed that Yorkshire Fog *(Holcus lanatus* L.) suppressed ryegrass *(Lolium perenne* L.) to a much larger extent when both were mycorrhizal than when they were uninfected. However, such experiments are difficult to interpret fully because the mechanisms of root competition are currently little understood.[19] It seems likely that mathematical models of competition such as that of Baldwin[104] could usefully be extended to take into account the influence of mycorrhizae.

C. Genotypic Variation in Hosts

Studies on wheat[105,106] have revealed large differences between individual genotypes in responsiveness to mycorrhizal infection, but the basis of these differences is not known. It is unfortunate that such studies have not been pursued further, because it is not generally appreciated that there may be considerable range of genotypes of a given species in a circumscribed area of natural vegetation. Burdon[107] found that in a small population of clover *(Trifolium repens* L.) there were as many as 50 distinct clones, and electrophoretic techniques subsequently revealed up to 45 clones per square meter.[115] The degree of variability for a range of plant characters was similar to that found in comparisons between populations from distinctly different environments.[108] There was, for example, a significant variation in resistance of individual members to two common foliar fungal pathogens. In an extension of this work, Freeman[115] studied VA mycorrhizal development in ten genotypes of clover. Variations in infection between genotypes were not significant, but there was a strong effect of microtopography. When each genotype was transplanted into soil derived from around the roots of each of the other nine, there was a significant but complex interaction between genotype and source of inoculum. Unexpectedly, the greatest infection in a genotype did not necessarily occur on soil from its parental site. This preliminary work suggests that studies on mycorrhizae/genotype interactions should be pursued further.

VII. CONCLUSIONS

An extensive understanding of the effects of mycorrhizae on plant nutrition has been gained through numerous pot experiments on sterilized or partly sterilized media, but it is dangerous to draw ecological inferences from such work. The general plea by Pigott[92] for more experiments in the field and less work on comparative physiology of plants in isolation applies with particular force to the study of mycorrhizae in plant ecology. It is, however, undeniable that to infer the role of natural mycorrhizae through appropriate experimentation is extraordinarily difficult. Soil fumigation with methyl bromide was used by Yost and Fox[109] in an ingenious series of experiments in which it was observed that growth of certain species at low concentrations of soil P was depressed by fumigation, whereas Chinese cabbage, immune to infection, was always stimulated. The conclusion was that fumigation suppressed growth of mycotrophic species by eliminating indigenous mycorrhizal fungi, but that there was concomitant increase in soil N. In several other experiments a depression of growth following fumigation in the field has been noted and attributed to elimination of natural mycorrhizae.[63,70,110-113] However, Buwalda et al.[56] have pointed out that this technique has severe limitations for species such as wheat which do not appear to be highly dependent on mycorrhizae. The manifold effects of fumigation on soil chemistry render it difficult to interpret the results. This problem was noted also by Sparling and Tinker[84] in their work on upland grasses. The observations of severely diminished plant uptake of P following fumigation constitute strong evidence that natural mycorrhizae are important organs for nutrient absorption in the field, but such information is far from unequivocal.

The general conclusion must be that although one may speculate freely on the basis of a large number of experiments under controlled conditions, there is a dearth of information from the field. Difficult though fieldwork may be, it is a task which must not be evaded if the function of mycorrhizae in vegetation is to be understood. It is the opinion of this author that our present knowledge of the role of mycorrhizae in the nutrition of plants in their natural environment unfortunately exemplifies the dictum of Mark Twain: "such a wholesale return of conjecture for a trifling investment of fact."

REFERENCES

1. Cooper, K. M., Physiology of VA mycorrhizal associations, in *VA Mycorrhiza,* Powell, C. Ll. and Bagyaraj, D. J., Eds., CRC Press, Boca Raton, Fla., 1984, 155.
2. Gianinazzi-Pearson, V. and Gianinazzi, S., The physiology of vesicular-arbuscular mycorrhizal roots, *Plant Soil,* 71, 197, 1983.
3. Hayman, D. S., Influence of soils and fertility on activity and survival of vesicular-arbuscular mycorrhizal fungi, *Phytopathology,* 72, 1119, 1982.
4. Hayman, D. S., The physiology of vesicular-arbuscular endomycorrhizal symbiosis, *Can. J. Bot.,* 61, 944, 1983.
5. Smith, S. E., Mycorrhizas of autotrophic higher plants, *Biol. Rev.,* 55, 475, 1980.
6. Tinker, P. B., Effects of vesicular-arbuscular mycorrhizas on plant nutrition and plant growth, *Physiol. Veg.,* 16, 743, 1978.
7. Tinker, P. B., The role of rhizosphere organisms in phosphorus uptake by plants, in *The Role of Phosphorus in Agriculture,* Kwasaneh, F. and Sample, E., Eds., American Society of Agronomy, Madison, Wisc., 1980, 617.

8. Tinker, P. B. and Gildon, A., Mycorrhizal fungi and ion uptake, in *Metals and Micronutrients,* Robb, D. A. and Pierpoint, W. S., Eds., Academic Press, London, 1983, 21.
9. Reisenauer, H. M., Mineral nutrients in soil solution, in *Environmental Biology,* Altman, P. L. and Dittmer, D. S., Eds., Federation of the American Society for Experimental Biology, Bethesda, Md., 1966, 507.
10. Itoh, S. and Barber, S. A., Phosphorus uptake by six plant species as related to root hairs, *Agron. J.,* 75, 457, 1983.
11. Capaccio, L. C. M. and Callow, J. A., The enzymes of polyphosphate metabolism in vesicular-arbuscular mycorrhizas, *New Phytol.,* 91, 81, 1982.
12. Sanders, F. E. and Tinker, P. B., Phosphate flow into mycorrhizal roots, *Pestic. Sci.,* 4, 385, 1973.
13. Owusu-Bennoah, E. and Wild, A., Autoradiography of the depletion zone of phosphate around onion roots in the presence of vesicular-arbuscular mycorrhiza, *New Phytol.,* 82, 133, 1979.
14. Bolan, N. S., Robson, A. D., Barrow, N. J., and Aylmore, L. A. G., Specific activity of phosphorus in mycorrhizal and non-mycorrhizal plants in relation to the availability of phosphorus to plants, *Soil Biol. Biochem.,* 16, 299, 1984.
15. Tate, K. R., The biological transformation of P in soil, in *Biological Processes and Soil Fertility,* Tinsley, J. and Darbyshire, J. F., Eds., Martinus Nijhoff, The Hague, 1984, 245.
16. Cooper, K. M. and Tinker, P. B., Translocation and transfer of nutrients in vesicular-arbuscular mycorrhizas. II. Uptake and translocation of phosphorus, zinc and sulphur, *New Phytol.,* 81, 43, 1978.
17. Rhodes, L. H. and Gerdemann, J. W., Hyphal translocation and uptake of sulphur by vesicular-arbuscular mycorrhizae of onion, *Soil Biol. Biochem.,* 70, 355, 1978.
18. Ames, R. N., Reid, C. P. P., Porter, L. K., and Cambardella, C., Hyphal uptake and transport of nitrogen from two ^{15}N-labelled sources by *Glomus mosseae,* a vesicular-arbuscular mycorrhizal fungus, *New Phytol.,* 95, 381, 1983.
19. Nye, P. H. and Tinker, P. B., *Solute Movement in the Soil-Root System,* Blackwell Scientific, Oxford, 1977.
20. Gildon, A. and Tinker, P. B., Interactions of vesicular-arbuscular mycorrhizal infections and heavy metals in plants. II. The effects of infection on uptake of copper, *New Phytol.,* 95, 263, 1983.
21. Gildon, A. and Tinker, P. B., Interactions of vesicular-arbuscular mycorrhizal infection and heavy metals in plants. I. The effects of heavy metals on the development of vesicular-arbuscular mycorrhizas, *New Phytol.,* 95, 247, 1983.
22. Bradley, R., Burt, A. J., and Read, D. J., Mycorrhizal infection and resistance to heavy metal toxicity in *Calluna vulgaris, Nature (London),* 292, 335, 1981.
23. Brown, M. T. and Wilkins, D. A., Zinc tolerance of mycorrhizal *Betula, New Phytol.,* 99, 101, 1985.
24. Killham, K. and Firestone, M. K., Vesicular-arbuscular mycorrhizal mediation of grass response to acidic and heavy metal deposition, *Plant Soil,* 72, 39, 1983.
25. Allen, M. F., Influence of vesicular-arbuscular mycorrhizae on water movement through *Bouteloua gracilis* (H.B.K.) Lag ex Steud, *New Phytol.,* 91, 191, 1982.
26. Allen, M. F. and Boosalis, M. G., Effects of two species of VA mycorrhizal fungi on drought tolerance of winter wheat, *New Phytol.,* 93, 67, 1983.
27. Hardie, K. and Leyton, L., The influence of VA mycorrhiza on growth and water relations of red clover. I. In phosphate deficient soil, *New Phytol.,* 89, 599, 1981.
28. Levy, Y. and Krikun, J., Effect of vesicular-arbuscular mycorrhizae on *Citrus jambhiri* water relations, *New Phytol.,* 85, 25, 1980.
29. Levy, Y., Syvertsen, J. P., and Nemec, S., Effect of drought stress and vesicular-arbuscular mycorrhiza on citrus transpiration and hydraulic conductivity of roots, *New Phytol.,* 93, 61, 1983.
30. Nelsen, C. E. and Safir, G. R., The water relations of well-watered, mycorrhizal, and non-mycorrhizal onion plants, *J. Am. Soc. Hortic. Sci.,* 107, 271, 1982.
31. Safir, G., Boyer, J. S., and Gerdemann, J. W., Nutrient status and mycorrhizal enhancement of water transport in soybean, *Plant Physiol.,* 49, 700, 1972.
32. Safir, G. R. and Nelsen, C. E., Water and nutrient uptake by vesicular-arbuscular mycorrhizal plants, in Role of Mycorrhizal Associations in Crop Production, Proc. of a Colloq. at Rutgers University, New Brunswick, N.J., 1980.
33. Snellgrove, R. C., Splittstoesser, W. E., Stribley, D. P., and Tinker, P. B., The distribution of carbon and the demand of the fungal symbiont in leek plants with vesicular-arbuscular mycorrhiza, *New Phytol.,* 92, 75, 1982.
34. Nelson, C. E. and Safir, G. R., Increased drought tolerance of mycorrhizal onion plants caused by improved phosphorus nutrition, *Planta,* 154, 407, 1982.
35. Fitter, A. H., Functioning of vesicular-arbuscular mycorrhizas under field conditions, *New Phytol.,* 99, 257, 1985.
36. Biermann, B. J. and Linderman, R. G., Increased geranium growth using pretransplant inoculation with a mycorrhizal fungus, *J. Am. Soc. Hortic. Sci.,* 109, 972, 1983.

37. Buwalda, J. G., The Development of Vesicular-Arbuscular Mycorrhizal Root Systems and Their Role in the Growth of Cereals, Ph.D. thesis, University of London, London, 1983.
38. Smith, S. E. and Smith, F. A., Could mycorrhizas be involved in changes in soil pH?, in *Proc. of the 7th Australian Legume Nodulation Conf. (AIBS Occasional Publ. No. 12)*, Kennedy, I. R. and Copeland, L., Eds., Australian Institute of Biological Science, Sydney, 1984, 125.
39. Bledsoe, C. S. and Zasoski, R. J., Effects of ammonium and nitrate on growth and nitrogen uptake by mycorrhizal Douglas-fir seedlings, *Plant Soil*, 71, 445, 1983.
40. Clement, A., Garbaye, J., and LeTacon, F., Importance des ectomycorhizes dans la resistance au calcaire du Pin noir *(Pinus nigra* Arn. ssp. *nigricans* Host), *Ecol. Plant*, 12, 111, 1977.
41. Lapeyrie, F. F. and Chilvers, G. A., An endomycorrhiza-ectomycorrhiza succession associated with enhanced growth of *Eucalyptus dumosa* seedlings planted in a calcareous soil, *New Phytol.*, 100, 93, 1985.
42. St. John, T. V. and Coleman, D. C., The role of mycorrhizae in plant ecology, *Can. J. Bot.*, 61, 1005, 1983.
43. Miller, D. D., Domoto, P. A., and Walker, C., Colonization and efficacy of different endomycorrhizal fungi with apple seedlings at two phosphorus levels, *New Phytol.*, 100, 393, 1985.
44. Miller, D. D., Domoto, P. A., and Walker, C., Mycorrhizal association and eighteen apple rootstock plantings in the United States, *New Phytol.*, 100, 379, 1985.
45. Black, R. and Tinker, P. B., The development of endomycorrhizal root systems. II. Effect of agronomic factors and soil conditions on the development of vesicular-arbuscular mycorrhizal infection in barley and on endophyte spore density, *New Phytol.*, 83, 401, 1979.
46. Strzemska, J., Mycorrhiza in farm crops grown in monoculture, in *Endomycorrhizas*, Sanders, F. E., Mosse, B., and Tinker, P. B., Eds., Academic Press, London, 1975, 527.
47. Wang, G. M., Soil pH and Vesicular-Arbuscular Mycorrhiza, Ph.D. thesis, University of London, London, 1984.
48. Wang, G. M., Stribley, D. P., Tinker, P. B., and Walker, C., Soil pH and vesicular-arbuscular mycorrhizas, in *Ecological Interactions in Soil: Plants, Microbes and Animals*, Fitter, A. H., Ed., British Ecological Society Special Publ. 4, Blackwell Scientific, Oxford, 1985.
49. Read, D. J., Koucheki, H. K., and Hodgson, J., Vesicular-arbuscular mycorrhiza in natural vegetation systems. I. The occurrence of infection, *New Phytol.*, 77, 641, 1976.
50. Sanders, F. E., Tinker, P. B., Black, R. L. B., and Palmerley, S. M., The development of endomycorrhizal root systems. I. Spread of infection and growth-promoting effects with four species of vesicular-arbuscular endophyte, *New Phytol.*, 78, 257, 1977.
51. Porter, W. M., The 'Most Probable Number' method for enumerating infective propagules in vesicular-arbuscular mycorrhizal fungi in soil, *Aust. J. Soil Res.*, 17, 515, 1979.
52. Wilson, J. M. and Trinick, M. J., Factors affecting the estimation of numbers of infective propagules of vesicular-arbuscular fungi by the most probable number method, *Aust. J. Soil Res.*, 21, 73, 1982.
53. Sanders, F. E. and Sheikh, N. A., The development of vesicular-arbuscular mycorrhizal infection in plant root systems, *Plant Soil*, 71, 223, 1983.
54. Abbott, L. K. and Robson, A. D., The role of vesicular-arbuscular mycorrhizal fungi in agriculture and the selection of fungi for inoculation, *Aust. J. Agric. Res.*, 33, 389, 1982.
55. Abbott, L. K., Robson, A. D., and Hall, I. R., Introduction of vesicular-arbuscular mycorrhizal fungi into agricultural soils, *Aust. J. Agric. Res.*, 34, 741, 1983.
56. Buwalda, J. G., Stribley, D. P., and Tinker, P. B., The vesicular-arbuscular mycorrhizal association in cereals grown in the field. I. Effects of vesicular-arbuscular mycorrhizal infection in first, second and third cereal crops, *J. Agric. Sci. Camb.*, 105, 631, 1985.
57. Hall, I. R., Growth of *Lotus pedunculatus* Cav. in an eroded soil containing soil pellets infested with endomycorrhizal fungi, *N.Z. J. Agric. Res.*, 23, 103, 1980.
58. Hall, I. R., Field trials assessing the effect of inoculating agricultural soils with endomycorrhizal fungi, *J. Agric. Sci. Camb.*, 102, 725, 1984.
59. Hayman, D. S. and Mosse, B., Improved growth of white clover in hill grasslands by mycorrhizal inoculation, *Ann. Appl. Biol.*, 93, 141, 1979.
60. Krikun, J., Spiegel, Y., Haas, J., Dodd, J., and Livescu, L., The influence of vesicular-arbuscular mycorrhiza on phosphorus nutrition of pepper seedlings in a loessial soil, *Hassadeh*, 62, 989, 1982.
61. Plenchette, C., Furlan, V., and Fortin, J. A., Growth stimulation of apple trees in unsterilised soil under field conditions with VA mycorrhizal inoculation, *Can. J. Bot.*, 59, 2003, 1981.
62. Plenchette, C., Fortin, J. A., and Furlan, V., Growth responses of several plant species to mycorrhizae in a soil of moderate P-fertility. I. Mycorrhizal dependency under field conditions, *Plant Soil*, 70, 199, 1983.
63. Powell, C. Ll., Inoculation of barley with efficient mycorrhizal fungi stimulates seed yield, *Plant Soil*, 59, 487, 1981.
64. Hall, I. R., Effect of inoculant endomycorrhizal fungi on white clover growth in soil cores, *J. Agric. Sci. Camb.*, 102, 719, 1984.

65. Powell, C. Ll. and Daniel, J., Growth of white clover in undisturbed soils after inoculation with efficient mycorrhizal fungi, *N.Z. J. Agric. Res.*, 21, 675, 1978.
66. Walker, C., Mize, C. W., and McNabb, H. S., Jr., Populations of endogonaceous fungi at two locations in central Iowa, *Can. J. Bot.*, 60, 2518, 1982.
67. Abbott, L. K., Comparative anatomy of vesicular-arbuscular mycorrhizas formed on subterranean clover, *Aust. J. Bot.*, 30, 485, 1982.
68. Abbott, L. K. and Robson, A. D., Infectivity and effectiveness of five endomycorrhizal fungi: competition with indigenous fungi in field soils, *Aust. J. Agric. Res.*, 32, 621, 1981.
69. Aldwell, F. E. B., Hall, I. R., and Smith, J. M. B., Enzyme linked immunosorbent assay (ELISA) to identify endomycorrhizal fungi, *Soil Biol. Biochem.*, 15, 377, 1983.
70. Wilson, J. M., Trinick, M. J., and Parker, C. A., The identification of vesicular-arbuscular mycorrhizas using immunofluorescence, *Soil Biol. Biochem.*, 15, 439, 1983.
71. Graham, J. H., Linderman, R. G., and Menge, J. A., Development of external hyphae by different isolates of mycorrhizal *Glomus* sp. in relation to root colonization and growth of Troyer citrange, *New Phytol.*, 91, 183, 1982.
72. Abbott, L. K. and Robson, A. D., Infectivity and effectiveness of five endomycorrhizal fungi: effect of inoculum type, *Aust. J. Agric. Res.*, 32, 631, 1981.
73. Dodd, J., Krikun, J., and Haas, J., Relative effectiveness of indigenous populations of vesicular-arbuscular mycorrhizal fungi from four sites in the Negev, *Isr. J. Bot.*, 32, 10, 1983.
74. Plenchette, C., Furlan, V., and Fortin, J. A., Effects of different endomycorrhizal fungi on five host plants grown on calcined montmorillonite clay, *J. Am. Soc. Hortic. Sci.*, 107, 535, 1982.
75. Powell, C. Ll., Selection of efficient VA mycorrhizal fungi, *Plant Soil*, 68, 3, 1982.
76. Haas, J. H. and Krikun, J., Efficacy of endomycorrhizal-fungus isolates and inoculum quantities required for growth response, *New Phytol.*, 100, 613, 1985.
77. Burnett, J. H., Presidential address-speciation in fungi, *Trans. Br. Mycol. Soc.*, 81, 1, 1983.
78. McIlveen, W. D. and Cole, H., Jr., Spore dispersal of Endogonaceae by worms, ants, wasps, and birds, *Can. J. Bot.*, 54, 1486, 1976.
79. Lambert, D. H., Cole, H., Jr., and Baker, D. E., Adaption of vesicular-arbuscular mycorrhizae to edaphic factors, *New Phytol.*, 85, 513, 1980.
80. Fitter, A. H., Influence of mycorrhizal infection on competition for phosphorus and potassium by two grasses, *New Phytol.*, 79, 119, 1977.
81. Cawse, P. A., Microbiology and biochemistry of irradiated soils, in *Soil Biochemistry*, Vol. 3, Paul, E. A. and McLaren, A. D., Eds., Marcel Dekker, New York, 1975, 213.
82. Jakobsen, I. and Andersen, A. J., Vesicular-arbuscular mycorrhiza and growth in barley: effects of irradiation and heating of soil, *Soil Biol. Biochem.*, 14, 171, 1982.
83. Rovira, A. D., Studies on soil fumigation. I. Effects on ammonium, nitrate and phosphate in soil and on the growth, nutrition and yield of wheat, *Soil Biol. Biochem.*, 8, 241, 1976.
84. Sparling, G. P. and Tinker, P. B., Mycorrhizal infection in Pennine grassland. II. Effects of mycorrhizal infection on the growth of some upland grasses on Y-irradiated soils, *J. Appl. Ecol.*, 15, 951, 1978.
85. Baylis, G. T. S., Root hairs and phycomycetous mycorrhizas in phosphorus-deficient soil, *Plant Soil*, 33, 713, 1970.
85a. Larsen, S., The effect of steam treatment of soil on phosphate availability, *Acta Agric. Scand.*, 16, 208, 1966.
86. Warnock, A. J., Fitter, A. H., and Usher, M. B., The influence of a springtail *Folsomia candida* (Insecta, Collembola) on the mycorrhizal association of leek *Allium porrum* and the vesicular-arbuscular mycorrhizal endophyte *Glomus fasciculatus*, *New Phytol.*, 90, 285, 1982.
87. Finlay, R. D., Interaction between soil micro-arthopods and endomycorrhizal associations of higher plants, in *Ecological Interactions in Soil: Plants, Microbes and Animals*, Fitter, A. H., Ed., British Ecological Society Special Publ. 4, Blackwell Scientific, Oxford, 1985, 318.
88. Bååth, E. and Hayman, D. S., Effect of soil volume and plant density on mycorrhizal infection and growth response, *Plant Soil*, 77, 373, 1984.
89. Pairunan, A. K., Robson, A. D., and Abbott, L. K., The effectiveness of vesicular-arbuscular mycorrhizas in increasing growth and phosphorus uptake of subterranean clover from phosphorus sources of different solubilities, *New Phytol.*, 84, 327, 1980.
90. Chapin, F. S., The mineral nutrition of wild plants, *Annu. Rev. Ecol. Syst.*, 11, 233, 1980.
91. Grime, J. P., Competition and diversity in herbaceous vegetation — a reply, *Nature (London)*, 244, 310, 1973.
92. Pigott, C. D., The experimental study of vegetation, *New Phytol.*, 90, 389, 1982.
93. Warner, A. and Mosse, B., Factors affecting the spread of vesicular mycorrhizal fungi in soil. I. Root density, *New Phytol.*, 90, 529, 1982.
94. Mosse, B., Stribley, D. P., and LeTacon, F., The ecology of mycorrhizae and mycorrhizal fungi, *Adv. Microb. Ecol.*, 5, 137, 1981.

95. Chiarello, N., Hickman, J. C., and Mooney, H. A., Endomycorrhizal role for interspecific transfer of phosphorus in a community of annual plants, *Science,* 217, 941, 1982.
96. Heap, A. J. and Newman, E. I., Links between roots by hyphae of vesicular-arbuscular mycorrhizas, *New Phytol.,* 85, 169, 1980.
97. Whittingham, J. and Read, D. J., Vesicular-arbuscular mycorrhiza in natural vegetation systems. III. Nutrient transfer between plants with mycorrhizal interconnections, *New Phytol.,* 90, 277, 1982.
98. Hirrel, M. C. and Gerdemann, J. W., Carbon transfer between onions infected with a vesicular-arbuscular mycorrhizal fungus, *New Phytol.,* 83, 731, 1979.
99. Heap, A. J. and Newman, E. I., The influence of vesicular-arbuscular mycorrhizas on phosphorus transfer between plants, *New Phytol.,* 85, 173, 1980.
100. Ritz, K. and Newman, E. I., Movement of ^{32}P between intact grassland plants of the same age, *Oikos,* 44, 138, 1984.
101. Newman, E. I., Interactions between plants, in *Physiological Plant Ecology III,* Encyclopedia of Plant Physiology, New Series 12C, Lange, O. L., Nobel, P. S., Osmond, C. B., and Ziegler, H., Eds., Springer-Verlag, Berlin, 1983, 679.
102. Buwalda, J. G., Growth of a clover-ryegrass association with vesicular-arbuscular mycorrhizas, *N.Z. J. Agric. Res.,* 23, 379, 1980.
103. Hall, I. R., Effects of endomycorrhizas on the competitive ability of white clover, *N.Z. J. Agric. Res.,* 21, 509, 1978.
104. Baldwin, J. P., Competition for plant nutrients in soil; a theoretical approach, *J. Agric. Sci. Camb.,* 87, 341, 1976.
105. Azcon, R. and Ocampo, J. A., Factors affecting the vesicular-arbuscular infection and mycorrhizal dependency of thirteen wheat cultivars, *New Phytol.,* 87, 677, 1981.
106. Bertheau, Y., Gianinazzi-Pearson, V., and Gianinazzi, S., Developpement et expression de l'association endomycorrhizienne chez le ble. I. Mise en evidence d'un effet varietal, *Ann. Amelior. Plantes,* 30, 77, 1980.
107. Burdon, J. J., Intra-specific diversity in a natural population of *Trifolium repens, J. Ecol.,* 68, 717, 1980.
108. Burdon, J. J., Variation in disease-resistance within a population of *Trifolium repens, J. Ecol.,* 68, 737, 1980.
109. Yost, R. S. and Fox, R. L., Contribution of mycorrhizae to P nutrition of crops growing on an oxisol, *Agron. J.,* 71, 903, 1979.
110. Kleinschmidt, G. D. and Gerdemann, J. W., Stunting of citrus seedlings in fumigated nursery soils related to the absence of mycorrhizae, *Phytopathology,* 62, 1447, 1972.
111. Martin, J. P., Baines, R. C., and Page, A. L., Observations on the occasional temporary growth inhibition of citrus seedlings following heat or fumigation treatment of the soil, *Soil Sci.,* 95, 175, 1963.
112. Plenchette, C., Fortin, J. A., and Furlan, V., Growth responses of several plant species to mycorrhizae in a soil of moderate P-fertility. II. Soil fumigation induced stunting of plants corrected by reintroduction of the wild endomycorrhizal flora, *Plant Soil,* 70, 211, 1983.
113. Timmer, L. W. and Leyden, R. F., Stunting of citrus seedlings in Texas and its correction by phosphorus fertilization and inoculation with mycorrhizal fungi, *J. Am. Soc. Hortic. Sci.,* 103, 533, 1978.
114. Read, D. J. and Birch, C. P. D., personal communication.
115. Freeman, A., personal communication.

Chapter 5

THE WATER RELATIONS OF VESICULAR-ARBUSCULAR MYCORRHIZAL SYSTEMS

Charles E. Nelsen

TABLE OF CONTENTS

I. INTRODUCTION

A. Mycorrhizal Interaction

The vesicular-arbuscular (VA) mycorrhizal system is a symbiotic interaction between one or more species of coenocytic fungi and the roots of higher plants. In the late 1950s and early 1960s interest in this interaction began to grow, and early work concentrated on describing the visual interaction and growth stimulation of the plant when the plant roots were infected by the VA mycorrhizal fungus.[1-3] Improved mineral nutrition (especially phosphorus) of the host plant was established as the reason for the host plant growth stimulation, and studies dealing with mineral and carbohydrate exchange dominated the mycorrhizal literature in the 1960s and 1970s.[4,5] The importance and universality of the mycorrhizal interaction became increasingly apparent due to this research, and, more recently, additional characteristics of the mycorrhizal system have begun to be studied. Changes in host plant hormone content[6] and mycorrhizal effects on host plant photosynthesis[7,8] have been reported. The water relations of the VA mycorrhizal system have also recently begun to be studied more intensely, and the advances in this field are the subject of this review. I hope to incorporate much of the information reported to date and cast a critical view on the interpretation of the reported data. Finally, I hope to offer enough speculation to stimulate the reader to ask his own critical questions, with the result that new ideas and reports will add to our current, limited knowledge on the water relations of the mycorrhizal system.

B. Plant Water Relations

An argument can be presented that water is the single most important factor limiting plant growth and crop yield.[9-11] Because of the importance of water to plant growth, there have been literally thousands of reports published dealing with water relations, water use, and aspects of drought resistance of higher plants. A subset of this literature deals with the interaction between plant nutrition and plant water use or drought resistance. Lahiri has reported that under conditions of cyclic drought stress, higher fertilization rates improve crop yield.[12] Begg and Turner discussed the reduction of nitrogen and phosphorus uptake by plants during water stress and suggested that nutrient deficiency may be partly involved in reduced plant growth during water stress.[10] This nutrient deficiency could be due to either lowered mineral availability as the soil moisture content declines, or to lowered mineral uptake rates by the plant as roots are exposed to soil water deficits; in relation to phosphorus, both situations have been reported.[13,14] Because of this interaction between water stress and phosphorus (as well as other mineral nutrients), it is not surprising that the mycorrhizal system influences and is influenced by water availability and drought stress. However, it is only recently that a sufficient number of studies dealing with the interaction between water use and VA mycorrhizae have been published to warrant a review of the literature. Finally, as will become evident, mycorrhizal researchers have only scratched the surface in the investigation of this interaction. Many different abiotic and biotic factors influence the water relations of VA mycorrhizal systems; in most cases even cause and effect remain nebulous.

C. Plant Water Relations — Description of Terms

The field of plant water relations is a complicated one, and for a comprehensive review the reader is directed to one or more of the referenced books and reviews.[9-11,15-18] However, for the benefit of the layman to plant water relations, a few simplified definitions will be presented here.

1. Water Potential

Plant water potential (Ψ_w) is a relative measure of the water status of a plant or plant tissue and is usually assigned units of pressure. Values for plant water potential are determined relative to the water potential of pure water at a standard temperature and pressure, which is arbitrarily assigned the value of 0 (zero). Because plant tissue within the normal range of atmospheric temperatures and pressures has a lower water potential than pure water, plant water potential values are generally reported as negative numbers.

In its simplest conception, plant water potential can be partitioned into three major components:

$$\Psi_w = \Psi_\pi + \Psi_m + \Psi_p \quad (1)$$

where Ψ_π is the the solute potential, Ψ_m is the matric potential, and Ψ_p is the pressure potential or turgor pressure. Solute potential is a function of the concentration of all water-soluble components in plant tissue and has a negative value. Matric potential is a function of water adsorption to nonsoluble components and hydration of macromolecules in plant tissue. Matric potential has a negative value and is often considered negligible within the physiological range of plant water status. Pressure potential is related to the turgor of each plant cell, is the attribute that gives rigidity to nonlignified plant tissues, and is generally given a positive value.

Mesophytic plants or plant tissues with ample available water growing in a relatively mild environment will often have water potential values of about −0.1 to −0.5 MegaPascals (MPa). Water potentials of mildly water-stressed tissue will often be in the range of −0.5 to −1.5 MPa, and more severely stressed tissue which will often sustain visibly assessable damage even upon rewatering will be in the range of −1.5 to −3.0 MPa. Water potential values below −3.0 or −4.0 MPa will often result in the death of tissue in mesophytes. Conversion of various water potential units used in the literature is easy because 1 MPa = 10 bars and 1 bar ≅ 1 atm.

2. Transpiration

Transpiration is the emission of water vapor from inside the plant structure (generally the leaf) to the atmosphere. All living plants transpire some water during their life cycle. Generally, transpiration is reported as a specific rate (mass or volume of water per unit time per unit leaf area). Transpiration rates are expressed using a number of units such as g H_2O dm^{-2} leaf area h^{-1}, $\mu g\ cm^{-2}\ s^{-1}$, and $\mu mol\ cm^{-2}\ m^{-1}$. Conversion for comparison, while necessary, is not obvious.

3. Water Use Efficiency

Water Use Efficiency (WUE) is the ratio of the weight or mass of dry plant matter produced, divided by the weight, mass, or volume of water lost to the atmosphere by transpiration through the plant and evaporation from the soil surface. WUE is expressed in many ways, and caution must be used in comparing values quoted by different authors. An older but similar term is transpiration ratio. Transpiration ratio (mass of water transpired, divided by the mass of dry matter accumulated) is approximately equal to the reciprocal of WUE, except WUE includes water evaporated from the soil surface as well as that transpired by the plant. WUE is most useful when discussing plants growing under ample moisture conditions. When moisture becomes limiting, water use and dry matter production are both influenced. Many authors discuss the more general term of drought resistance.

4. Drought Resistance

Drought resistance is a generic term that most authors find necessary to define for a given set of conditions. Levitt[19] may offer one of the best descriptions of the various aspects of drought resistance, but again caution must be used when comparing levels of drought resistance described by different authors. In general, a specific plant is said to be more drought resistant when it performs better than another plant in some specifically defined way when exposed to a *similar level* of water stress.

5. Hydraulic Conductivity

In plant water relations, hydraulic conductivity is a description of the relative ease or difficulty with which liquid water moves through the plant. Conductivity values are generally calculated after determining a pressure difference in and volume flow of water through a plant or plant tissue. Conductivity values are expressed in terms of volume water flow per unit cross-sectional area per unit pressure per unit time or linear flow per unit pressure per unit time ($m\ell\ cm^{-2}\ MPa^{-1}\ s^{-1}$ or $cm\ MPa^{-1}\ s^{-1}$). Liquid water flows "more easily" through a plant with a high level of hydraulic conductivity than through one with a low level.

6. Leaf Resistance

Leaf resistance (often equated with stomatal resistance) is a measure of the resistance to water vapor transfer from inside plant tissue to the atmosphere. This resistance is due to the presence of the surface layers of the plant, including the epidermis, cuticle, and stomata. There is generally a large gradient of water vapor concentration from the inside of the leaf to the atmosphere. Loss of water vapor concentration from the plant tissue is partially controlled by this leaf resistance. Leaf resistance values are generally expressed in seconds per centimeter or seconds per meter. A plant with a high level of leaf resistance will transpire less water per unit leaf area than a plant with a low level of leaf resistance.

II. EFFECTS OF MYCORRHIZAL INFECTION ON HOST PLANT WATER USE

A. Changes in Plant Water Relations

The investigation of the water relations of the mycorrhizal association is a new field of research. Most of the literature dates from 1980 and virtually all have been published since 1970. The first review dealing with water stress and mycorrhizae was published in 1979 and dealt mainly with ecto-mycorrhizae.[20] The author was quite speculative on the subject of VA mycorrhizae because of the lack of information available at that time. More recently, Safir and Nelsen published a short review on VA mycorrhizae and plant water use, but it dealt primarily with unpublished data.[21]

In 1971, Safir et al.[22] were the first to describe changes in plant water relations due to mycorrhizal inoculation. They inoculated pots of autoclaved soil with a mixture of soil, mycorrhizal maize roots, and spores of the mycorrhizal fungus *Glomus mosseae,* and planted soybean in these pots. Nonmycorrhizal controls were soybeans planted in autoclaved soil containing autoclaved inoculum plus a sieved water filtrate of unsterilized inoculum. This final filtrate addition ensured that other microorganisms which might affect plant growth were present in the control pots. They then allowed soybean plants of various ages to wilt by withholding water and monitored water uptake after the addition of water to the dry soil. Twenty-day-old mycorrhizal plants had no measurable growth stimulation over controls and there were no differences in water uptake. However, between day 30 and day 40, both mycorrhizal growth stimulation and differences in rates of water uptake (i.e., recovery from mild water stress) were evident. Thirty-day-old mycorrhizal soybeans had hydraulic conductivity values that were approximately 70% higher than nonmycorrhizal control plants ($1.4 \times 10^{-5}\ cm\ s^{-1}\ MPa^{-1}$

vs. 0.8×10^{-5} cm s^{-1} MPa^{-1}). The change in hydraulic conductivity was determined using two different methods and was reproducible. They presented four hypotheses as to how the presence of the mycorrhizae could alter the hydraulic conductivities. First, the external hyphae might increase the total surface area of the root system much as would an increased number of root hairs. Second, the hyphae which penetrate the root cortex to the endodermis could provide a low-resistance pathway for water movement across the root. Third, the hyphae could enhance nutrient uptake, which in turn could decrease the resistance to water transport within the roots. Finally, mycorrhizal infection might increase root growth so that there is a larger root system. They were able to discount only one of the four hypotheses. The root systems of the two types of soybeans were not significantly different in size, so it was unlikely that the changes in hydraulic conductivities were due to a stimulation of the root system.

In a later report, these same authors elaborated on their results using the soybean/*G. mosseae* system.[23] By measuring hydraulic conductivities of whole plants and detached plant tops (stem plus leaves), they were able to demonstrate that most or all of the differences in hydraulic conductivity were associated with the plant roots. Stem plus leaf conductivities were high and similar for both mycorrhizal and nonmycorrhizal soybean plants. Root conductivities were lower and the differences reported for the whole plant[22] were reflected by similar differences in the roots.

In order to elucidate the mechanism of the alteration in hydraulic conductivity, Safir et al.[23] treated a number of mycorrhizal and nonmycorrhizal plants with additional fertilizer in order to improve the nutritional status of the plants. The addition of 400 mℓ of a complete nutrient fertilizer increased the hydraulic conductivity of the nonmycorrhizal plants so that it was not different from that of the mycorrhizal plants. Additional mineral nutrition had no effect on the mycorrhizal soybeans. These results support the hypothesis that the mycorrhizal fungus improved the nutritional status of the host, which in turn increased the hydraulic conductivity of the roots.

To investigate the possible direct involvement of the fungus on water transport, a third experiment was described.[23] The fungicide Pentachloronitrobenzene (PCNB) was added to the soil of mycorrhizal and nonmycorrhizal soybeans for 48 hr prior to the determination of hydraulic conductivities. It had previously been determined that PCNB reduced mycorrhizal enhancement of nutrient uptake,[24] and Safir and co-workers suggested that if the fungus was actively assisting in water uptake, treatment with PCNB would reduce fungal activity and reduce hydraulic conductivity levels of the mycorrhizal plants to the level of the nonmycorrhizal plants. Treatment with PCNB had no effect on water transport of either mycorrhizal or nonmycorrhizal plants, so this argued against the fungus acting as an active, low-resistance pathway for water movement within the roots.

Levy and Krikun reported a number of changes in the water relations of VA mycorrhizal citrus trees.[7] They determined leaf water potentials, leaf resistances, leaf proline levels, and photosynthesis rates, and calculated transpiration rates before, during, and after a short period (3 days) of water stress on 8-month-old seedlings. Water stress was imposed on the plants by withholding water and stress symptoms developed after the first day. Leaf water potentials of mycorrhizal and nonmycorrhizal plants of similar size were virtually the same throughout the 3 days of stress development and 4 days of stress relief (by rewatering). Stomatal resistance values were lower and calculated transpiration rates were slightly higher in the mycorrhizal plants during stress development, but no significant differences were reported for the experiment. During the 4-day recovery period, leaf resistance values appeared lower and photosynthesis rates and calculated transpiration rates appeared higher for the mycorrhizal plants, but few conclusions can be drawn since it was not reported that the experiment was either replicated or repeated.

Table 1
REPORTED VALUES FOR TRANSPIRATION RATES OF VARIOUS PLANT SPECIES

Plant type	Typical transpiration rate[a] (g $H_2O \times dm^{-2} \times hr^{-1}$)	Ref.
Citrus	0.07—0.14	7
Citrus	1.08—1.80	25
Citrus	0.36—0.72	26
Onion	0.50—1.00	27
Various	0.72—3.60	28
Citrus	14.40—21.60	29

[a] Reported transpiration rates were converted to the same units to simplify comparison. Methodology used to determine transpiration rates was not the same in all reports.

Levy and Krikun speculated that in their system, alterations in root/shoot hormone balances due to the presence of mycorrhizal infection may be influencing the citrus water relations, rather than a change in root hydraulic conductivity as reported by Safir et al.[23] An alternative hypothesis is that the hydraulic conductivity analysis from their study was done on mycorrhizal and nonmycorrhizal plants of similar size, with the similarity in size being obtained by growing the plants under a high soil nutritional regime (daily waterings with a 0.1% solution of 20:20:20,N:P:K commercial fertilizer). In essence, their experiment was similar to the high fertilization treatment of Safir et al.[23] Transpiration rates were quite low (Table 1) relative to other reports and it would be informative to repeat this work under conditions which would result in more typical transpirational responses.

In a thorough and comprehensive report, Hardie and Leyton further addressed the effects of VA mycorrhizae on the water relations of the host plant.[30] They investigated the changes in growth, transpiration rates, leaf resistances, and rates of recovery from water stress for mycorrhizal and nonmycorrhizal clover plants grown in low phosphorus soils containing three levels of phosphorus fertilizer. They also went a step further and attempted to define the mechanism by which the changes occurred. They measured tissue phosphorus contents of roots and shoots as a measure of plant nutrition and determined a number of root characteristics including fresh and dry weight, root length, fresh weight per unit root length, water flow rate per plant, and water flow rate per unit root length. Based on their results, they drew a series of conclusions:

1. The hydraulic conductivities of clover roots were much higher in mycorrhizal than nonmycorrhizal clover plants. They attributed part of the higher conductivity to the greater length and diameter of mycorrhizal roots. Because the hydraulic conductivities of mycorrhizal roots were two to three times higher than nonmycorrhizal roots based on a per unit length of root, they also suggested that these increases might be attributed to translocation of water in or on hyphae extending out into the soil.
2. When soil water contents were adequate, transpiration rates of mycorrhizal plants were much higher than uninfected plants, due to higher root conductivities, larger leaf surface areas, and lower leaf diffusion resistances.
3. When water became limiting, leaf resistances of mycorrhizal plants increased to values approximately equal to nonmycorrhizal plants, and transpiration rates on

a unit leaf area basis were below that of nonmycorrhizal plants. However, because of the much greater total leaf area of the mycorrhizal plants, total water throughput was still greater for the mycorrhizal plants on a per plant basis. This resulted in the mycorrhizal plants wilting more rapidly.

4. Mycorrhizal plants extracted soil moisture content down to lower water potentials (≥1MPa), possibly due to the higher root conductivities coupled with lower leaf water potentials (not measured).
5. Stressed mycorrhizal plants recovered more rapidly than did nonmycorrhizal plants when soil was rewatered, again probably due to the higher levels of root conductivity. This is similar to earlier results reported by Safir et al.[22]

This paper is important because of the relatively large numbers of plant parameters monitored during the course of the experiment (Table 2). The information presented allows the investigation of a number of potential mechanisms by which mycorrhizal infection alters plant water relations. Hardie and Leyton demonstrated that the root conductivities were higher based on the whole root system and on a per unit length of the root system. They suggested that the two- to three-times increase in root conductivity on a per unit root length basis could not be accounted for by the reported increase in root diameter or the per unit root length increase in root surface area, and that the fungal hyphae extending out into the soil might account for the increased ability in water uptake. While this hypothesis is consistent with the data, the idea of hyphal translocation of water has been suggested as minimal by others,[31,32] and a second hypothesis might also be presented. Improvement in plant nutrition is also likely to improve water conductivity, especially if plant nutrition (phosphorus, in this case) is limiting and at deficient levels in the nonmycorrhizal plants. Phosphorus nutrition is higher in the mycorrhizal plants at all three soil levels tested in this study (Table 2), and it is suggested that the tissue phosphorus levels of the nonmycorrhizal plants (range = 0.07 to 0.13%, dry weight) were at or below critical levels (reported to be 0.10 to 0.13%) that will cause stunting and have other physiological effects.[33] The data presented by Hardie and Leyton (summarized in Table 2) demonstrated that for the nonmycorrhizal plants, increasing soil phosphorus levels resulted in an increase in root hydraulic conductivity as well as a number of other parameters. Mycorrhizal fungi are well known to improve plant phosphorus nutrition and appeared to be more efficient at increasing tissue phosphorus levels in this study than did added phosphorus fertilization (Table 2). This improved phosphorus nutrition due to the mycorrhizal infection could also be hypothesized to have a direct effect on membrane resistance to water flow, an area where resistance to water flow is by far the highest.[34]

Mycorrhizal infection resulted in higher transpirational flux on both a whole plant basis (5 to 20 times greater) and on a unit leaf area basis (approximately 10 to 30% higher) when adequate soil moisture was present (Table 2). The mycorrhizal plants were considerably larger in all three soil phosphorus treatments, which explained the greater water flux per plant; the greater transpiration rate per unit leaf area was not due to different plant sizes. Phosphorus deficiency of the nonmycorrhizal plants at all three soil phosphorus levels could have a direct effect on water uptake (i.e., lower root hydraulic conductivity) or on stomatal control. Alternatively, poor nutritional status could result in a decreased capability to maintain a favorable water status,[27] with the result that stomates would be partially closed and transpiration rates would be reduced. While leaf water potentials were not determined, leaf diffusion resistance was lower in mycorrhizal plants, supporting the hypothesis that the nonmycorrhizal plants were suffering from mild water stress, even under conditions of adequate moisture.

When water was limiting, the mycorrhizal plants wilted first, probably due to their larger size and more rapid depletion of soil water. After wilting, mycorrhizal plants

Table 2
SUMMARY OF PLANT GROWTH, PHOSPHORUS CONTENT, AND WATER RELATIONS VALUES REPORTED BY HARDIE AND LEYTON[30]

Plant parameter[a]	Soil P level[b]	Nonmycorrhizal	Mycorrhizal
Plant dry wt	Low	84	714
(mg)	Med	245	1090
	High	479	925
% P; root, shoot	Low	0.13, 0.08	0.21, 0.15
(% dry wt)	Med	0.12, 0.07	0.22, 0.15
	High	0.12, 0.09	0.23, 0.19
Root fresh wt	Low	0.17	0.37
(g)	Med	0.28	0.49
	High	0.40	0.53
Root fresh wt	Low	6.5	9.1
per unit length	Med	7.2	9.4
($g\ cm^{-1} \times 10^{-4}$)	High	7.8	9.1
Water flow rate	Low	21.6	93.8
per plant	Med	50.0	136.0
($cm^3\ s^{-1} \times 10^{-5}$)	High	57.3	155.3
Water flow rate per	Low	8.2	24.7
unit root length	Med	12.0	25.0
($cm^3\ s^{-1}\ cm^{-1} \times 10^{-8}$)	High	10.6	23.6
Transpiration per plant[c]	Low	0.6	11.9
($g\ day^{-1}$)	Med	1.3	15.9
	High	3.2	14.3
Transpiration rate[c]	Low	90	101
($mg\ cm^{-2}\ day^{-1}$)	Med	75	100
	High	75	100
Leaf diffusion resistance[c]	Low[d]	13.0	2.9
($s\ cm^{-1}$)	Med[d]	4.7	3.2
	High[d]	3.7	2.4
Transpiration per plant[e]	Low	0.2	4.3
($g\ day^{-1}$)	Med	0.9	6.4
	High	1.1	6.2
Transpiration rate[e]	Low[d]	42	23
($mg\ cm^{-2}\ day^{-1}$)	Med	45	27
	High	37	24
Leaf diffusion resistance[e]	Low	14.0	17.0
($s\ cm^{-1}$)	Med	14.0	16.0
	High	17.0	14.0

[a] Each value is the mean of 12 replicate samples.
[b] Soil phosphorus (P) base level was 5 mg kg^{-1} bicarbonate extractable P; low = 10 mg kg^{-1}, med = 25 mg kg^{-1}, high = 40 mg kg^{-1} added P as calcium phosphate.
[c] With adequate soil moisture available.
[d] Each value is the mean of six replicate samples.
[e] Water stressed, 24 hr before visible wilting.

had a lower transpiration rate on a per unit leaf area basis, but not on a whole plant basis due to their larger total size. Final values of leaf diffusion resistances appeared similar (after wilting) for all treatments.

Final soil water potential values were lower for the mycorrhizal plants (range = -1.0 to -1.2 MPa). The smaller roots of nonmycorrhizal plants may not have fully explored the soil volume, thus leaving pockets of moisture and resulting in a higher, bulk soil water potential. Visual inspection confirmed that the soil volume was less filled by the roots of the nonmycorrhizal plants. Additionally, the mycorrhizal plants may better have been able to develop or survive lower leaf water potentials, thus drying out the soil more efficiently. Leaf turgor can be maintained during stress by increasing osmotic concentrations in the vacuole and cytoplasm, but no evidence for or against this possibility was presented.

Finally, upon rewatering, mycorrhizal plants recovered more quickly than did nonmycorrhizal plants, as determined by time to return to a condition of low leaf diffusion resistance. This suggests that mycorrhizal plants can recover more quickly from periods of water stress, and the more rapid recovery is probably due to the lower root resistances of the mycorrhizal plants. As such, mycorrhizae probably are advantageous to plants which are subjected to *brief* periods of water stress. However, it must be pointed out that during extended periods of water stress, mycorrhizal plants could suffer more due to their larger size and more rapid depletion of soil water. (This will be addressed more fully in Section II.B.) The authors speculated that fungal alteration of root/shoot hormone levels may have some effect on one or more of the described responses. This possibility has been previously suggested.[6] While possible, the "cause and effect" relationships of plant hormones in vivo are difficult to establish, and the alternate hypothesis of phosphorus nutrition changes the authors presented seems to have been much more firmly established. Adequate mineral nutrition is likely to affect a number of plant parameters (including water relations and hormone levels) and the probable phosphorus deficiencies in the nonmycorrhizal plants appear to have been established (Table 2).

Working with the prairie grass *Bouteloua gracilis,* Allen et al.[8] reported a number of differences in the water relations of 5-month-old greenhouse-grown mycorrhizal and nonmycorrhizal plants. In their study, the mycorrhizal plants had higher transpiration rates, lower leaf resistances, and higher hydraulic conductivities over a wide range of soil moistures. These differences occurred during soil drying and did not differ significantly between mycorrhizal and nonmycorrhizal treatments. Photosynthesis also was greater for mycorrhizal plants as determined by an increase in CO_2 uptake in the light. This difference in CO_2 uptake was only partially explained by a difference in leaf resistance to gas diffusion, and they attributed two thirds of the increase to a reduction in the liquid-phase resistance of the mesophyll cells. None of the plants were fertilized in their study, and tissue phosphorus levels were higher in mycorrhizal plants in both leaves and roots (Table 3). The increased levels of phosphorus nutrition might explain the reported differences in water relations and photosynthesis, except that all levels were above typical phosphorus concentrations assumed to be adequated for growth (Table 3 and Epstein[33]). Nonetheless, the improved phosphorus nutrition of the mycorrhizal plants was established and it is known that improved plant nutrition will improve water use and drought resistance.[12]

Resistances to water fluxes (hydraulic conductivities) presented by Allen et al.[8] were calculated using the Ohm's law analogy for water transport under steady-state condition.[36] In order to use this equation one must know both soil and leaf water potentials, and soil water potential must be *less* negative than leaf water potential. Their Figure 1A shows that as the soil dried, leaf water potentials became less negative than soil water potentials. This would lead to negative values for hydraulic conductivity if the Ohm's law analogy was used implicitly, and probably indicates steady-state conditions were not achieved in their study. This may help explain the apparent increase in hydraulic conductivity as the soil dried, an event different from that reported by others.[37,38] The periods when soil water potentials were less negative than leaf water potentials would also indicate that "transpiration" rates shown in Figure 1C are more likely water loss due to leaf drying unless soil and/or leaf water potential values are in error.

Table 3
PHOSPHORUS CONCENTRATIONS OF MYCORRHIZAL AND NONMYCORRHIZAL *BOUTELOUA GRACILIS* COMPARED WITH OTHER PLANT SPECIES AS REPORTED IN THE LITERATURE

Treatment	Phosphorus as phosphate (mmol kg^{-1} fresh wt)	Phosphorus[a] (% dry wt)	Plant type	Ref.
Mycorrhizal (leaf)	49.8	1.03	*Bouteloua gracilis*	8
Mycorrhizal (root)	11.5	0.24	*B. gracilis*	8
Nonmycorrhizal (leaf)	29.4	0.55	*B. gracilis*	8
Nonmycorrhizal (root)	7.8	0.16	*B. gracilis*	8
Phosphorus deficient	—	0.1	Onion	35
Not phosphorus deficient	—	0.3—0.6	Onion	35
Phosphorus deficient	—	<0.10	Various	33
Typical adequate phosphorus levels	—	0.13—0.60	Various	33

[a] Conversion made from phosphorus as phosphate (mmol kg^{-1} fresh weight) assuming phosphate is measured as the $H_2PO_4^+$ ion and dry weight is 15% of fresh weight.

If so, the leaf drying would also indicate that steady-state conditions had not been attained.

Finally, it was reported that " . . . even though midday transpiration rates were higher in mycorrhizal than in nonmycorrhizal plants, the rate of soil water depletion was not significantly different. This suggests that mycorrhizal plants are more efficient in obtaining the available moisture . . . " Since they reported that their plants were not different in size, the implications of this statement are not possible. If the soil moisture depletion rates are similar, the water loss rates must be similar *over time.* If the integrated transpiration rates are different, the soil moisture depletion rates must be different, especially in pots where total available soil water is finite. If mycorrhizal plants are transpiring water faster from a limited soil moisture pool, they must deplete that pool faster.[29] Their data, however, still suggest that the mycorrhizal relationship is a useful adaptation in the dry habitats where *B. gracilis* is found.

In a later report using the same system,[39] Allen again demonstrated that mycorrhizal *B. gracilis* had higher transpiration rates, higher hydraulic conductivities, and lower leaf resistances than nonmycorrhizal plants at similar leaf water potentials. He attempted to define the mechanism by which mycorrhizal infection increased water uptake by intensely studying the roots of both mycorrhizal and nonmycorrhizal plants. Individual and total root hair lengths were found to be higher in mycorrhizal plants, and levels of infection in the mycorrhizal plants were carefully documented.

Based on differences in water uptake between mycorrhizal and nonmycorrhizal plants and on the number of mycorrhizal entry points per unit root length of the mycorrhizal roots, Allen calculated a fungal water uptake and transport to the root which would account for the difference in water uptake (100 nℓ hr^{-1} per hyphal entry point). This is a value similar to that reported for evapotranspiration in the coenocytic fungus *Phycomyces blakesleeanus* (131 nℓ hr^{-1} per sporangiophore).[40] Based on this calculation, Allen suggested that the increased water uptake is directly attributable to fungal uptake and transport (i.e., a predominantly direct flow mechanism) which has been suggested earlier.[30]

While this theory is possible, some assumptions were made that may not be supported and other evidence suggests an alternative mechanism. First, Allen made the assumption that the rate of water uptake is similar between mycorrhizal and nonmycorrhizal plants per *unit root* surface area. There is no valid reason to make this as-

Table 4
CHANGES IN PLANT WATER RELATIONS OF WELL-WATERED, MYCORRHIZAL, OR PHOSPHORUS FERTILIZED ONION PLANTS RELATIVE TO NONMYCORRHIZAL, NONFERTILIZED PLANTS[27]

Treatment	Leaf water[a] potential (MPa)	Hydraulic conductivity (cm MPa^{-1} s^{-1} × 10^{-6})	Leaf resistance (s cm^{-1})	Transpiration rate (g dm^{-2} hr^{-1})
Nonmycorrhizal nonfertilized	6.5 z	1.9 z	5.0 z	0.48 z
Mycorrhizal nonfertilized	3.4 y	7.9 y	1.2 y	1.00 y
Nonmycorrhizal plus fertilizer	3.2 y	7.8 y	1.3 y	0.90 y

[a] Each value is the mean of four replications. These are the results from one of four experiments with similar results. Values in each column followed by different letters are significantly different at the 5% level by the Duncan's Multiple Range test.

sumption, particularly if differences in nutrition (i.e., phosphorus) might have occurred. It is likely that the mycorrhizal plants had increased phosphorus nutrition (a typical mycorrhizal response), especially since the plants used were more than 2 months of age and were never fertilized. Improved plant nutrition is likely to alter plant water use.[12,13] Tissue phosphorus levels were not measured in this report. Secondly, the water uptake rates reported for *P. blakesleeanus*[40] were under vapor-driving gradients (atmospheric water activities, W_a) of between 0.98 and 0.70 W_a, which are equivalent to water potential values of −1.0 and −15.7 MPa. The driving gradient from the soil to the plant root cortex (mycorrhizal hyphae do not penetrate through the endodermis) is likely to be less than −1.0 MPa and *far* less than −7.0 to −15.7 MPa, the driving gradients for which 131 nℓ hr^{-1} water uptake was reported by Cowan et al.[40] Finally, Sanders and Tinker made similar calculations to try to attribute extra water uptake by mycorrhizal plants to the hyphae. Their calculated value (2.9 × 10^{-2} nℓ s^{-1} per hyphal entry point) was equal to 104 nℓ hr^{-1}, a value they suggested was "ridiculously high".[31]

An alternative hypothesis, implicating improved phosphorus nutrition, could also explain changes in plant water use, photosynthetic rates, and in vivo plant hormone levels. The transport of phosphorus to the plant by external mycorrhizal hyphae and improved nutrition of mycorrhizal plants have been well established and seem the simplest explanation of the reported phenomena.

Nelsen and Safir[27] investigated the water relations of mycorrhizal onion plants under conditions of an ample soil moisture supply. Their work included the use of two sets of nonmycorrhizal controls, one set grown at a soil phosphorus level which was low and equal to that used for the mycorrhizal plants (10 ppm P), and one set grown at a high soil phosphorus level (base soil plus 50 ppm P) to stimulate nonmycorrhizal plants to grow to a similar size as mycorrhizal plants. The use of both types of controls allowed them to investigate the water relations of the mycorrhizal plants compared to nonmycorrhizal plants of similar size and similar soil phosphorus conditions. With one exception,[23] the use of the double set of controls had not previously been used in water relations investigations relating to mycorrhizae.[7,8,22,30,39] Mycorrhizal infection resulted in more favorable water relations when compared to nonmycorrhizal plants grown under low soil phosphorus conditions (Table 4). Added soil phosphorus (50 ppm) improved all the water relations parameters such that they were not different from the mycorrhizal plants (Table 4), strongly supporting the theory that changes in plant nutrition are implicated in the changes that occur in plant water relations when plants are infected with a mycorrhizal fungus. The reported results were confirmed in four separate experiments (three at 60% relative humidity and one at 40% relative humidity); repetition of this type is needed in future mycorrhizal reports.

Nelsen and Safir also suggested that the reported differences in leaf water potentials, transpiration rates, and leaf resistances were due to the reported differences in hydraulic conductivities. For a given evaporative demand, a decrease in hydraulic conductivity would lead to a lower leaf water potential. A plant can counter this decrease in leaf water potential by increasing leaf resistance by partial or cyclic closing of the stomates, thus allowing partial (or total) recovery of leaf water status. This increase in leaf resistance to vapor transfer would then reduce the transpiration rate (Table 4). Their results suggested that under conditions of high water and phosphorus availability, mycorrhizal infection may not have major effects on plant-water relationships. While under low soil phosphorus conditions, mycorrhizae would improve both plant nutrition and plant-water relations.

B. Mycorrhizae and Drought Resistance in Plants

The effects of mycorrhizal infection on the drought resistance of plants are a separate subject from changes in plant-water relations due to mycorrhizae (Section II.A). Implications or suggestions that higher hydraulic conductivities, increased water uptake, increased transpiration rates, and/or lowered stomatal (leaf) resistances somehow improve drought resistance in plants can often result in a contradiction. The need for drought resistance implies that the test plants are subjected to drought stress imposed in some manner by a limited supply or limited rate of delivery of water. Any increased use of water, when water is limiting, with *all* other factors being equal, will result in plants that will suffer more severely from stress and, thus, would be termed less drought resistant. Logic suggests that, when water is limiting, bigger plants or plants that transpire more water will use the available water faster and suffer greater stress. When discussing drought resistance it is important to integrate all the plant responses to the limitation of water and ascertain the results over time, since drought resistance implies a favorable response (as defined by man) to limited water over an exposure time.

1. Short Exposure to Drought Stress

Mesophytic plants have evolved under conditions of generally ample moisture supplies. As such, mesophytic species have developed a "genetic memory" of the water availability patterns of this environment in which they evolved.[11] In general, any drought stress events that are likely to occur will be of short duration, and selection will probably develop plants that can rapidly recover upon rehydration of the soil. Mycorrhizal infection will aid recovery from short (and isolated) drought stress events through the increasing of the hydraulic conductivity of the plants. This was shown by Safir et al.[22] with soybean, and Hardie and Leyton[30] using red clover. In both reports the test plants were exposed to a single, relatively mild drought stress, and plants were then rehydrated by watering the soil. Rate of recovery was monitored (in different ways) and mycorrhizal plants recovered more quickly than did the nonmycorrhizal controls; that is, the leaves of the mycorrhizal plants rehydrated faster than did the nonmycorrhizal plants, resulting in quicker recovery of leaf water potential[22] or faster recovery of leaf turgor.[30] From the results reported by Nelsen and Safir[27] that added phosphorus fertilizer also increased hydraulic conductivity, one might predict that any way of improving plant nutrition would result in a similar response to a single, short episode of water stress. It is likely that this at least partially explains the results of Levy and Krikun.[7] They reported no differences in recovery of leaf water potential for well-fertilized citrus trees when water stressed once.

Table 5
LEVELS OF PHOSPHORUS IN PLANT TISSUE OF MYCORRHIZAL AND NONMYCORRHIZAL PLANTS WHEN EXPOSED TO DIFFERENT LEVELS OF SOIL MOISTURE[42]

Treatment[a]	Phosphorus content (% dry wt)	Phosphorus content (mg/plant)
Mycorrhizal, well watered	0.22	1.6
Mycorrhizal drought stressed	0.28	0.7
Nonmycorrhizal, well watered	0.26	1.3
Nonmycorrhizal drought stressed	0.12[b]	0.09[c]

[a] The mycorrhizal plants were grown in low phosphorus soil, the nonmycorrhizal plants in the same soil with 60 ppm phosphorus added as KH_2PO_4.

[b] Onions containing this level of phosphorus are stunted and show symptoms of phosphorus deficiency.[35]

[c] Onion seeds used in this experiment contained an average of 0.024 mg phosphorus per seed.

2. *Exposure to Severe or Long Periods of Drought Stress*

While any drought stress event is likely to reduce plant productivity,[41] severe, extended, or repeated drought stress events will more severely stress a plant, are of greater interest to man, and are more directly relevant to agriculture in dry land areas.[9-11] By definition, investigations of this type are time consuming, difficult, and seldom done.

The effects of mycorrhizal infection on the drought resistance of onion were recently described by Nelsen and Safir.[42] Inoculation of onion with the mycorrhizal fungus *G. etunicatum* improved the drought resistance of cyclicly water stressed plants as determined by greater fresh- and dry weight increases over a 12-week period. Nonmycorrhizal plants were grown in high-phosphorus soils (6 ppm phosphorus plus 30 ppm added phosphorus). Mycorrhizal plants were grown in low-phosphorus soils (6 ppm phosphorus). One half of the mycorrhizal plants and the nonmycorrhizal controls were grown with ample soil moisture, and one half were repeatedly stressed by withholding water until the monitored soil moisture dropped to −1.0 MPa, at which time the pot would be rewatered. The drought cycle occurred an average of seven times for each stressed plant between week 4 and week 12.

All stressed plants were exposed to similar levels of water stress, as determined by measurements of leaf water potential and transpiration rates. During the 8 weeks of stress, the mycorrhizal plants added more fresh and dry weight such that they were four times as large as the stressed, nonmycorrhizal plants.

As mentioned, leaf water potentials did not differ among the water stressed treatments, nor did soil phosphorus levels change during the course of the experiments. The nonmycorrhizal plants were grown in soil with the highest available phosphorus, but were the smallest plants. Nonetheless, when tissue phosphorus was determined on a percent dry weight basis or on a total phosphorus content basis, only the nonmycorrhizal, stressed plants had low (and deficient[35]) levels of phosphorus (Table 5). Nelsen and Safir hypothesized that, under conditions of soil drought, phosphorus was less mobile,[13] and the nonmycorrhizal plants were both drought stressed and phosphorus deficient. The mycorrhizal plants were supplied with adequate phosphorus due to the mycorrhizal infection and were drought stressed, but not nutrient deficient. The healthier state of the plant resulted in a more optimum exploitation of the available soil moisture (better WUE) and thus a greater increase in biomass. These experiments were conducted three times with similar results.

Previous to this report, plant growth stimulations due to mycorrhizal infection could be duplicated by increases in soil phosphorus by fertilization.[43] In this report, increases of soil phosphorus which would increase growth under well-watered conditions did not result in a similar growth increase under conditions of drought stress. Increased drought tolerance resulted from improved phosphorus nutrition due to the mycorrhizal infection and suggested that mycorrhizae may be even more important to plant growth under dry conditions than when soil moisture is plentiful. These results were supported by what may be the only field study involving mycorrhizae and drought stress. Bolgiano et al.[44] determined onion yield and root infection by mycorrhizae in a Michigan organic soil field. The onions were grown at three phosphorus levels and two watering regimes. When soil moisture was kept high by irrigation, soil phosphorus levels above 15 $\mu g\ cm^{-3}$ greatly reduced root infection by mycorrhizal fungi. In the nonirrigated (rain-fed) plots, root infection was not reduced until the soil phosphorus levels exceeded 30 $\mu g\ cm^{-3}$. Irrigation did not increase plant yield. These results, coupled with those of Nelsen and Safir,[42] would support the hypothesis that mycorrhizae are more important to onion growth under dry conditions and that mycorrhizae are implicated in improved phosphorus nutrition under conditions of limited soil moisture.

Working with citrus seedlings, Levy et al.[29] performed an experiment involving the cyclic water-stressing of mycorrhizal and nonmycorrhizal plants. They determined leaf water potentials of stressed and nonstressed, mycorrhizal and nonmycorrhizal plants after three cycles of water stress, each lasting 5 to 7 days. Unlike Nelsen and Safir[42] who found that all the stressed treatments had similar leaf water potentials, Levy et al. reported that the stressed, mycorrhizal plants had significantly lower (more negative) leaf water potentials. They concluded that the greater total root length and higher transpiration rates of the mycorrhizal plants may have more quickly depleted the available soil moisture and resulted in the mycorrhizal plants being exposed to more severe episodes of water stress than did the nonmycorrhizal, stressed plants. This conclusion is supported by the lower leaf water potentials of the mycorrhizal, stressed plants.

They also determined transpiration rates of well-watered plants that had been continuously well watered or previously stressed and hydraulic conductivities of stressed and nonstressed mycorrhizal and nonmycorrhizal plants. Supplied with ample soil moisture, mycorrhizal plants (both continuously well watered and previously drought stressed) had generally (and significantly) higher transpiration rates than did nonmycorrhizal plants. Drought stress reduced the hydraulic conductivities of both inoculated and noninoculated plants, with mycorrhizal plants having the lowest conductivity of the four treatments. Since Kramer has previously shown that severe stress can decrease root hydraulic conductivity,[45] the lowered hydraulic conductivity of the mycorrhizal, stressed plants is likely due to the more severe levels of water stress to which these plants were exposed. In turn, this reduced conductivity reduced the ability of the mycorrhizal, stressed plants to recover from stress during the night.

It is important to note that Levy et al.[29] did measure leaf tissue phosphorus levels and that both mycorrhizal and nonmycorrhizal plants (the latter which had been fertilized with 210 $mg\ kg^{-1}$ superphosphate) had leaf tissue phosphorus levels considered to be adequate for growth of citrus.[46] Therefore, the differences reported are not likely to be nutritional in nature unless root tissue phosphorus levels were considerably lower than those found in the leaves.

The faster reduction in soil moisture may have been due to the slightly (but not significantly) larger size of the mycorrhizal plants. Levy et al. also suggested that small pot size may have aggravated the difference in treatments and that the result may be different in either the field or larger pots. They did conclude, however, that mycorrhizal infection enabled the root systems to more effectively deplete soil moisture, even to the point of causing *greater* water stress in mycorrhizal plants when grown in pots.

Whether this increases or decreases drought resistance depends on many other factors such as total and temporal water supply, crop density, crop type, yield factor, and evaporative demand and needs to be investigated in the field or simulated field conditions.

Allen and Boosalis described the effects of two different species of mycorrhizal fungi *(G. fasciculatum* and *G. mosseae)* on wheat leaf water potentials and stomatal resistances under well-watered conditions and during exposures to decreasing soil moisture.[47] Using the winter wheat variety "Centurk" grown in pots, they exposed 2-month-old vegetative plants to a continuous supply of soil moisture and drought stress by withholding water until visual wilting, at which time pots were rewatered. Within a watering regime, nonmycorrhizal and *G. fasciculatum*-infected plants were of similar size, while *G. mosseae*-infected plants were significantly smaller. Water stress reduced the weight of plants in all three fungal treatments. Since tissue phosphorus levels were not reported, it cannot be determined if the size differences and subsequent water relations differences were nutritionally related.

Under conditions of ample soil moisture, infection with either fungus reduced daytime stomatal resistance from 146 to about 60 s m^{-1}, but did not change leaf water potentials. This could indicate that stomatal resistances were directly affected by infection in some way, or that differences in hydraulic conductivities to liquid water flow caused subtle changes in leaf water potentials that altered stomatal resistances in the nonmycorrhizal plants, allowing recovery of leaf water potentials to similar levels.[27] Feedback mechanisms involving stomata and leaf water potential levels are well-known.[48]

Under water stress conditions, *G. fasciculatum*-infected plants had the lowest leaf water potentials after 7 days due to their larger size (relative to *G. mosseae*-infected plants) and their more open stomates (relative to the nonmycorrhizal controls), which resulted in a more rapid depletion of the limited available soil moisture in the pots. *G. mosseae*-infected plants had the highest (least negative) leaf water potentials after 7 days of stress, probably due to their much smaller size, although pot weight loss or total water loss was not reported.

Unlike the plants grown under well-watered conditions, plants of all three fungal treatments which had grown under cyclic drought stress had similar levels of both stomatal resistance and leaf water potential at the beginning of a drought cycle. During stress development, stomates of nonmycorrhizal and *G. mosseae*-infected plants appeared to close at leaf water potential values between about −1.5 and −1.8 MPa, while *G. fasciculatum*-infected plants did not show stomatal closure until leaf water potentials dropped below −2.0 MPa. This was probably due to a lower osmotic potential (−2.9 MPa) in the *G. fasciculatum*-infected plants (vs. −1.9 to −2.1 MPa for the *G. mosseae*-infected and nonmycorrhizal plants, respectively). The ability of the *G. fasciculatum*-infected plants to reduce the osmotic potential and maintain turgor under stress conditions is important and needs to be investigated further. Furthermore, whether this is a direct effect of mycorrhizae on the plant or an indirect effect due to improved nutrition should also be determined.

In an attempt to simulate drought stress under field conditions, Nelsen and Maiti[49] grew spring wheat plants (cv. Anza) singly to maturity in the greenhouse in large PVC tubes (1.5 m tall by 15 cm diameter). Each tube was filled with 35 kg of dry soil in which no viable mycorrhizal spores could be found. At tube filling, water was added with the soil such that the initial stored soil moisture was 10 cm of available water throughout the soil column. A 2 × 2 × 3 × 3 factorial design was set up as follows: soil level, dry or wet; mycorrhizal inoculum added *(G. etunicatum),* no or yes; nitrogen added equivalent to 0, 50, or 100 kg ha^{-1}; and phosphorus added equivalent to 0, 30, or 60 kg ha^{-1}. There were three replicate tubes per treatment and the experiment was performed twice.

Table 6
EFFECTS OF DROUGHT STRESS AND MYCORRHIZAL INOCULATION OF THE SOIL, ON WHEAT BIOMASS, YIELD, AND YIELD COMPONENTS; EXPERIMENT 1[49]

	Treatment			
	Wet (40 cm available moisture)		Dry[a] (10 cm available moisture)	
Measured plant parameter	Mycorrhizal[b]	Nonmycorrhizal	Mycorrhizal	Nonmycorrhizal
Total biomass (g)	23.11	24.41	4.19	4.15
Grain yield (g)	12.25	13.02	2.08	2.16
Heads per plant	10.55	11.00	3.14	3.14
Grains per plant	343.25	365.59	63.00	63.25
Grains per head	32.53	33.23	20.06	20.14
Av grain wt (mg)	35.88	35.82	32.43	34.77
Harvest index	0.52	0.52	0.47	0.51

[a] Dry soil significantly reduced biomass, yield, head number per plant, grains per plant, and average grain weight (5%).
[b] Mycorrhizal inoculation significantly reduced yield and grains per plant (5%).

The dry tubes had no additional water added through the season. Wet tubes had water added at 2-week intervals for the first 8 weeks of growth (1300 mℓ/2 weeks) such that the total available moisture was 40 cm of water. Since the tube surface area was 177 cm^2, the net seeding rate was equivalent to 20 to 25 kg ha^{-1}. After senescence, plants were harvested, air dried, and data collected on above-ground biomass, grain yield, heads per plant, total grain number per plant, grain number per head, average grain weight, and harvest index. Analysis of variance was performed on the data and showed that water availability was the major factor in determining total above-ground biomass and grain yield. Decreasing water availability significantly reduced biomass 82% and grain yield 83% in experiment 1 (Table 6). The decrease in water availability also significantly reduced the number of heads per plant, number of grains per plant, and average grain weight. Fertilizer level had no effect in these experiments. In experiment 1, the presence of the mycorrhizal inoculum resulted in a small but significant (5%) reduction in grain yield (6.0 and 3.7%, respectively, in the wet and dry treatments). No such effect was seen in experiment 2.

The decreases in yield reported in these experiments due to water stress are typical. Anza, the wheat cultivar used in these experiments, had previously been reported to respond positively to mycorrhizal inoculation under well-watered conditions,[50] although cereals are generally less responsive to mycorrhizae than other crops. The decrease in yield found in the presence of the mycorrhizal inoculum was unusual but not unique, however, this response under less than optimal water conditions is in opposition to other reports where mycorrhizae increased drought resistance.[42] While the effects of the mycorrhizal inoculation were small and inconsistent relative to the effects of drought, the possible decrease in yield shows that there is a continued need for future research on the effects of mycorrhizae under conditions of environmental stress.

III. EFFECTS OF SOIL MOISTURE CHANGES ON MYCORRHIZAL FUNGI

Up to this point, this review has dealt entirely with effects of the mycorrhizal fungus on plant water relations. Certainly there must also be direct or indirect effects of

Table 7
EFFECTS OF AVAILABLE MOISTURE ON PLANT GROWTH AND MYCORRHIZAL SPORE PRODUCTION FOR INOCULATED *KHAYA GRANDIFOLIOLA* PLANTS[51]

Watering regime	Plant dry wt (g)	Mycorrhizal spore no. per 25 mℓ soil
Weekly	23	7
Twice weekly	31	16
Every 2nd day	62	27
Daily	57	62
Twice daily	45	17
L.S.D. 5%	—	33.3

changing water availability on the fungus. This subject has been sorely ignored thus far, no doubt for two reasons: this field of research is itself rather new, and it is easier to measure plant parameters than fungal parameters. Nonetheless, there is a limited number of reports available.

Redhead[51] investigated the effect of three environmental parameters on plant growth, root infection by mycorrhizae, and the mycorrhizal spore number in the soil. He varied the light intensity, nutrient supply, and water regimes for pot-grown *Khaya grandifoliola* plants in separate experiments. Water regime was changed by watering weekly, twice weekly, every second day, daily, or twice daily, and plants were grown for 35 weeks.

Of the environmental factors tested, available moisture had the greatest effect on root infection; both plant and fungus showed a water level optimum as determined by maximum plant weight and fungal spore number in the soil (Table 7). The amount of water which was optimal for plant growth also resulted in the greatest production of fungal spores. This direct relationship between plant biomass and fungal spore number suggests that the effect on the fungus is a secondary effect due to changes in plant size and, presumably, available photosynthate. As the plant increases in size, more carbohydrates are available for fungal growth and reproduction; as overwatering occurs, plant growth decreases and fungal reproduction drops (Table 7).

This hypothesis is supported by more recent evidence of Nelsen and Safir.[42] They exposed mycorrhizal onion plants to well-watered and cyclic drought-stressed conditions for 8 weeks in soils of two phosphorus levels. Both drought stress and phosphorus fertilization reduced spore production by the fungus *G. etunicatum* (Table 8). Plant fresh weight was reduced 68 and 67% by drought stress at the low and high levels of soil phosphorus, and spore production was reduced 61 and 57% by stress in the same soil phosphorus conditions. These changes in spore production occurred while root infection was not greatly altered by drought. The changes in spore production are likely due to the changes in plant size, since fungal activity (of other species of phycomycetous fungi) is not greatly reduced by soil water potentials incurred in the experiments of Nelsen and Safir.[42,52]

Many more controlled experiments investigating the effects of drought (and other environmental stresses) on mycorrhizal fungi are sorely needed.

IV. CONCLUSIONS AND SPECULATIONS

There seems little doubt that, like plant growth changes, mycorrhizal fungi can cause changes in plant water relations and may, at least in some cases, improve drought

Table 8
PLANT SIZE AND MYCORRHIZAL SPORE PRODUCTION AFTER 8 WEEKS OF EXPOSURE TO WELL-WATERED OR CYCLIC, WATER-STRESSED CONDITIONS[42]

Watering regime	Soil phosphorus[a]	Plant fresh wt (g)	Spore no.[b] (g^{-1} soil)
Well watered	High	5.8	18.6 ± 1.8
Cyclic drought	High	1.8	8.0 ± 2.6
Well watered	Low	5.6	28.7 ± 2.1
Cyclic drought	Low	1.9	11.3 ± 1.4

[a] Low = base soil, no fertilizer, 6 ppm available P; high = base soil plus fertilized with 30 ppm P as KH_2PO_4.
[b] Mean of four replicate determinations ± standard error of the mean.

resistance. However, with the limited amount of information available at the present time, few other conclusions can be drawn, and additional comments are better labeled as speculations or presented as hypotheses.

I would suggest that most changes in water relations reported thus far are likely to be secondary responses due to improved nutrition. In those cases where plant nutrition is investigated along with water relations changes, the nonmycorrhizal plants are often at or near tissue phosphorus concentrations considered to be deficient. Often, in papers where hyphal translocation of water is invoked or unspecified hormone changes are suggested, tissue phosphorus levels are not measured. It would seem logical when there is such a large body of evidence relating mycorrhizae to improved plant phosphorus nutrition, that improved nutrition might be involved in altered water relations and/or drought resistance.

One report, that of Levy et al.,[29] demonstrated differences in transpiration rates between mycorrhizal and nonmycorrhizal citrus plants which did not appear to be due to differences in tissue phosphorus levels. This report and the small number of reports in general, certainly leave open to argument other effects of mycorrhizae besides improved nutrition, but "Ockham's razor" would suggest that tissue phosphorus level determinations or phosphorus fertilized controls should be included in experiments investigating mycorrhizal effects on plants.

Mycorrhizae may indeed improve plant drought resistance or tolerance.[42] Again, however, in many cases this may be due to improved plant nutrition.[42] There is a limited number of early papers that are referenced as evidence that mycorrhizae improve survival under periods of water stress.[53-55] Careful inspection of these reports indicates that nutrition is as likely or more likely to be involved in the described response. Indeed, in one case referenced as an example of mycorrhizae improving the survival of plants exposed to a period of water stress, the plants were water stressed, grown on a coal mine spoil, covered by sediment deposits, and flooded for a period of 30 hr![54] If reports such as these are used as examples from the available literature, this indicates a *great* need for continued and increased levels of research on mycorrhizae and plant water use!

More generally, mycorrhizal investigations relating to plant water use (and any mycorrhizal response) need to go beyond description of phenomena to more reliable, replicated, and repeated experiments. In most cases, it appears that reported experiments are only done one time and in some cases it is impossible to tell if there is even more than one replicate plant per treatment. Experiments dealing with mycorrhizae are difficult to perform and time consuming, but until the reported results are repeatable, they are phenomena, not facts. Nonrepetition of experiments could be the biggest omission of mycorrhizal research to date.

Finally, extension to the field is essentially nonexistent in mycorrhizal plant water relations studies. Controlled field or simulated field experiments have not been done, with two exceptions, that of Bolgiano et al.[44] and Nelsen and Maiti.[49] Potential uses of mycorrhizae in agriculture will remain as potential, until more field work is done.

The increased interest in mycorrhizae in general and their interaction with plant water relations is exciting. Improved instrumentation for controlled environment work is becoming increasingly available. With the addition of more replicated and repeated experiments, much can be learned about mycorrhizal and plant water use since we have only "scratched the surface" in this intriguing field of research.

The author would also like to bring attention to a paper by Sieverding which was found after the manuscript preparation.[56] He reported that VA mycorrhizae improved plant growth and water use of two species of host plant under water stress. Soil type, soil phosphorus levels and availability, and inherent mycorrhizae dependency of the host plant all were involved in the mycorrhizal response. The reported data support the nutritional aspects of the improved drought resistance by the host plant when infected by VA mycorrhizae.

REFERENCES

1. Mosse, B., Fructifications of an *Endogone* species causing endotrophic mycorrhiza in fruit plants, *Ann. Bot. (London)*, 20, 349, 1956.
2. Baylis, G. T. S., Effect of vesicular-arbuscular mycorrhizas on growth of *Griselinia littoralis* (Cornaceae), *New Phytol.*, 58, 274, 1959.
3. Gerdemann, J. W., The effect of mycorrhiza on the growth of maize, *Mycologia*, 3, 342, 1964.
4. Mosse, B., Advances in the study of vesicular-arbuscular mycorrhiza, *Annu. Rev. Phytopathol.*, 11, 171, 1973.
5. Tinker, P. B. H., Effects of vesicular-arbuscular mycorrhizas on higher plants, *Symp. Soc. Exp. Biol.*, 29, 325, 1975.
6. Allen, M. F., Moore, T. S., Jr., and Christensen, M., Phytohormone changes in *Bouteloua gracilis* infected by vesicular-arbuscular mycorrhizae. I. Cytokinin increases in the host plant, *Can. J. Bot.*, 58, 371, 1980.
7. Levy, Y. and Krikun, J., Effects of vesicular-arbuscular mycorrhiza on *Citrus jambhiri* water relations, *New Phytol.*, 85, 25, 1980.
8. Allen, M. F., Smith, W. K., Moore, T. S., Jr., and Christensen, M., Comparative water relations and photosynthesis of mycorrhizal and nonmycorrhizal *Bouteloua gracilis*, *New Phytol.*, 88, 683, 1981.
9. Boyer, J. S. and McPherson, H. G., Physiology of water deficits in cereal crops, *Adv. Agron.*, 27, 1, 1975.
10. Begg, J. E. and Turner, N. C., Crop water deficits, *Adv. Agron.*, 28, 161, 1976.
11. Hanson, A. D. and Nelsen, C. E., Water: adaptation of crops to drought prone environments, in *The Biology of Crop Productivity*, Carlson, P. S., Ed., Academic Press, New York, 1980, 77.
12. Lahiri, A. N., Interaction of water stress and mineral nutrition on growth and yield, in *Adaptation of Plants to Water and High Temperature Stress*, Turner, N. C. and Kramer, P. J., Eds., John Wiley & Sons, New York, 1980, 341.
13. Viets, F. G., Water deficits and nutrient availability, in *Water Deficits and Plant Growth*, Vol. 3, Kozlowski, T. T., Ed., Academic Press, New York, 1972, 217.
14. Greenway, H., Hughes, P. G., and Klepper, B., Effects of water deficit on phosphorus nutrition of tomato plants, *Physiol. Plant*, 22, 199, 1969.
15. Kozolowski, T. T., *Water Deficits and Plant Growth*, Vols. 1 to 6, Academic Press, New York, 1968 to 1981.
16. Mussell, H. and Staples, R. C., *Stress Physiology in Crop Plants*, John Wiley & Sons, New York, 1979, 1.

17. Turner, N. C. and Kramer, P. J., *Adaptation of Plants to Water and High Temperature Stress*, John Wiley & Sons, New York, 1980, 1.
18. Kramer, P. J., *Water Relations of Plants*, Academic Press, New York, 1983, 1.
19. Levitt, J., *Responses of Plants to Environmental Stresses*, Academic Press, New York, 1972, 1.
20. Reid, C. P., Mycorrhizae and water stress, in *Plant Physiology and Symbiosis*, Vol. 6, Riedacher, A. and Gagnaire-Michard, J., Eds., CNRF, Nancy, France, 1979, 392.
21. Safir, G. R. and Nelsen, C. E., Water and nutrient uptake by vesicular-arbuscular mycorrhizal plants, in *Mycorrhizal Associations and Crop Production*, Myers, R., Ed., New Jersey Agric. Exp. Stn. Res. Rep. RO4400-01-81, 1981, 25.
22. Safir, G. R., Boyer, J. S., and Gerdemann, J. W., Mycorrhizal enhancement of water transport in soybean, *Science*, 172, 581, 1971.
23. Safir, G. R., Boyer, J. S., and Gerdemann, J. W., Nutrient status and mycorrhizal enhancement of water transport in soybean, *Plant Physiol.*, 49, 700, 1972.
24. Gray, L. E. and Gerdemann, J. W., Uptake of phosphorus-32 by vesicular-arbuscular mycorrhizae, *Plant Soil*, 30, 415, 1969.
25. Camacho-B, S. E., Kaufmann, M. R., and Hall, A. E., Leaf water potential response to transpiration by Citrus, *Physiol. Plant*, 31, 101, 1974.
26. Kaufmann, M. R. and Levy, Y., Stomatal response of *Citrus jambhiri* to water stress and humidity, *Physiol. Plant*, 38, 105, 1976.
27. Nelsen, C. E. and Safir, G. R., The water relations of well watered, mycorrhizal and nonmycorrhizal onion plants, *J. Am. Soc. Hortic. Sci.*, 107, 271, 1982.
28. Nobel, P. S., *Biophysical Plant Physiology*, W. H. Freeman, San Francisco, 1974, 323.
29. Levy, Y., Syvertsen, J. P., and Nemec, S., Effect of drought stress and vesicular-arbuscular mycorrhiza on Citrus transpiration and hydraulic conductivity of roots, *New Phytol.*, 93, 61, 1983.
30. Hardie, K. and Leyton, L., The influence of vesicular-arbuscular mycorrhiza on growth and water relations of red clover, *New Phytol.*, 89, 599, 1981.
31. Sanders, F. E. and Tinker, P. B., Phosphate flow into mycorrhizal roots, *Pestic. Sci.*, 4, 385, 1973.
32. Cooper, K. M. and Tinker, P. B., Translocation and transfer of nutrients in vesicular-arbuscular mycorrhizas. IV. Effects of environmental variables on movement of phosphorus, *New Phytol.*, 88, 327, 1981.
33. Epstein, E., *Mineral Nutrition of Plants: Principles and Perspectives*, John Wiley & Sons, New York, 1972, 70.
34. Nobel, P. S., *Biophysical Plant Physiology*, W. H. Freeman, San Francisco, 1974, 395.
35. Stribley, D. P., Tinker, P. B., and Rayner, J. H., Relation of internal phosphorus concentration and plant weight in plants infected by vesicular-arbuscular mycorrhizas, *New Phytol.*, 86, 261, 1980.
36. van den Honert, T. H., Water transport in plants as a catenary process, *Discuss. Faraday Soc.*, 3, 146, 1948.
37. Boyer, J. S., Recovery of photosynthesis in sunflower after a period of low leaf water potential, *Plant Physiol.*, 47, 816, 1971.
38. Nye, P. G. and Tinker, P. B., *Solute Movement in the Soil-Root System*, University of California Press, Berkeley, 1977, 26.
39. Allen, M. F., Influence of vesicular-arbuscular mycorrhizae on water movement through *Bouteloua gracilis*, *New Phytol.*, 91, 191, 1982.
40. Cowan, M. C., Lewis, B. G., and Thain, T. F., Uptake of potassium by the developing sporangiophore of *Phycomyces blakesleeanus*, *Trans. Br. Mycol. Soc.*, 58, 113, 1972.
41. Tazaki, T., Ishihara, K., and Ushijima, T., Influence of water stress on the photosynthesis and productivity of plants in humid areas, in *Adaptation of Plants to Water and High Temperature Stress*, Turner, N. C. and Kramer, P. J., Eds., John Wiley & Sons, New York, 1980, 309.
42. Nelsen, C. E. and Safir, G. R., Increased drought tolerance of mycorrhizal onion plants caused by improved phosphorus nutrition, *Planta*, 154, 407, 1982.
43. Rhodes, L. H. and Gerdemann, J. W., Nutrient translocation in vesicular-arbuscular mycorrhizae, in *Cellular Interactions in Symbiosis and Parasitism*, Cook, C. B., Pappas, P. W., and Rudolph, E. D., Eds., Ohio State University Press, Columbus, 1980, 173.
44. Bolgiano, N. C., Safir, G. R., and Warncke, D. D., Mycorrhizal infection and growth of onion in the field in relation to phosphorus and water availability, *J. Am. Soc. Hortic. Sci.*, 108, 819, 1983.
45. Kramer, P. J., Effects of wilting on the subsequent intake of water by plants, *Am. J. Bot.*, 37, 280, 1950.
46. Chapman, H. D., The mineral nutrition of citrus, in *The Citrus Industry*, Vol. 2, Reuther, W., Batchelor, L., and Webber, H., Eds., University of California Press, Riverside, 1968, 127.
47. Allen, M. F. and Boosalis, M. G., Effects of two species of VA mycorrhizal fungi on drought tolerance of winter wheat, *New Phytol.*, 93, 67, 1983.
48. Lange, O. L., Nobel, P. S., Osmond, C. B., and Ziegler, H., *Physiological Plant Ecology II, Encyclopedia of Plant Physiology*, Vol. 12B, Springer-Verlag, Berlin, 1983, 1.

49. Nelsen, C. E. and Maiti, T. C., The effects of drought stress, mycorrhizal innoculation and soil nutrition on wheat yield, *Phytopathology,* 73, 841, 1983.
50. Azcon, R. and Ocampo, V. A., Factors affecting the vesicular-arbuscular infection and mycorrhizal dependency of 13 wheat cultivars, *New Phytol.,* 87, 677, 1981.
51. Redhead, J. F., Endotrophic mycorrhizas in Nigeria: some aspects of the ecology of the endotrophic mycorrhizal association of *Khaya grandifoliola* C. DC., in *Endomycorrhizas,* Sanders, F. E., Mosse, B., and Tinker, P. B., Eds., Academic Press, New York, 1975, 447.
52. Duniway, J. M., Water relations of water molds, *Annu. Rev. Phytopathol.,* 17, 431, 1979.
53. Mosse, B. and Hayman, D. S., Plant growth responses to vesicular-arbuscular mycorrhiza, *New Phytol.,* 70, 29, 1971.
54. Aldon, E. F., Endomycorrhizae enhance survival and growth of four-wing saltbush on coal mine spoils, *USDA Forest Res. Note RM-294,* 1975, 1.
55. Janos, D. P., Vesicular-arbuscular mycorrhizae affect lowland tropical rain forest plant growth, *Ecology,* 61, 151, 1980.
56. Sieverding, E., Influence of soil water regimes on VA mycorrhizae, *Z. Acker-und Pflanzenban,* 150, 400, 1981.

Chapter 6

CARBON REQUIREMENTS OF VESICULAR-ARBUSCULAR MYCORRHIZAE

David Harris and Elder A. Paul

TABLE OF CONTENTS

I. INTRODUCTION

The importance of the supply of carbon (C) from the host plant to the mycorrhizal fungus was recognized by Franck in the earliest days of mycorrhiza research. All available evidence suggests that vesicular-arbuscular (VA) mycorrhizal fungi have little or no capacity for independent growth and they must be considered obligate biotrophs deriving all C compounds from the plant. The subjects of C nutrition and transfer in mycorrhizae have recently been reviewed by Harley and Smith.[1] We will concentrate on recent measurements of C flow in VA mycorrhizal plants and examine some theories of the transfer of C between host and fungus in relation to these measurements.

Most attention has been given to the stimulation of plant growth by VA mycorrhizal symbiosis, but there are also examples where infection reduces plant growth.[2,3] Consideration of the energy requirements of the fungal endophyte relative to the effects of symbiosis on host-plant photosynthesis offers an explanation of the balance between mutualism and parasitism in the mycorrhizal association.

II. CARBON TRANSFER

The dependence of VA mycorrhizal infection on host photosynthesis has been demonstrated by manipulation of light intensity, photoperiod, or ozone concentration.[4-7] Direct evidence for the transfer of photosynthate from host to fungus was found by Ho and Trappe,[8] who detected ^{14}C in extramatrical VA-fungal hyphae and spores several weeks after exposing mycorrhizal *Festuca* plants to $^{14}CO_2$. These findings were confirmed over much shorter time periods by Cox et al.,[9] who showed by autoradiography accumulation of ^{14}C in intraradicle hyphae, arbuscular cells, and vesicles within 27 hr of pulse labeling the host plants with $^{14}CO_2$. Similarly, Bevege et al.[10] found ^{14}C in extramatrical hyphae by scintillation counting of the tracer.

A. Mechanisms of Carbon Transfer

The movement of C to the fungus is presumed to occur at the arbuscule, though it is possible that C compounds are also taken up by unspecialized hyphae within the root cortex. Suggested mechanisms of transfer of materials between symbionts range from the simple uptake of materials normally present in the apoplast or intrafacial matrix of the arbuscule, to fairly elaborate specific transport systems.[1,11-13] Rates of transfer may be controlled by the concentration gradient between the host and fungal cytoplasm, by the area and permeability of the interface, or by the activity of specific uptake processes.

1. Carbon Sinks

Sink-mediated transfer of C may require no special properties for the host-fungus interface. Sucrose, unloaded from the phloem into the apoplast, or sugars and organic and amino acids that have diffused from plant cells into the apoplast or the interfacial matrix of the arbuscule may be taken up by the fungus in competition with plant cells and other microorganisms. Transfer may be accelerated by increasing the concentration gradient between the host and fungal cytoplasm, by rapid catabolic fungal metabolism, or by conversion of the diffusate into compounds not utilized by the host and preferably not susceptible to diffusion back across the fungal membrane. Lewis and Harley[14] suggest that in ectomycorrhizae, carbohydrates such as trehalose and mannitol could perform such a function, since they are not normally metabolized by the host. These compounds were not detected in VA mycorrhizae[4,10] and the carbohydrate spectrum of VA mycorrhizal roots was similar to that of uninfected roots.

The accumulation of lipid in hyphae, vesicles, and spores of VA mycorrhiza[15,16] is often pronounced and lipid has been suggested to be a sink for C within the fungus. The conversion of plant carbohydrate to fungal lipid could maintain a carbohydrate concentration gradient favoring movement of C to the fungus. Lipid globules could be carried by protoplasmic streaming[17] along a concentration gradient towards the external hyphae. The autoradiographs of Cox et al.[9] show by far the greatest concentrations of plant-derived ^{14}C in vesicles where large lipid accumulations are also found. However, the experiments of Lösel and Cooper[18] failed to show evidence of specific C-sink compounds in lipid or other fractions extracted from mycorrhizal roots after labeling onions with $^{14}CO_2$.

The accumulation of soluble C compounds within roots increases the potential for efflux from the plant. Bowen[19] found that the exudation of amino compounds from P-deficient pine roots was associated with high concentrations within the plant, rather than to effects on membrane permeability. Enhanced mycorrhizal infection of P-deficient clover has been attributed to the accumulation of carbohydrate in the roots.[20,21]

2. Leaky Membranes and Diversion of Cell Wall Precursors

An inverse relationship between root exudation and P concentration in citrus and sudangrass has been observed. The reducing sugar fraction of root exudates was correlated with mycorrhizal infection.[11,12] The increased exudation in P-deficient plants was not associated with higher concentrations of reducing sugars or amino acids in roots, but was attributed to increased "leakage" by membranes. It was proposed that increased exudation by P-deficient plants stimulated mycorrhizal infection until root P concentrations were increased sufficiently to reduce leakage of the host plasmalemma. The consequent reduction in the supply of C as exudates, in turn, curtailed mycorrhizal infection. Graham et al.[12] also suggested that exudates could support the postinfection development of the fungus.

Schwab et al.[22] estimated the total loss of carbohydrate, amino acids, and carboxylic acids from nonmycorrhizal, P-deficient sudangrass roots to be 15 mg g^{-1} day^{-1}. Fungal growth in equivalent mycorrhizal plants was estimated to require 20 to 40 mg substrate day^{-1}. They concluded that it was unlikely that the mycorrhizal fungus captured all the compounds exuded from P-deficient roots and that the measured rates of exudation were insufficient to support the development of the mycorrhizae.

Schwab et al.[22] suggest that the concentrations of metabolites in the apoplast may be higher than indicated by measurements of exudation at the root surface. An increased efflux might be achieved due to efficient scavenging by the fungus. It is also possible that some metabolite released by the fungus could further affect the permeability of the host membrane. There are many examples of such alterations in membrane permeability resulting from pathogenic infection,[23,24] but none have been demonstrated in mycorrhizae.

Harley[1,25] has suggested that mycorrhizal fungi could interfere with the normal synthesis of plant cell walls and divert cell wall precursors that are secreted into the interfacial matrix of the arbuscule. Electron micrographs, showing high numbers of vesicles containing fibrillar carbohydrate in the interfacial matrix of mycorrhizal onions,[26] lend some support to this hypothesis. Further evidence is needed before it is possible to determine whether such a mechanism could transfer enough C for the development of mycorrhizal infection.

3. Active Transport

Woolhouse[13] proposed an elaborate carrier-mediated transfer system for both carbohydrates and Pi at the arbuscular interface. In this scheme, carbohydrates are passively released through the host plasmalemma and then actively taken up by the endophyte by an energy-linked sugar transport system at the fungal membrane. Conversely,

Pi is released by the hydrolysis of polyphosphate in the arbuscule, leaked into the interfacial matrix, and actively transported across the host plasmalemma. This dual carrier system has circumstantial support both from ultrastructural studies of the arbuscule and by analogy with the haustoria of powdery mildew which are structurally similar. Studies of ATPase activity in the plasma membranes of mildew haustoria have shown "domains" of low activity in the host plasmalemma surrounding the haustorium and high activity in the opposed fungal membrane.[27] It is suggested that the invaginated region of the host plasmalemma is unable to control the efflux of sucrose from the cell into the interfacial matrix where it is actively taken up (probably after inversion) by the fungal plasmalemma which shows high ATPase activity. The efflux has been estimated as 2.4 ng sucrose 100 mm^{-2} sec^{-1}, which is three orders of magnitude higher than effluxes estimated for other plant cells.[28,29] The situation in arbuscules is more complex in that active transport, though of different materials, is presumed to occur in both host and fungal membranes. Consequently, the high ATPase activity found in the invaginated regions of the host plasmalemma at the arbuscule[30] may be associated with the uptake of Pi and does not necessarily discount the possibility of low activity with respect to sucrose.

If the rate of C transfer per unit interfacial area (flux density) in arbuscules is similar to that in mildew haustoria and C uptake by the fungus occurs only in the arbuscule, then transfer of 5 mg C g^{-1} root day^{-1} to the fungus would require an interfacial surface area of 7×10^3 mm^2g^{-1} root. Cox et al.[9] found that cortical cells of leek with arbuscules of *Glomus mosseae* contained 3.1 times more plasmalemma than uninfected cells. In spherical cells of 80 μm diameter, this would correspond to an increase in plasmalemma of 4×10^4 μm^2 $cell^{-1}$. If this represents the arbuscule interface, the transfer of 5 mg C g^{-1} root day^{-1} would require 1.6×10^5 arbuscules per gram dry weight root. The average dry weight of pea root cells[31] (9 mm from tip) of 5×10^{-9} g gives 1.4×10^8 cells g^{-1} root. Therefore, at the assumed flux density, about 0.1% of root cells must contain active arbuscules to transfer C at 5 mg g^{-1} root day^{-1}. Schwab et al.[32] estimated that during the most rapid phase of mycorrhizal development in sudangrass when 19.4% of the root length was colonized, 2.5% of the total root volume consisted of arbuscules. Thus, if arbuscules are functionally similar to mildew haustoria, potential transfer rates may exceed 100 mg C g^{-1} root day^{-1}.

III. FUNGAL CARBON REQUIREMENTS

It is useful to consider how much C might be required to support the growth and metabolism of mycorrhizal fungi, since direct measurement of the C flux across the host-fungus interface awaits direct evidence of where and how this occurs.

A. Fungal Biomass

Most measurements of VA mycorrhizal infection have been made by assessing the frequency of fungal structures in stained root segments; these do not give good measurements of fungal weight or volume. The development of the chitin method for the measurement of VAM fungal biomass[33] has enabled a number of recent estimates of the mass of fungal material in infected roots. These vary from 3% of root dry weight in sorghum, to 16% in soybeans (Table 1). The chitin method shares some of the problems of the staining techniques. It does not distinguish living or active fungal material from dead or inactive and it is not well adapted to the measurement of extramatrical hyphal material in soil. It also requires the assumption that the ratio of chitin to fungal biomass within the root is similar to that of external hyphae. The ratio of extraradicle to intraradicle fungal material in a soybean-*Glomus fasciculatum* association has been found to vary with age, from 7:1 at 4 weeks to 0.4:1 at 14 weeks.[37] Thus,

Table 1
BIOMASS OF VA MYCORRHIZA IN THE ROOTS OF SYMBIOTIC ASSOCIATIONS

Fungus	Host	Root dw (%)	Method	Infection (%)	Ref.
G. fasciculatum	*Centrosema*	14	Chitin	95	33
G. mosseae	*Centrosema*	7	Chitin	100	33
A. laevis	*Centrosema*	5	Chitin	95	33
G. mosseae	*Vicia faba*	6	Microscopy	60	34
G. fasciculatum	*Glycine max*	16	Chitin	70	35
G. fasciculatum	*Sorghum*	3	Chitin	50	36

Table 2
INCREMENTS IN VA MYCORRHIZAL FUNGAL BIOMASS C AND ESTIMATED C REQUIREMENTS FOR GROWTH AND MAINTENANCE IN A SOYBEAN *G. FASCICULATUM* SYMBIOSIS (CALCULATED FROM DATA OF BETHLENFALVAY ET AL.[37])

			Respiration			
Time (days)	Root dw (g)	Fungal C[a] (mg)	ΔFC[b]	Growth[c] (mg C day^{-1})	Maintenance[d] (mg C day^{-1})	Total (b + c + d)
28	0.7	6.4	0.4	0.3	10.2	10.9
42	1.1	26.5	1.7	1.1	42.4	45.2
56	2.3	44.7	2.9	1.9	71.5	76.3
70	2.3	95.2	6.3	4.2	91.4	101.9

[a] Fungal C = 0.4 fungal dry weight.
[b] $\Delta FC = \mu x$.
[c] Growth respiration $G_R = \mu x / Y_{max} \cdot (1 - Y_{max})$.
[d] Maintenance respiration = ax/Y_{max} where μ = specific growth rate (day^{-1}); x = fungal C (mg); Y_{max} = yield coefficient, 0.6 (mg C mg^{-1} C); and a = specific maintenance rate (day^{-1}).

the assumption that extraradicle hyphae constitute roughly 50% of the total fungal biomass, which will be made in the absence of direct measurements, should be treated with caution.

B. Fungal Growth and Respiration

The kinetics of fungal growth are frequently complex because of the apical nature of hyphal elongation. However, fungal biomass in mycorrhizae can increase exponentially (Table 2), at least during part of the growth curve. During this phase Monod kinetics can be applied and growth described by a specific growth rate μ ($time^{-1}$):

$$\mu = \frac{(\ln x_t - \ln x_0)}{t} \tag{1}$$

where x_0 = biomass C at zero time, x_t = biomass at time t.

The rate of substrate utilization is

$$\frac{-ds}{dt} = \frac{\mu x}{Y} \tag{2}$$

where s = substrate (g C) and Y = the efficiency of conversion of substrate C into biomass C (g C g^{-1} C).

Y deviates from the maximum, Y_{max}, depending on the proportion of substrate used for maintenance of the standing biomass. In steady-state growth, the maintenance requirement is proportional to the standing biomass, and the rate of substrate utilization then becomes

$$\frac{-ds}{dt} = \frac{\mu x}{Y_{max}} + mx \tag{3}$$

where m = the maintenance coefficient (g C g^{-1}C $time^{-1}$).

The maintenance coefficient m is related to a mass-independent parameter, the specific maintenance rate a ($time^{-1}$) by

$$m = \frac{a}{Y_{max}} \tag{4}$$

The respiration associated with growth G_R (g C g^{-1}C $time^{-1}$) at Y_{max} is

$$G_R = \frac{\mu x}{Y_{max}} \cdot (1 - Y_{max}) \tag{5}$$

Total substrate used for growth and respiration is substrate = Δbiomass + growth respiration + maintenance respiration

$$\frac{-ds}{dt} = \mu x + \frac{\mu x}{Y_{max}} \cdot (1 - Y_{max}) + \frac{ax}{Y_{max}} \tag{6}$$

The data of Bethlenfalvay et al.[37] show an exponential increase of fungal biomass in mycorrhizal soybeans between 4 and 10 weeks. From this a specific growth rate (μ) of 0.064 hr^{-1} and a doubling time of about 11 days can be calculated. A maximum growth yield (Y_{max}) of 0.6 (g biomass C per g substrate C utilized) has been found for many aerobic heterotrophs growing on a variety of substrates.[39] Estimation of a specific maintenance rate (a) is more difficult since published values for the maintenance requirements of soil biomass[40-43] are frequently one or two orders of magnitude smaller than those determined experimentally in the chemostat,[38] and these in turn vary according to organism and growth conditions.

The high turnover rate of arbuscules ($t_{0.5}$ = 2 to 8 days)[1] and the energy costs for active transport and cytoplasmic streaming in mycorrhizal fungi suggest that maintenance requirements may be fairly high. The C requirements for the development of mycorrhizal fungi can be calculated if the specific maintenance rate is known. If we provisionally accept the value of 0.04 hr^{-1} for (a) given by Barber and Lynch[44] for rhizosphere populations, the resulting estimates of maintenance respiration (Table 2) suggest that the overall growth yield for *G. fasciculatum* is less than 0.05. Comparison of the calculated C requirements with net rates of photosynthesis in soybeans grown under similar conditions[45] indicates that mycorrhizal respiration would consume 30 to 40% of total photosynthate. It must be concluded that the maintenance component is overestimated, either by an inappropriate value for (a), or because only a small proportion of the fungal thallus is alive. If a large proportion of mycorrhizal hyphae are dead, this would greatly affect the understanding and analysis of nutrient transport

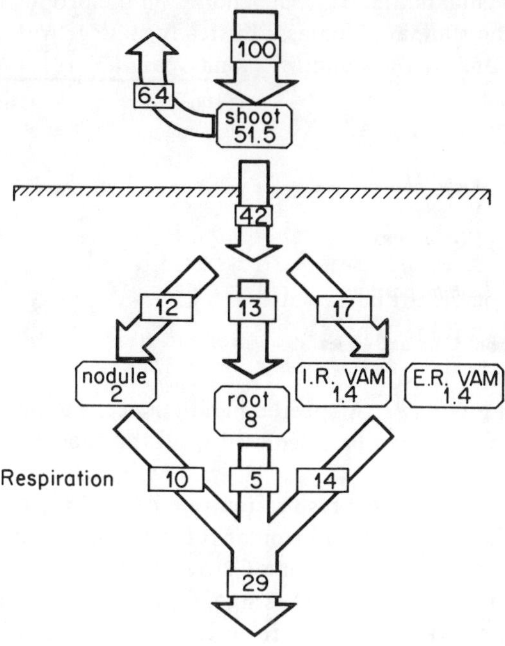

FIGURE 1. ^{14}C allocations in a soybean-*Rhizobium-Glomus* association. ^{14}C allocations shown as percentages of total C uptake. IR VAM = intraradicle VA mycorrhizal biomass, ER VAM = extraradicle VA mycorrhizal biomass, C allocation to extraradicle mycorrhizal biomass assumes external biomass = internal biomass.

and infection development. Measurements of the proportion of living fungal material in mycorrhizae are therefore important and urgent. Direct determination of maintenance respiration in VAM fungi will probably depend on growth in pure culture, but provisional estimates are being made using whole-plant C-flow measurements.[45]

The calculations of mycorrhizal respiration from C allocation experiments are made by subtraction of $^{14}CO_2$ evolution rates for nonmycorrhizal roots (and rhizosphere microflora) from those of roots with mycorrhizae. It is possible that basal plant respiration in mycorrhizal roots is altered in response to infection. Cox et al.[9] noted increased membrane and cytoplasmic contents of cells with arbuscules and suggested that part of the additional respiration was of plant origin. Conversely, there is evidence to suggest that mycorrhiza reduce root exudation and that extraradicle mycorrhizal biomass is formed at the expense of the soil biomass.[36] Therefore, although the increased respiration of mycorrhizal roots can be considered as part of the C cost of the symbiosis, calculation of fungal respiration may be subject to error, most probably of overestimation.

A comparison of respiration rates calculated from fungal growth rates and experimentally derived rates for respiration and C accumulation in fungal biomass has been made.[45] In 42-day-old soybeans infected with *G. fasciculatum,* the increment in fungal C per plant (3.7 mg C day^{-1}), calculated from biomass measurements, agreed well with the value of 3.6 mg C day^{-1} obtained from ^{14}C measurements (Figure 1). Fungal maintenance respiration, calculated as total ^{14}C allocated to the mycorrhiza, minus C used

Table 3
EFFECTS OF VA MYCORRHIZA ON C ALLOCATION AND PHOTOSYNTHESIS

Fungus	Host	Control (%)		Ref.
		C allocation to symbiont	C uptake[a] (net photosynthesis)	
G. mosseae	*Vicia faba*	+10	+21	55
G. mosseae	*V. faba*	+5	+8	34
G. fasciculatum	*Glycine max*	+14	+20	45
G. fasciculatum	*Sorghum bicolor*	+4	−21	35
G. mosseae	*Allium porrum*	+12	+13	47

[a] Photosynthetic rate as C uptake g^{-1} leaf dry weight.

in growth was 14.8 mg C day^{-1}. A specific maintenance rate (a) of 0.008 hr^{-1} and an overall growth yield of 0.18 can be calculated if all the fungal biomass is assumed to have been alive.

Harley et al.[46] reported a fungal respiration rate of 11.5 mg CO_2 g^{-1} hyphae h^{-1} for an ectomycorrhizal sheath, and a very similar rate was estimated by Snellgrove et al.[47] for *G. mosseae* infection in leeks (11.0 mg CO_2 g^{-1} hr^{-1}). The rate for *G. fasciculatum* in soybeans[45] was almost identical at 11.7 mg CO_2 g^{-1} hr^{-1}. The extremely close correspondence between these values must be regarded as coincidental.

IV. CARBON UPTAKE AND ALLOCATION IN HOST PLANTS

A. Assimilate Partitioning

Alterations in source-sink relationships through modifications of their relative strengths influence translocation rates of sucrose from source to sink regions.[48] Translocation rates have been correlated with the activity of sink regions,[49] leaf sucrose levels, and sucrose-P synthetase activity.[50] Sucrose-P synthetase can in turn be regulated by sucrose concentration and the exchange of Pi with triose-P at the chloroplast membrane.[51] Regulation of sucrose export depends, therefore, on a number of factors. These include the relative strength of source and sinks; leaf Pi concentrations acting via effects on the triose-P translocation, sucrose-P synthetase, and sucrose-P phosphatase activities.[52] In addition, hormones either of microbial[53] or plant origin[54] may be involved.

Most of the attempts to measure the C cost of VA mycorrhizal symbiosis have used pulse-chase labeling with $^{14}CO_2$ to measure the uptake and distribution of photosynthate in plants with or without mycorrhiza. All of these measurements have shown increased C allocation to mycorrhizal roots, varying from 4 to 14% of total photosynthate (Table 3). This range includes legumes, leeks and sorghum infected with *Glomus mosseae* or *G. fasciculatum* at root infection percentages of 50 to 70%. Carbon allocation in 6-week-old soybeans which were both mycorrhizal and nodulated (Figure 1) shows that the three components of the plant-microbial association each consumed about 1/3 of the ^{14}C translocated to the root system. Intraradicle fungal biomass accounted for 1.4% of total photosynthate and a similar amount was attributed to extraradicle hyphae. The respiration of mycorrhiza and nodules together represented 70% of the total below-ground $^{14}CO_2$ evolution.[45]

B. Host Photosynthesis

Losel and Cooper[18] found increased translocation of ^{14}C-labeled photosynthate to the roots of mycorrhizal onion and suggested that this reflected increased photosynthesis in mycorrhizal plants. Allen et al.[56] noted that photosynthesis in mycorrhizal *Bouteloua gracilis* was increased compared to nonmycorrhizal controls without affecting plant dry weight, and they suggested that the additional photosynthate had been utilized by the mycorrhiza. The ^{14}C pulse labeling experiments summarized in Table 3 show increased translocation of photosynthate from the plant shoot accompanied by increased photosynthesis of 8 to 21%. In these experiments, mycorrhizal plants are compared to uninfected controls which have been supplied with P fertilizer to match the growth curve of the mycorrhizal plants as closely as possible. This is done to enable comparison between mycorrhizal and uninfected plants that are of similar size, development, and growth rate. If this aim is achieved, the photosynthetic rate of mycorrhizal plants must inevitably exceed that of nonmycorrhizal plants because the mycorrhizal fungi use host-derived C. The lower rates of photosynthesis of mycorrhizal sorghum[36] compared to fertilized plants reflect the failure of the fertilizer treatments to exactly emulate the growth curve of the mycorrhizal plants. The phenomenon of "compensation"[34] is of interest because a number of different host-plant responses to the C demands of mycorrhizal and *Rhizobium* symbioses has been found.[45] These enable the plant to increase the overall efficiency of photosynthesis for a given mass of leaf.

Kucey and Paul[34] found that in mycorrhizal faba beans the fungus used 3.5% of the total photosynthate, and C fixation rates per unit shoot dry weight were 8% higher in mycorrhizal than in control plants. Very similar results were reported by Snellgrove et al.[47] for mycorrhizal leeks, where on a dry weight basis the photosynthetic rates of mycorrhizal plants were 13% greater than the uninfected controls. However, on a fresh weight or leaf area basis, rates for mycorrhizal and P-compensated leeks were similar because mycorrhizal plants had smaller dry matter contents and greater specific leaf areas than controls. Snellgrove et al. concluded that photosynthetic compensation was due to greater hydration in the mycorrhizal plants. This enabled the net assimilation rate of mycorrhizal plants to equal that of nonmycorrhizal plants, but required less shoot dry matter. Experiments with soybeans having both mycorrhiza and *Rhizobium* have confirmed these conclusions with regard to the effects of mycorrhizal infection on photosynthesis and specific leaf area.[45] However, other effects are apparent in the triple symbiosis. Plants with *Rhizobium* alone showed greater rates of C fixation than controls. Specific leaf area in nodulated plants was only slightly increased and photosynthetic rates were greater on both a leaf area and leaf dry weight basis. Infection of soybean by *G. fasciculatum* and *Rhizobium* resulted in a further increase in photosynthetic rate per unit leaf dry weight, but not per unit leaf area. The addition of the mycorrhiza increased specific leaf area, N and P concentrations and reduced the amount of starch in leaves by 50%.

The above results indicate three levels of response of photosynthesis to the C requirements of the symbioses.

1. Removal of Sink Limitation

The growth of the control plants was P limited, and it is suggested that photosynthesis was "sink" limited due to the accumulation of end products which could not be used for growth. Low sink demand may result in the accumulation of sucrose-6-P in leaf mesophyll cells.[49] This would have the effect of binding Pi in inactive pools, thereby reducing the exchange of Pi and triose phosphate across the chloroplast membrane.[57] Starch would ultimately accumulate in the chloroplast; this has been associated with decreased rates of photosynthesis.[58,59] A low Pi concentration in the chloroplast may also directly affect the C-reduction cycle via the stromal ADP/ATP ratio.[60]

Rhizobium symbiosis in this experiment consumed *circa* 13% of the total photosynthate (Figure 1), and it is suggested that the associated increase in photosynthetic rate was primarily due to the increased below-ground sink for C. Removal of sink limitation allowed nodulated plants to achieve maximal photosynthetic rates per unit leaf area under these conditions.

2. Morphological Adaptation

Triple symbionts were able to further increase net photosynthesis by increasing specific leaf area (SLA). This would appear to be the primary effect of mycorrhiza on photosynthetic rate. It is not yet known whether mycorrhiza can affect sink limitation when present as the only symbiont. In leek, the increased SLA was associated with an increase in leaf hydration; this would seem to be the simplest mechanism for leaf expansion. In the soybean-*Rhizobium*-VAM association the increase in SLA could be partly explained by the decrease in starch concentration, but no effect on leaf water content was found. Other effects which may have influenced SLA include changes in the ratio of primary (main stem) and secondary (lateral branch) leaves in mycorrhizal soybeans.[61]

3. Source Limitation

Increased N and P concentrations in leaves of VA mycorrhizal, nodulated plants were accompanied by decreased starch content. Increased mobilization of starch may have resulted from the sink demand of the dually infected root system and higher leaf Pi concentrations. It is also possible that the growth of these plants was C limited, due to the strong sink for C generated by the combined effects of both mycorrhizal and rhizobial symbioses and, thus, were unable to maintain the level of starch reserves found in plants without symbionts. Carbon allocation to the microbial symbionts accounted for 28% of total photosynthate.

V. MYCORRHIZAL C/P INTERACTIONS

Mycotrophic growth of plants is most frequently attributed to enhanced P uptake. Mycorrhizal enhancement of plant growth usually declines as P concentrations in hosts are increased.[62-64] The enhanced uptake of P is gained at the cost of host photosynthate utilized by the mycorrhiza; in P-limited plants growth is stimulated because C is abundant. However, if P uptake becomes nonlimiting and photosynthate utilization by the mycorrhiza is not curtailed, the mycorrhizal plant may become C limited, leading to a reduction in plant growth compared to nonmycorrhizal controls. This is particularly likely if photosynthesis is limited, for example, by low light intensity,[66] or other sinks for C such as *Rhizobium* are supported. Host growth depression during early growth may later be replaced by growth stimulation as the supply of soil P is diminished.[35,67] The balance of effects of mycorrhiza on P uptake, C fixation, and allocation can be seen as a dynamic process leading to increased concentrations of P in mycorrhizal plants.[68,69] Comparison of mycorrhizal C utilization and the mycorrhiza-enhanced uptake of P in soybean gives a ratio of 140 C/1 P (g atom/g atom) during mycotrophic growth,[45] but this ratio would be expected to vary according to the P status of the host.

The available evidence suggests that increased photosynthetic rates resulting from mycorrhizal infection are primarily due to morphological adaptations that increase specific leaf area.[45,47] If plants are grown in a sward or closed canopy where increases in leaf area cannot increase photosynthesis, the possibility of growth depression should increase since the expansion of leaves cannot compensate for C utilization by the mycorrhiza. Conversely, increased light intensity or CO_2 concentration should extend the range of tissue P concentrations which allow mycotrophic growth.

This review has shown that enough information is now available to initiate attempts to quantify the C relationships of symbiotic plant-fungal associations. This has required extrapolation of data from one association to another, and the studies performed have been restricted to a limited number of VA mycorrhizal species and to host plants grown in pots. However, there is no evidence to suggest that the information gained cannot be related to field conditions.

Tracer techniques provide a powerful tool for the measurement of C allocation, but the need for uninoculated controls of similar size and growth rate relative to mycorrhizal plants is likely to restrict the range of associations and conditions that can be studied. The evaluation of the C requirements of mycorrhizal fungi has highlighted several areas where specific measurements are needed. These include respiration rates and maintenance requirements of mycorrhizal fungi and the proportion of living to dead or inactive hyphae in mycorrhizae. Little is known of the mechanisms of transfer of C and P, their controls and interactions. An understanding of these mechanisms would aid the interpretation of C/P interactions at the whole plant level.

REFERENCES

1. Harley, J. L. and Smith, S. E., *Mycorrhizal Symbioses,* Academic Press, London, 1983.
2. Buwalda, J. G. and Goh, K. M., Host fungus competition for carbon as a cause of growth depressions in vesicular-arbuscular mycorrhizal ryegrass, *Soil Biol. Biochem.,* 14, 103, 1982.
3. Bethlenfalvay, G. J., Brown, M. S., and Pacovsky, R. S., Parasitic and mutualistic associations between a mycorrhizal fungus and soybean: development of the host plant, *Phytopathology,* 72, 889, 1982.
4. Hayman, D. S., Plant growth responses to vesicular-arbuscular mycorrhiza. VI. Effect of light and temperature, *New Phytol.,* 73, 71, 1974.
5. Tolle, R., Üntersuchungen über die Psuedomykorrhiza von Gramineen, *Arch. Mikrobiol.,* 30, 285, 1958.
6. Johnson, C. R., Menge, J. A., Schwab, S., and Ting, I. P., Interaction of photoperiod and vesicular arbuscular mycorrhizae on growth and metabolism of sweet orange, *New Phytol.,* 90, 665, 1982.
7. McCool, P. M., Effect of Ozone Stress on Development of Vesicular-Arbuscular Mycorrhiza and Response of Tomato and Citrus, Ph.D. thesis, University of California, Riverside, 1981.
8. Ho, I. and Trappe, J. M., Translocation of ^{14}C from *Festuca* plants to their endomycorrhizal fungi, *Nature (London),* 244, 30, 1973.
9. Cox, G., Sanders, F. E., Tinker, P. B., and Wild, J. A., Ultrastructural evidence relating to host-endophyte transfer in a vesicular-arbuscular mycorrhiza, in *Endomycorrhizas,* Sanders, F. E., Mosse, B., and Tinker, P. B., Eds., Academic Press, London, 1975, 297.
10. Bevege, D. I., Bowen, G. D., and Skinner, M. F., Comparative carbohydrate physiology of ecto- and endomycorrhizas, in *Endomycorrhizas,* Sanders, F. E., Mosse, B., and Tinker, P. B., Eds., Academic Press, London, 1975, 149.
11. Ratnayake, R. T., Leonard, R. T., and Menge, J. A., Root exudation in relation to supply of phosphorus and its possible relevance to mycorrhizal formation, *New Phytol.,* 81, 543, 1978.
12. Graham, J. T., Leonard, R. T., and Menge, J. A., Membrane-mediated decrease in root exudation responsible for phosphorus inhibition of vesicular-arbuscular mycorrhiza formation, *Plant Physiol.,* 68, 548, 1981.
13. Woolhouse, H. W., Phosphate physiology of vesicular-arbuscular mycorrhizas, in *Endomycorrhizas,* Sanders, F. E., Mosse, B., and Tinker, P. B., Eds., Academic Press, London, 1975, 241.
14. Lewis, D. H. and Harley, J. L., Carbohydrate physiology of mycorrhizal roots of Beech. III. Movement of sugars between host and fungus, *New Phytol.,* 64, 256, 1965.
15. Mosse, B., Advances in the study of vesicular-arbuscular mycorrhiza, *Annu. Rev. Phytopathol.,* 11, 171, 1973.
16. Cooper, K. M. and Losel, D. M., Lipid physiology of vesicular arbuscular mycorrhiza. I. Composition of lipids in roots of onion, clover and ryegrass infected with *Glomus mosseae, New Phytol.,* 80, 143, 1978.
17. Rhodes, L. H. and Gerdemann, J. W., Nutrient translocation in vesicular-arbuscular mycorrhiza, in *Cellular Interactions in Symbiosis and Parasitism,* Cooks, C. B., Pappas, P. W., and Rudolph, E. D., Eds., Ohio State University Press, Columbus, 1980, 173.

18. Lösel, D. M. and Cooper, K. M., Incorporation of ^{14}C-labelled substrates by uninfected and VA mycorrhizal roots of onion, *New Phytol.*, 83, 415, 1979.
19. Bowen, G. D., Nutrient status affects loss of amides and amino acids from pine roots, *Plant Soil*, 30, 139, 1969.
20. Same, B. I., Robson, A. D., and Abbott, L. K., Phosphorus, soluble carbohydrates and endomycorrhizal infection, *Soil Biol. Biochem.*, 15, 593, 1983.
21. Jasper, D. A., Robson, A. D., and Abbott, L. K., Phosphorus and the formation of vesicular arbuscular mycorrhizas, *Soil Biol. Biochem.*, 11, 501, 1979.
22. Schwab, S. M., Menge, J. A., and Leonard, R. T., Quantitative and qualitative effects of phosphorus on extracts and exudates of sudangrass in relation to vesicular-arbuscular mycorrhiza formation, *Plant Physiol.*, 73, 761, 1983.
23. Wheeler, H., Disease alterations in permeability and membranes, in *Plant Disease*, Vol. 3, Horsfall, J. G. and Cowling, E. B., Eds., Academic Press, New York, 1978, chap. 15.
24. Scott Russel, R., *Plant Root Systems*, McGraw-Hill, London, 1977, 113.
25. Harley, J. L., *The Biology of Mycorrhiza*, 2nd ed., Leonard Hill, London, 1969.
26. Dexheimer, J., Gianinazzi, S., and Gianinazzi-Pearson, V., Ultrastructural cytochemistry of the host fungus interface in the endomycorrhizal association *Glomus mosseae/Allium cepa*, *Z. Pflanzenphysiol.*, 92, 191, 1982.
27. Spencer-Phillips, P. T. N. and Gay, J. L., Domains of ATPase in plasma membranes and transport through infected plant cells, *New Phytol.*, 89, 393, 1981.
28. Edelman, J., Schoolar, A. I., and Bonner, W. B., Permeability of sugar cane chloroplasts to sucrose, *J. Exp. Bot.*, 22, 534, 1971.
29. Humphreys, T. E., Sucrose transport at the tonoplast, *Phytochemistry*, 12, 1211, 1973.
30. Marx, C., Dexheimer, J., Gianinazzi-Pearson, V., and Gianinazzi, S., Enzymatic studies on the metabolism of vesicular-arbuscular mycorrhizas. IV. Ultracytoenzymological evidence (ATPase) for active transfer processes in the host-arbuscule interface, *New Phytol.*, 90, 37, 1982.
31. Brown, R. and Broadbent, D., The development of cells in the growing zones of the root, *J. Exp. Bot.*, 1, 249, 1950.
32. Schwab, S. M., Menge, J. A., and Leonard, R. T., Comparison of stages of vesicular-arbuscular mycorrhiza formation in sudangrass grown at two levels of phosphorus nutrition, *Am. J. Bot.*, 70, 1225, 1983.
33. Hepper, C. M., A colorimetric method for estimating vesicular-arbuscular mycorrhizal infection in roots, *Soil Biol. Biochem.*, 9, 15, 1977.
34. Kucey, R. M. N. and Paul, E. A., Carbon flow, photosynthesis, and N_2 foxation in mycorrhizal and nodulated faba beans, *(Vicia faba L.)*, *Soil Biol. Biochem.*, 14, 407, 1982.
35. Bethlenfalvay, G. J., Pacovsky, R. M., Brown, M. S., and Fuller, G., Mycotrophic growth and mutualistic development of host plant and fungal endophyte in an endomycorrhizal symbiosis, *Plant Soil*, 68, 43, 1982.
36. Harris, D., Pacovsky, R. S., and Paul, E. A., Carbon flow and N_2 fixation in Sorghum, *Azospirillum, Glomus* associations, in *Proc. 6th N.A.C.O.M. Meeting, Bend, Oregon*, University of Oregon, Corvallis, 1984, 374 pp.
37. Bethlenfalvay, G. J., Brown, M. S., and Pacovsky, R. S., Relationships between host and endophyte development in mycorrhizal soybeans, *New Phytol.*, 90, 537, 1982.
38. Pirt, J. S., *Principles of Microbe and Cell Cultivation*, John Wiley & Sons, New York, 1975, chap. 2 and 8.
39. Payne, W. J., Energy and growth of heterotrophs, *Ann. Rev. Microbiol.*, 24, 17, 1970.
40. Babuik, L. A. and Paul, E. A., The use of fluorescein isothiocyanate in the determination of microbial biomass of grassland soil, *Can. J. Microbiol.*, 16, 57, 1970.
41. Behra, B. and Wagner, G. H., Microbial growth rate in glucose ammended soil, *Proc. Soil Sci. Soc. Am.*, 38, 591, 1974.
42. Shields, J. A., Paul, E. A., Lowe, W. E., and Parkinson, D., Turnover of microbial tissue in soil under field conditions, *Soil Biol. Biochem.*, 6, 31, 1973.
43. Chapman, J. S. and Gray, T. R. G., Endogenous metabolism and macromolecular composition of *Arthrobacter globiformis*, *Soil Biol. Biochem.*, 13, 11, 1981.
44. Barber, D. A. and Lynch, J. M., Microbial growth in the rhizosphere, *Soil Biol. Biochem.*, 9, 305, 1977.
45. Harris, D., Pacovsky, R. S., and Paul, E. A., Carbon economy of soybean, *Rhizobium, Glomus* associations, *New Phytol.*, 101, 427, 1985.
46. Harley, J. L., McCready, C. C., and Jennings, D. H., Salt respiration of excised beech mycorrhizas. II, *New Phytol.*, 55, 489, 1956.
47. Snellgrove, R. C., Splittstoesser, W. E., Stribley, D. P., and Tinker, P. B., The distribution of carbon and the demand of the fungal symbiont in leek plants with vesicular-arbuscular mycorrhizas, *New Phytol.*, 92, 75, 1982.

48. Gifford, R. M. and Evans, L. T., Photosynthesis, carbon partitioning and yield, *Ann. Rev. Plant Physiol.*, 32, 485, 1981.
49. Hurewitz, J. and Janes, H. W., Effect of altering the root zone temperature on growth, translocation, carbon exchange rate, and leaf starch accumulation in the tomato, *Plant Physiol.*, 73, 46, 1983.
50. Silvius, J. E., Kremer, R. E., and Lee, R. E., Carbon assimilation and translocation in soybean leaves at different stages of development, *Plant Physiol.*, 62, 54, 1978.
51. Priess, J., Starch, sucrose biosynthesis and partition of carbon in plants are regulated by orthophosphate and triosephosphates, *TIBS*, January, 24, 1984.
52. Salerno, G. L. and Pontis, H. G., Regulation of sucrose levels in plant cells, in *Mechanisms of Polysaccharide Polymerization and Depolymerization*, Marshall, J. J., Ed., Academic Press, New York, 1980, 31.
53. Slankis, V., Hormonal relationships in mycorrhiza, in *Ectomycorrhyzae: Their Ecology and Physiology*, Marks, G. C. and Kozlowski, T. T., Eds., Academic Press, New York, 1973, 231.
54. Allen, M. F., Moore, T. S., Jr., and Christensen, M., Phytohormone changes in *Bouteloua gracilis* infected by vesicular-arbuscular mycorrhizae. II. Altered levels of gibberellin-like substances and absisic acid in the host plant, *Can. J. Bot.*, 60, 468, 1982.
55. Pang, P. C. and Paul, E. A., Effects of vesicular-arbuscular mycorrhiza on ^{14}C and ^{15}N distribution in nodulated fababeans, *Can. J. Soil Sci.*, 60, 241, 1980.
56. Allen, M. F., Smith, W. K., Moore, T. S., Jr., and Christensen, M., Comparative water relations and photosynthesis of mycorrhizal and non-mycorrhizal *Bouteloua gracilis* H. B. K. Lag ex Steud., *New Phytol.*, 88, 683, 1981.
57. Giaquinta, R. T., Translocation of sucrose and oligosaccharides, in *The Biochemistry of Plants*, Vol. 3, Academic Press, New York, 1980, 271.
58. Thorne, J. H. and Koller, H. R., Influence of assimilate demand on photosynthesis, diffusive resistance, translocation and carbohydrate levels in soybean leaves, *Plant Physiol.*, 54, 201, 1974.
59. Chatterton, N. J., Carlson, G. E., Hungerford, D. R., and Lee, D. R., Effect of tillering and cool nights on photosynthesis and chloroplast starch in Pangola, *Crop Sci.*, 12, 206, 1972.
60. Walker, D. A. and Robinson, S. P., Regulation of photosynthetic carbon assimilation, in *Photosynthetic Carbon Assimilation*, Siegelman, H. W. and Hinds, G., Eds., Plenum Press, New York, 1978, 43.
61. Pacovsky, R. S., Paul, E. A., and Bethlenfalvay, G. J., Response of mycorrhizal and phosphorus fertilized soybean to nodulation by *Rhizofium* or combined nitrogen application, *Plant Physiol. Crop Sci.*, 26, 145, 1986.
62. Daft, M. J. and Nicolson, T. H., Effect of *Endogone* mycorrhiza on plant growth. II. Influence of soluble phosphate on endophyte and host in maize, *New Phytol.*, 72, 127, 1969.
63. Mosse, B., Plant growth responses to vesicular-arbuscular mycorrhiza. IV. In soil given additional phosphate, *New Phytol.*, 72, 127, 1973.
64. Sanders, F. E., The effect of foliar-applied phosphate on the mycorrhizal infection of onion roots, in *Endomycorrhizas*, Sanders, F. E., Mosse, B., and Tinker, P. B., Eds., Academic Press, London, 1975, 261.
65. Menge, J. A., Steirle, D., Bagyaraj, D. J., Johnson, E. L. V., and Leonard, R. T., Phosphorus concentration in plant responsible for inhibition of mycorrhizal infection, *New Phytol.*, 80, 575, 1978.
66. Bethlenfalvay, G. J. and Pacovsky, R. S., Light effects in mycorrhizal soybeans, *Plant Physiol.*, 73, 963, 1983.
67. Cooper, K. M., Growth responses to the formation of endotrophic mycorrhizas in *Solanum, Leptospermum* and New Zealand ferns, in *Endomycorrhizas*, Sanders, F. E., Mosse, B., and Tinker, P. B., Eds., Academic Press, London, 1975, 391.
68. Stribley, D. P., Tinker, P. B., and Snellgrove, R. C., Effect of vesicular-arbuscular mycorrhizal fungi on the relations of plant growth, internal phosphorus concentration and soil phosphate analyses, *J. Soil Sci.*, 31, 655, 1980.
69. Stribley, D. P., Tinker, P. B., and Rayner, J. H., Relation of internal phosphorus concentration and plant weight in plants infected by vesicular-arbuscular mycorrhizas, *New Phytol.*, 86, 261, 1980.

Chapter 7

VA MYCORRHIZAS IN HUMID TROPICAL ECOSYSTEMS

David P. Janos

TABLE OF CONTENTS

I. INTRODUCTION

Relatively little specific information is available concerning the responses of vesicular-arbuscular (VA) mycorrhizal fungi to tropical environments; discussion of the ecophysiology of VA mycorrhizas in the tropics seems to require a temperate/tropical comparison and extrapolation of data from temperate-zone fungi. Although the cosmopolitan occurrence of genera and some species of VA mycorrhizal fungi might legitimize such extrapolation, it potentially is misleading because of the probable adaptation of VA mycorrhizal fungi to local edaphic conditions.[1] In this chapter I assume that tropical VA mycorrhizal fungi are adapted to different soil and root environments than are mycorrhizal fungi in other regions; I characterize tropical soil and root environments, and I speculate upon aspects of the ecophysiology of VA mycorrhizas that may be peculiar to the tropics.

I focus on VA mycorrhizas in the lowland humid tropics, because more information is available for humid than for dry areas[2-6] and I know of only one paper that mentions highland mycorrhizas.[5] I consider data concerning plants growing in the lowland humid tropics, whether or not they are strictly tropical species, and also include data from the subtropics on species with tropical origins. I review all of the information regarding humid tropical forest species of which I am aware, but do not comprehensively treat agricultural literature.

A. Tropical Mycorrhiza Research

Although one of the earliest studies of the occurrence of mycorrhizas was conducted in Java in 1896,[7] mycorrhizas in the tropics subsequently have received much less attention than in the temperate zones. Surveys of mycorrhizal hosts and fungi in native tropical vegetation have been sporadic, and recent experimental investigations have dealt primarily with effects on crop growth. Most information on the physiology of mycorrhizas comes from temperate-zone studies.[8] This reflects both the paucity of trained mycorrhiza researchers in tropical countries,[9] and the needs of those nations for research applied to practical problems. The mycorrhizas of tropical forest trees and especially of important timber species are virtually unstudied;[10] lack of study belies the ecological and potential economic importance of tropical forest mycorrhizas.

Most tropical tree species that have been examined form VA mycorrhizas[10] which can be indispensable for seedling survival and growth.[11] Notwithstanding the consequent prevalence of VA mycorrhizal fungus mycelia, unusually few spores of these fungi are found in tropical forest soils[10] which may be degraded by loss of the fungi after forest clearing.[12] Because tropical forests may disappear by the middle of the next century,[13] in order to insure the productivity of forestry and agriculture it is imperative to catalog, understand, and preserve tropical VA mycorrhizas.

B. Lowland Humid Tropical Ecosystems

The lowland humid tropics may be defined as those regions of the earth's surface below approximately 1500 m elevation where mean annual biotemperature exceeds 24°C (biotemperature scores temperatures below 0 or above 30°C as 0), and annual rainfall equals or exceeds potential evaporative and transpirational loss of water to the atmosphere.[14] These regions are free of frost and have similar annual and diurnal temperature variation; if seasons occur, they are determined by changes in rainfall. Humid tropical lowlands annually receive more than 1500 mm of rain, and usually have no more than two dry months (less than 100 mm rainfall per month) per year.

The lowland humid tropics predominantly lie between the Tropics of Cancer and Capricorn in three main geographical regions: South and Central America and the Caribbean (the Neotropics), Africa, and Asia-Australia-Oceania (the Paleotropics).

The natural vegetation of the humid lowlands is a closed-canopy forest, and togehter these regions represent a potential forested area of almost 16 million km^2 (approximately 10% of the world's land and 31% of the tropics) distributed as 8.03 million km^2 in Latin America, 3.87 in Asia, and 3.62 in Africa.[13] At least half of this area has already been deforested, however, and the remaining forest will be converted to other types of vegetation within 50 years at present rates of deforestation. Subsistence farming, fuelwood gathering, pasture establishment (especially in the Neotropics), and timber extraction (especially in Asia) are major causes of forest loss.

Within the lowland humid tropics, forest composition and structure are influenced by phytogeography, average biotemperature, ratio of potential evapotranspiration to precipitation, and soil. Species diversity, primary productivity, canopy height, forest structural complexity, and epiphyte abundance tend to increase with precipitation and to decrease with elevation.[14] Lowland tropical forests are the richest in species on earth, typically containing 60 to 400 tree species per square kilometer with all species in low abundance. Areas of especially poor soils and hydric associations, however, have low diversity. Forests on impoverished, sandy soils can be dominated by a single species of tree which may form ectomycorrhizas.[10] Brackish-water mangrove forests and freshwater swamps often contain five or fewer tree species; in freshwater these are often palms. I know of no published information on the mycorrhizal status of mangroves, but one temperate-zone swamp tree species has VA mycorrhizas,[15] and VA mycorrhizas recently have been observed on a common palm species in the seasonally inundated Venezuelan llanos.[16]

A typical, diverse, lowland humid tropical forest has a closed but uneven canopy at 35 to 45 m height (with emergents to 50 m) of mesophyllic evergreen trees with 30 to 40 individuals per hectare. Subordinate vertical strata include 80 to 160 subcanopy trees per hectare at 30 m height, an understory of 400 to 500 shrubs and small trees per hectare at 5 to 20 m including many palms, and a forb layer dominated by aroids, ferns, and fern-allies. Canopy trees have relatively thin, smooth-barked, columnar boles and are often buttressed. Woody climbers (lianas) and epiphytic aroids, bromeliads, ferns, orchids, mosses, and lichens are abundant. Various forms of cauliflory (literally "stemflowering") are common, and I speculate that those few species that flower on specialized branches which lie on the ground surface[17] might carry VA mycorrhizal fungus propagules on or in their fruits or seeds. Although the vertical distributions of major roots of tropical trees can be comparable to those of temperate-zone trees, the major proportion of feeder roots occurs in the top 20 cm of the soil.[18]

II. TROPICAL SOIL ENVIRONMENTS

Tropical soils are diverse, encompassing highly infertile acid sands, acid clays of volcanic origin, and limited extents of soils developed from limestone.[14] Generalizations in this chapter, however, principally pertain to the most common types of tropical soils, Oxisols and Ultisols. I discuss the support of both native vegetation and crops by tropical soils.

Although often stereotyped as infertile, not all tropical soils are deficient when judged by standards of crop production. Mollisols, usually associated with limestone parent materials, and Alfisols are fertile and sustain productive agriculture. These soils with which centers of population co-occur[19] constitute one third of Asian tropical soils, but are much less abundant in Africa and the Neotropics.[14] They have high base contents, and acidity is seldom a problem, except in West African Alfisols which are gravelly and have a low buffering capacity. Nitrogen is the nutrient that most often limits crop production. Phosphorus deficiencies may occur, but are easily corrected by fertilization and liming.[20]

Oxisols and Ultisols account for two thirds of world tropical soils by area. These two soil orders have similar properties and include most of the "red" soils of the tropics. They are clays with good physical characteristics that are usually well drained, but deficient chemically.[14] They are acid, have high potential aluminum toxicity, low cation exchange capacities indicative of high leaching potential, and low available phosphorus, potassium, calcium, magnesium, sulfur, zinc, and additional micronutrients.

Although highly infertile acid sands (Spodosols and Psamments) comprise only about 7% of tropical soils,[14] they must be mentioned in the context of tropical mycorrhizas. It was for these soils that the involvement of mycorrhizal fungi in a "direct mineral cycle" was first proposed,[21] and they have received scientific attention disproportionate to their areal extent. Direct mineral cycling, most often interpreted as involving enzymatic ability to mineralize litter on the part of mycorrhizal fungi, never has been demonstrated nor is it likely to apply to VA mycorrhizal fungi.[10]

A. Mineral Nutrients

Year-round high temperatures and rainfall in excess of potential evapotranspiration markedly influence biological and physicochemical activity in tropical soils; decomposition is about five times more rapid than in temperate-zone forests, and potential mineral nutrient loss through leaching is great.[19,22] Bases are replaced on soil colloid exchange surfaces by abundant aluminum ions and by dissociated protons from carbonic acid formed as a consequence of the respiration of soil organisms. Soil parent materials are usually highly weathered, such that, with the exception of nitrogen from the atmosphere, there is no ready resupply of mineral nutrients that are lost from the ecosystem. The possibility of mineral loss with little replenishment necessitates efficient mineral recycling mechanisms.[10,17,21,23]

1. Organic Matter

Although decomposition rates are generally higher in tropical than temperate-zone forests, the equilibrium organic matter contents of their soils are similar because the productivity of tropical forests exceeds that of temperate-zone forests.[19] The mean organic matter content of the top 30 cm of several hundred Hawaiian, Puerto Rican, and East African soils ranged from 3.36 to 3.75%.[24] Lowland humid tropical acid sands (Spodosols and Psamments) are exceptional, however, having as much as 30% of their total biomass in humus.[25] This accumulation probably reflects slow decomposition[26] because of the inhibitory effects of very high phenolic content of litter[27,28] on decomposers and nitrifying[29] organisms.

The similar soil organic matter contents of temperate-zone and tropical forests suggest that because of the high standing-crop of tropical forests and the low cation exchange capacities and base saturations of Oxisols and Ultisols, most essential mineral nutrient reserves are stored in vegetation. This is true for potassium, calcium, and magnesium, but more than 70% of nitrogen and phosphorus reserves can be found in the topsoil.[14] Moreover, throughfall and stemflow, which carry mineral nutrients leached from leaves, add more potassium, sulfur, and sodium to the soil solution than the respective inputs of these elements in litterfall.[30] Retention of the mineral nutrient capital of tropical forests demands efficient uptake from the soil, notwithstanding the emphasis placed on recycling from litter by the direct mineral cycling hypothesis. Nevertheless, fine roots and fungus hyphae in close association with organic matter are in an advantageous position to absorb nutrients released by the activities of decomposer organisms.

St. John and co-workers have examined the tendency of roots in a Brazilian tropical forest and of VA mycorrhizal fungus hyphae in the laboratory to associate with organic matter.[31-33] They demonstrated that unbranched roots and hyphae grow in random directions until organic matter is encountered, then branch to colonize the organic substrate. An active microflora in organic matter was implicated in the stimulation of root branching.[34] Such positioning of roots and hyphae is especially beneficial for uptake of sorbed or poorly mobile ions such as phosphate, sulfur, zinc, and to some extent, potassium and ammonium.[8,35]

2. Phosphorus

In a survey of published data from 62 tropical forests, Vitousek[36] concluded that many tropical forests cycle phosphorus more efficiently than most temperate-zone forests, and that litterfall in such tropical forests is limited by phosphorus supply. Crop production in the tropics, however, is most often limited by nitrogen deficiency.[24] Phosphorus and sulfur follow nitrogen in importance to crops, with phosphorus deficiency especially common in Oxisols and Ultisols. Phosphorus and calcium supplements are almost always needed to prevent broken bones in range cattle in Latin America[19] where Oxisols and Ultisols predominate.

Oxisols and Ultisols are known as "phosphate-fixing" soils because of their high capacity to convert soluble phosphates to insoluble forms by combination with exchangeable aluminum, iron, and aluminum and hydrous oxides.[14,24] This tendency increases with acidity and clay content, and although phosphorus thus stabilized is in equilibrium with the soluble pool of soil phosphorus, the equilibrium is poorly defined[37] and favors the insoluble forms. Phosphorus in organic forms such as phytin and nucleic acids also tends to be immobilized in such soils by occlusion in organomineral aggregates.[38,39] Immobilization, lack of unweathered parent materials, and leaching when vegetation and normal uptake processes are disrupted conspire to impose phosphorus limitations on both forest and crop growth. Small phosphatic fertilizer additions to crops are ineffective in alleviating such phosphorus limitation, although the addition of phosphorus or soluble silica can improve cation exchange capacity by precipitating exchangeable aluminum.[24,40]

The phosphorus immobilization capacity of most lowland humid tropical soils establishes conditions under which the well-known beneficial effects of mycorrhizas on uptake of solution phosphorus may be maximized; solution phosphorus is low and reserve phosphorus is present (in immobilized forms; sometimes in very large quantities[24]).[35,39] Localized phosphorus-depletion zones may quickly develop around absorbing organs, but for VA mycorrhizal roots and hyphae in contrast with uninfected roots: (1) phosphorus will be withdrawn from a large volume of soil,[41] (2) phosphorus depletion zones of hyphae will overlap less rapidly than those of roots or root hairs[41] which are thereby less efficient than hyphae,[12] (3) longevity of functional mycorrhizas in excess of that of uninfected roots will enhance the uptake of relatively unavailable phosphates as they are solubilized or desorbed,[37,39] and (4) VA mycorrhizas or fungus hyphae may have a higher affinity for phosphorus and, hence, be capable of its uptake from lower solution concentrations than are uninfected roots.[41,42] Any one of these phenomena would improve the phosphorus nutrition of VA mycorrhizal plants in comparison with uninfected plants in almost all unamended tropical soils.

Other possibilities exist for the enhancement of uptake of complex phosphates by VA mycorrhizas, although all involve increased solubilization or desorption of stable phosphates, and their release to the soil solution from which they may be absorbed by uninfected roots as well as by mycorrhizas. Acid phosphatase produced by VA mycorrhizas may more actively hydrolyze organic phosphates such as phytates than do the phosphatases produced by uninfected roots,[41] provided that the phosphatase can reach occluded organic sources. Organic acids, which have been demonstrated to be produced by some ectomycorrhizal fungi,[38] may disaggregate clay-organic complexes. Ligands that chelate metallic ions such as iron also might release immobilized organic and inorganic phosphorus; iron chelators are known to be produced by some ectomycorrhizal fungi.[39] Although there is no evidence that VA mycorrhizal fungi produce organic acids or metal-ion chelators, the utility of phosphatases which they do produce[41] could be increased by the proliferation of hyphae in the presence of nonoccluded, nonimmobilized organic phosphorus.[32] In addition, VA mycorrhizas might stimulate the growth of phosphorus-solubilizing bacteria in the rhizosphere,[43] but bacterial populations are generally low under the acid conditions of tropical soils.

Most research concerning phosphorus uptake by VA mycorrhizas that utilized tropical plants or soils has focused on growth responses to phosphatic fertilizers of low solubility such as rock phosphate. Such fertilizers have the advantages of low cost and little immobilization because of slow solubilization, but the extent to which they are utilized is uncertain. Isotopic labeling of three Brazilian soils showed that VA mycorrhizal individuals of two tropical grasses and a legume utilized the same soluble phosphorus pool as uninfected plants,[44] which is consistent with the results of similar experiments conducted with temperate-zone plants and soils.[41] Nevertheless, VA mycorrhizal cassava was able to extract phosphorus from solutions of lower concentration than could uninfected plants.[42] High affinity for phosphorus may result in enhanced uptake from relatively insoluble sources as they slowly solubilize or desorb.[37] Some experiments have shown, however, that overall phosphorus availability from insoluble sources was not increased by VA mycorrhizas, even though plant growth responded to fertilization, and that uptake of phosphorus from rock phosphates by mycorrhizas declined over time.[41] Additions of rock phosphates to tropical soils have either not affected[45-48] or have enhanced[46,47,49,50] plant growth responses to mycorrhizas. Enhancement seems to depend on the availability of soluble phosphorus which is influenced by the amount of rock phosphate added,[47,49] its solubility,[50] and by pH.[46]

3. Nitrogen

In contrast to phosphorus, nitrogen lost from lowland humid tropical ecosystems can be readily replaced by conversion of abundant atmospheric nitrogen to organic forms by asymbiotic and symbiotic nitrogen-fixing bacteria and cyanobacteria.[24,51] *Beijerinkia* and *Azotobacter* on leaf surfaces may contribute 12 to 40 kg per hectare per year,[24] and rain forest canopy lichens in which phycobionts are predominantly cyanobacteria may contribute an additional 1.5 to 8 kg N per hectare per year.[52] In soil, unlike *Azotobacter, Beijerinkia* can tolerate low pH and may add from 0 to 25 kg N per hectare per year.[24,51] Symbiotic fixation by nodulated legumes or actinorhizal plants[43] (depending on their abundance) contributes 34 to 68 kg N per hectare per year in forest.[24] Nitrogen loss from the plant/soil system is accounted for by leaching, especially of highly mobile nitrate ions, and denitrification during very active organic decomposition and temporary waterlogging. When forest is cleared for agriculture, nitrogen inputs decline[51] and losses may increase, contributing to commonly observed nitrogen deficiency for crop production. Only flooded rice soils have some immunity to nitrogen deficiency, because of high rates of N-fixation by cyanobacteria and heterotrophic soil bacteria associated with rice roots.[53]

Most lowland humid tropical forests are not nitrogen limited,[36] although those on Spodosols can be[22,36,54] because of inhibition of nitrification by very low pH and high tannin contents of litter.[54] Soil organic matter is the most important source of nitrogen for plants,[14,24] and nitrogen cycling is more closely linked to that of carbon than is phosphorus cycling.[39] Nitrogen is rapidly mineralized (a three-step process involving aminization, ammonification, and nitrification that is wholly mediated by the activities of microorganisms) under the prevailing conditions of uniform, relatively high temperatures and low C/N ratios (on the order of 10 to 15[55]) of most tropical forest soils.[24]

Moisture has the greatest influence on rates of N mineralization; alternate wetting and drying is most favorable for nitrification, because nitrifying organisms can remain active under low moisture conditions and then proliferate rapidly as more moisture becomes available, releasing a flush of nitrate.[55] The critical C/N ratio above which mineralization stops is lower under constant moisture conditions than with alternate wetting and drying.[24] Thus, an even rainfall distribution minimizes nitrate fluctuations in the soil[24,55] and probably reduces total nitrification. Temporary anaerobiosis during heavy rains can inhibit nitrification by the obligately aerobic nitrifying bacteria, thereby intensifying competition for ammonium among nitrifying bacteria, other microorganisms, and plant roots and further reducing nitrification.

The importance of mycorrhizas for nitrogen supply to host plants is likely to depend on the relative amounts of ammonium and nitrate available in the soil, and on soil moisture. If nitrification is inhibited, e.g., by abundant tannins and low pH as on Spodosols, or by uniformly high moisture, then ammonium tends to accumulate. Furthermore, nitrate retention by sandy soils such as Spodosols and Psamments is lower than by clays, because the latter hold water and dissolved ions in abundant micropores which are lacking in sandy soils.[24] In addition, moisture influences the diffusion of ions to roots; its lack can diminish the apparent diffusion coefficients of mobile ions by more than an order of magnitude.[8] Under well-watered conditions, enhancement of mineral nutrient uptake by mycorrhizas is not expected for ions such as nitrate that readily diffuse through soil to root surfaces. Ammonium, however, is somewhat less mobile than nitrate.

VA mycorrhizas can enhance the uptake of ammonium and nitrogen from organic sources[56] and may preferentially utilize ammonium.[41] They could potentially enhance nitrate uptake as well, if competition for mineral nutrients is intense, such as when nitrogen mineralization is inhibited or root densities are very high, as on Spodosols and in pastures. The presence of VA mycorrhizas has reduced the amount of nitrate leached from cores of a temperate-zone soil,[57] although it was not determined if this was attributable to enhanced uptake by mycorrhizas or to immobilization in microbes. A VA mycorrhizal fungus has been shown to produce nitrate reductase which is needed for utilization of nitrate.[58]

On Oxisols and Ultisols where nitrate is relatively uniformly available throughout the year, perhaps with a flush at the beginning of the rainy season, VA mycorrhizas could enhance nitrate uptake should it be limited by the low surface areas of prevalent magnolioid roots (see Section III.B.1). If the magnolioid root morphology accentuates the relative contribution of VA mycorrhizas to nitrate uptake, benefit in addition to that of improved nitrogen supply may accrue to the host plant. When roots assimilate nitrate directly, excess hydroxyl ions are produced and pH is regulated by the production of carboxylic anions that are stored together with cations in vacuoles; when VA mycorrhizal fungi assimilate nitrate they are likely to regulate pH by hydroxyl ion extrusion.[58] The latter might result in a saving of photosynthate by the host plant,[58] and an increase in rhizosphere pH which would favor phosphorus and sulfur solubilization.

VA mycorrhizas might enhance host nitrogen nutrition in ways besides inorganic ion uptake. Proliferation of hyphae in organic material[31-33] may facilitate uptake of simple organic nitrogen compounds by VA mycorrhizas.[58] VA mycorrhizas may indirectly increase nitrogen availability by stimulation of free-living nitrogen-fixing organisms in the mycorrhizosphere,[43] or by enhancement of nodulation of legumes and actinorhiza formation.[43]

4. Sulfur and Other Nutrients

Sulfur deficiencies for crop growth have been reported from all over the tropics.[24]

Sulfur deficiency is common in soils high in allophane, oxides, or sand, or with low organic matter content; burning vegetation residues during land clearing can volatilize 75% of the sulfur present in the plant/soil system.[24] Sulfate ions, like phosphate ions, tend to be sorbed by soils high in iron and aluminum oxides, such as many Oxisols and Ultisols.[24] Sorption increases as pH decreases, because hydroxyl ions tend to replace sulfate ions at high pH. Sorbed sulfur, however, is held much less tightly than phosphorus. In tropical forest soils, much sulfur is in organic form,[24] and sulfur cycling is intermediate to that of nitrogen and phosphorus.[39]

Oxisols and Ultisols are low base (calcium, magnesium, potassium, and sodium) status soils, and base deficiencies may appear on them in crops that have been relieved of nitrogen, phosphorus, and sulfur limitations,[24] although calcium and magnesium deficiencies can be masked by aluminum toxicity. Some Oxisols contain so little zinc that crops are deficient in this micronutrient, but zinc deficiency is most common on calcareous soils.

Sulfate and potassium are only slightly less mobile than nitrate in moist soil, but VA mycorrhizas have been shown to enhance uptake of both sulfur and potassium as well as that of calcium and zinc.[8,41,47,59-61] VA mycorrhizas improved potassium uptake by a temperate-zone tree with hairless, magnolioid roots by 23% when phosphorus did not limit the growth of uninfected plants, although mycorrhizas improved phosphorus uptake by 15% when supplemental phosphorus was not provided.[59] This improvement of potassium uptake of approximately one seventh the improvement of phosphorus uptake by 157% when supplemental phosphorus was not provided.[59] This improvement of potassium uptake of approximately one seventh the improvement of phosphorus uptake is similar to sulfur translocation about one seventh as great as that of phosphorus by VA mycorrhizas of clover.[62] In the latter experiment, the molar amount of zinc translocated to the host was only 1/35th that of phosphorus. In tropical soils, more soluble forms of zinc than those complexed with clay and organic matter are most likely to be taken up by mycorrhizas.[61] Effects of VA mycorrhizas on calcium supply to hosts are probably minor,[60] although calcium is likely an important nutrient for VA mycorrhizal fungi because it is a constituent of polyphosphate granules in hyphae and may function in phosphorus transfer to hosts by stimulating alkaline phosphatase production.[41]

Phosphorus supplementation can reduce the concentrations of other elements in plant tissue by suppressing mycorrhiza formation[60,63] and may sometimes cause the appearance of minor deficiency symptoms even in mycorrhizal plants.[63] Phosphorus supplementation reduced total cation concentration in wheat and barley whether or not mycorrhizal, possibly by a dilution effect of large plant size, but VA mycorrhizas enhanced total anion concentration, including that of sulfate, regardless of phosphorus availability.[64] The mechanism of the latter effect for chloride, bromide, and sulfate is uncertain, but seemed unlikely to be attributable to uptake and translocation by VA mycorrhizal fungus hyphae. Buwalda et al.[64] suggested that high local photosynthate demand by mycorrhizal fungi in cortical cells might shift the host metabolic pathway for pH regulation from production of malate and other organic acids sequestered in vacuoles to hydroxyl or bicarbonate efflux. This mechanism, similar to that probably used by VA mycorrhizal fungi during nitrate uptake and assimilation, requires uptake of counterbalancing anions to maintain electrical neutrality.

B. pH and Physical Conditions

Soil reaction, moisture, temperature, and aeration influence ion availability, mobility, and uptake; hence, they influence the importance of mycorrhizas to host plants. In addition, they may directly affect root development and the provision of sites for mycorrhizal infection. Aluminum toxicity of subsoil in the tropics is commonly a barrier against deep roots.[24] Waterlogging and the formation of hardened plinthite, however, are likely to impede root development in only 12 and 11% by area of acid low base status lowland humid tropical soils, respectively.[20] The danger of hardened "laterite" (plinthite) formation after land clearing in the tropics has been greatly exaggerated,[19] although 16% of acid low base soils has a compaction hazard upon tillage.[14,20] Surprisingly, the main physical constraint of Oxisols (and to a slightly lesser extent of Ultisols) is low soil water-holding capacity. Drought stress especially affects crops, but is even noticeable among rain forest trees.[65] In forest, notwithstanding potential drought stress, most feeder roots are superficial,[18] probably because of competition for mineral nutrients from organic matter. In cropland, deep rooting for moisture is probably limited by aluminum toxicity.

1. Acidity and Aluminum Toxicity

Most lowland humid tropical soils are acid, with pH decreasing with increased depth. Exchangeable aluminum, not hydrogen, is the dominant cation associated with soil acidity.[24] By area, 72% of acid low base status tropical soils has a pH lower than 5.0 or greater than 60% aluminum saturation in its top 50 cm.[20] Except at pH values below 4.2, in which case cation uptake may be stopped or even reversed, hydrogen ion concentration has no direct effect on plant growth; effects of low pH usually result from aluminum or manganese toxicity, and calcium or magnesium deficiency.[24] High aluminum concentrations may cause formation of thick, stubby roots with necrotic spots. Aluminum often accumulates in roots and impedes uptake and translocation of phosphorus and calcium to shoots; aluminum toxicity seems to operate by precipitating phosphorus in root cell walls and cytoplasmic membranes as aluminum phosphates.[24] Many strictly tropical crops are well adapted to acid soil conditions, however, and aluminum-tolerant varieties often are able to tolerate low levels of available phosphorus. Bowen[8] has suggested that there might be a mycorrhizal component to such tolerance. Although small quantities of manganese are needed for plant growth (usually 1 to 4 ppm in the soil solution), toxic levels of this cation are present in some tropical soils.

Soil reaction can affect mycorrhizas directly through effects on fungus spore germination and germ tube growth, and indirectly by influencing phosphorus availability to uninfected roots and their susceptibility to infection. VA mycorrhizas have been found in soils from pH 2.7 to 9.2, but different fungus species have different pH tolerances.[66] In acid tropical soils, aluminum and, possibly, manganese can reduce spore germination and germ tube growth, thereby reducing root infection.[66] Effects of pH on host response to mycorrhizas are the product of several interacting factors, so that it is difficult to draw overall generalizations.[67] As liming of acid tropical soils increases phosphorus availability, host response to mycorrhizas may be enhanced provided that phosphorus levels are not so high that infection is suppressed. That high levels of available phosphorus can suppress mycorrhizal infection, probably by affecting hosts instead of by directly affecting mycorrhizal fungi, is well-documented.[41] In contrast, when phosphorus is supplied as a calcium salt with highest availability at low pH, plant response to mycorrhizas can be greatest at low pH if uninfected plants cannot take up sufficient phosphorus for growth.[67,68]

2. Moisture and Aeration

Poor rooting because of aluminum toxicity can exacerbate moisture stress. By area, 56% of acid low base status tropical soils has less than 10% available moisture in its top 50 cm.[20] Oxisols act like sands at low water tensions, draining freely, but hold water like clays at high tensions resulting in a narrow range of water availability.[24]

Drought susceptibility completely out of proportion to clay and water contents is caused by most iron and aluminum polyphosphates and occluded organic matter aggregates falling in the sand-particle size range. Water infiltration rates consequently are high, so puddling during heavy rains is seldom a problem; aeration and root penetration are good.

In forest, the proportion of deciduous to evergreen species indicates the degree of water stress during the dry season.[65] During the rainy season, moisture stress can occur even in well-watered plants if transpiration is high.[24,69] Because of their exposure, canopy trees may have high transpiration rates. They are also likely to have high resistance to water movement because of long flow paths and lower amounts of absorbing surface per root mass. Hence, leaf water potentials of canopy trees may fall to low levels.[69] The funneling effect of large trees can partially compensate by stemflow keeping the soil around dominant individuals wetter than around subdominants. Understory plants have very low transpirational water loss even during dry periods if the overstory does not lose too much foliage. Crops have high transpiration rates and poorly developed root systems when young[70] or when suffering nutrient stress or toxicity.

Effects of VA mycorrhizas on plant water status may be especially important for canopy trees and crops because of their potentially high transpiration rates and the low moisture availability in many tropical soils. VA mycorrhizas have decreased susceptibility to wilting and transplant shock in tropical trees,[11,71,72] and they have improved water utilization by a tropical shrub[68,73,74] and tropical crops.[68,73,75] Although VA mycorrhizas occur in both aquatic habitats and deserts[41] and appear capable of tolerating very high and very low moisture availability, infection of temperate-zone plants is reduced by excessively high or low soil water potentials.[76,77] Lower infection of a tropical tree in the field during the dry season in an arid zone,[4] and maximum infection of seedlings of an African mahogany at intermediate moisture levels[78] are consistent with temperate-zone findings. In other experiments with tropical plants, however, correlation of infection with moisture availability was poor because of interactions of this factor with soil type, phosphorus availability, temperature, and host and fungus species.[68,73-75] Although VA mycorrhizas can be expected to improve host water utilization, the magnitude of improvement depends upon a complex of factors and is not predictable at present.

The mechanisms by which mycorrhizas improve host water relations are uncertain. Improved water utilization has resulted most often as a secondary effect of improved phosphorus nutrition;[41,77] improved water use may reflect decreased resistance to water flow in roots,[41] stems, and leaves.[41,72,79] Some small amount of water uptake and translocation occurs in VA mycorrhizal fungus hyphae,[41] and this is likely to be important when soil moisture availability is low. Drought resistance cannot always be duplicated in the absence of mycorrhizas by added phosphorus.[41] Water has low diffusivity at high tensions; with sparse rooting in sands[8] and Oxisols, hyphae ramifying in clay or organic matter aggregates might by-pass dry zones developed around actively absorbing roots and maintain continuity of water transfer to roots.[41] Such an effect could be especially important if hyphae can survive low moisture availabilities that root hairs cannot,[41] and for species lacking root hairs.

Most tropical soils are well-aerated because of the free drainage of their macropores, but if drainage is impeded, low oxygen tensions are potentially detrimental for mycorrhiza function. Soil oxygen concentrations of 12 or 16% maximized mycorrhizal infection, plant dry weight, and uptake of several mineral nutrients by a tropical composite species.[80] The common observation that most mycorrhizas are superficial in the soil[81,82] may be explained by decreased oxygen availability with increased depth in soil. Because of mass flow of mineral nutrients to roots coupled with diminished physiological activity of mycorrhizal fungi resulting from low oxygen availability under very high soil moisture regimes, mycorrhizas may negatively affect host growth. This possibility makes mycorrhizal physiology in seasonally flooded sites[15,16] especially interesting.

3. Temperature

Soil temperature can influence both moisture availability in the soil and mineral nutrient supply to plants; temperature primarily affects soil moisture indirectly by influencing transpiration, but can also affect the physiological activities of mycorrhizal fungi and decomposer microorganisms, thereby directly influencing water and mineral nutrient uptake and mineralization of organic material. Tropical soil temperatures closely approximate air temperatures and are similarly invariant diurnally and annually, changing by less than 5°C between mean summer and mean winter temperature at 50 cm depth.[24] Average lowland soil temperatures generally exceed 22°C. A bare soil in Zaire had a record 86°C surface temperature, but was a nearly normal 30°C at 10 cm; at the same site, surface soil temperature was 25°C in forest and 34°C in grassland.[24] Invariant high temperatures throughout the year can be somewhat disadvantageous for cultivation because there is no cool period to regulate the water demand of young crop plants; frost does not restore soil structure for water, air, and root penetration; and humus, on which structure and porosity mainly depend, decomposes rapidly.[70]

VA mycorrhizas of several tropical crops had their greatest effect on host growth and phosphorus uptake at 30°C soil temperature,[67,68] suggesting that single-peak temperature optima exist for mycorrhiza function. Even if inoculated plants were pretreated for 20 days at the optimum temperature before exposure to lower or higher temperatures, the disadvantages of nonoptimal temperatures were not eliminated, hence, temperature must have acted through effects on fungus metabolism instead of influencing the development of infection. An investigation of the combined effects of temperature and soil moisture similarly indicated that optimum mycorrhizal efficiency at 30°C was attributable to enhanced fungus physiological activity.[74] This accords with the finding of a temperature optimum (at 15 to 25°C) for phosphorus transport by hyphae of a VA mycorrhizal fungus in a split-plate experiment with white clover.[83]

In addition to influencing VA mycorrhizal fungus metabolic activity, temperature can also affect the development of infection. Infection by VA mycorrhizal fungi from southwestern Oregon was greatly reduced or prevented at soil temperatures above 29.5°C,[84] although a VA mycorrhizal fungus species common in the rhizosphere of summer crops in Florida had its optimum for infection near 30°C.[85] A comparison of six VA mycorrhizal fungus species from Florida demonstrated that temperature optima for infection are specific to fungus species.[86] Temperature may affect infection either through its direct effect on fungus metabolism or indirectly by influencing leakage of root metabolites necessary for fungus activity.[87] It is probable that indigenous lowland humid tropical VA mycorrhizal fungi are adapted to prevailing forest soil temperatures; their ability to tolerate the elevated soil temperatures at less than 30 cm depth attendant upon clearing is uncertain, especially in view of the low variation in soil temperature that the fungi normally experience.

C. Soil Changes after Clearing

When tropical soils are cleared of forest, topsoil temperatures may increase 7 to 11°C,[24] causing more rapid organic matter decomposition and sometimes having adverse effects on soil structure,[88] aeration porosity, and water percolation.[89] In general, soil structure tends to deteriorate under cultivation because the decline in organic matter lowers the average size of soil aggregates. This, however, might improve moisture availability in some soils and might reduce subsequent mineral loss by leaching.[24] Nevertheless, runoff and erosion on slopes can be increased by decreased infiltration, exacerbating fertility problems.

Clearing can alter soil reaction depending on whether or not slash is burned. Soil acidity increased slightly with increased exposure when slash was manually removed,[88] but with burning, acidity can markedly decrease.[90] Increased pH can make immobilized phosphorus more soluble, and might thereby increase its loss by leaching. Clearing and clearing with burning[90,91] increased leaching loss of phosphorus, nitrogen, potassium, calcium, and magnesium; burning volatilized carbon, nitrogen, and sulfur.[90] In another study,[92] clearing without burning, followed by abandonment to natural regeneration, resulted in immediate flushes followed by the usual declines of nitrate, potassium, calcium, magnesium, and manganese; because of the presence of large amounts of slowly decomposing woody litter, however, soil organic matter and associated phosphate, ammonium, and zinc increased during the first 8 years of regrowth. Exchangeable aluminum, iron, and hydrogen were markedly increased for 8 to 16 years, although pH and soil structure were highly stable.

The effects of forest clearing without burning are likely to enhance host response to VA mycorrhizas. The presence of slash moderates soil temperature and runoff, and as woody litter slowly decomposes, available phosphate and ammonium may increase, although probably not so quickly that mycorrhiza formation is suppressed. Leaching of nitrate and cations amplifies any value of mycorrhizas to hosts for nitrogen and cation uptake; elevated exchangeable aluminum heightens any beneficial effects of mycorrhizas in overcoming aluminum toxicity. Forest clearing without burning or slash removal is probably similar in effect on mycorrhizas to the normal regeneration process of lowland tropical forest which involves the opening of tree-fall canopy gaps, except for the influence of scale of disturbance on the arrival of seeds and rate of root mass reestablishment.[12]

Forest clearing followed by burning likely disfavors immediate plant responses to mycorrhizas, yet this practice is integral to the practice of shifting cultivation which accounts for a high proportion of tropical agricultural land use.[20] Burning of slash releases a transient flush to the soil of those nutrients that are not volatilized, and crop productivity is highly dependent on this nutrient flush. Phosphorus solubilization increases as pH rises after burning, and available phosphorus might reach sufficient levels to suppress mycorrhiza formation by mycorrhiza-independent crop species (see Section III.B). Burning of slash is likely to bare portions of the soil surface, increasing soil temperature. This too could inhibit mycorrhiza formation, although VA mycorrhizal fungus spores can survive temperatures that greatly reduce or totally inhibit mycorrhiza formation.[84] Subsequent mineral nutrient deficiencies and aluminum toxicity constitute the main soil constraints on shifting agriculture that cause abandonment of fields to forest regrowth or planting to pasture after 1 to 2 years of cropping.[20] Weeds and forest species can be as strong a constraint to crop production as diminished soil fertility,[20] however, perhaps because they are better able than crops to take advantage of mycorrhizas as fertility plummets.

III. ROOT ENVIRONMENTS

VA mycorrhizal fungi simultaneously occupy two different environments (the soil and roots) and their ecophysiology is influenced by both. Soil chemical and physical parameters determine the potential benefits of mycorrhizas to hosts, but host/fungus interactions influence realization of those benefits.

A. Plant Associates

Relatively few surveys of the occurrence of VA mycorrhizas in lowland humid tropical habitats or on tropical hosts have been conducted, but data are available from all three main tropical regions of the world (Asia,[7,93-99] Africa,[100,101] and the

Neotropics[11,82,102-108]). Unfortunately, most accounts do not provide full information concerning (1) how species were selected for examination; (2) whether or not seedlings, saplings, or adults were examined; (3) species stature (e.g., understory large herb, subcanopy tree, etc.); (4) number of samples examined per individual and species; (5) depth in soil of sampled roots and intensity of infection; and (6) habitat type and soil chemistry. Nevertheless, the pattern emerging from most of these studies is one of ubiquity of VA mycorrhizas in lowland humid tropical ecosystems, and infrequent occurrence or sparse infection in pioneer or colonizing species and in some crops.

Most infection surveys have underscored the regularity of VA mycorrhizal infection of tropical trees; exceptional records pertain to species that have other types of mycorrhizas or lack mycorrhizas. Of several typically tropical species growing in Japan (e.g., *Carica papaya, Terminalia cattapa, Psidium guajava)* only pineapple lacked VA mycorrhizas.[94] In India, 22 forest tree species were all infected by endogonaceous fungi, although some infections consisted of only vesicles, arbuscules, or hyphal pelotons (all variations of what I am loosely calling "VA mycorrhizas").[95] A survey of about 200 species of Malaysian forest trees which were not listed found that just *Durio zibethinus* lacked mycorrhizas.[97] Only betel nut palm and 3 species of Sapotaceae lacked mycorrhizas of 63 tree species belonging to 26 families in a lowland tropical forest aboretum in Sri Lanka.[99] From Brazilian *cerrado,* 56 species described as having "endotrophic mycorrhizae with septate mycelia"[104] probably had VA mycorrhizas, because some of the host taxa listed have been repeatedly reported to have VA mycorrhizas elsewhere. In 47 families collected primarily near Manaus, Brazil, 121 species had VA mycorrhizas.[105] In Cuba, 11 species of forest trees and 6 savanna forbs had VA mycorrhizas.[106] VA mycorrhizal fungus spores were extracted from the rhizospheres of 62 species of Brazilian *cerrado* plants in 32 families,[108] but finding spores in the rhizosphere of a plant need not imply that it is infected.

Several studies accord with ubiquity of infection in forest trees, but report less frequent or lower intensity infection in seral and herbaceous species. In a survey of 75 species, including mosses, ferns, gymnosperms, monocots, and dicots in a botanical garden with forest remnants in Java, all 46 forest tree species examined had VA mycorrhizas, but 6 (i.e., species of *Equisetum, Asplenium, Pollia, Paspalum, Musa,* and *Nepenthes*) of 29 herbaceous species lacked mycorrhizas.[7] In Trinidad, of 93 species surveyed, 13 forest trees all had VA mycorrhizas, and had the highest average intensity of infection when compared with weed, secondary bush, orchard savanna, and crop species, not all of which were infected.[102] In Nigeria, only very young plants growing in newly cleared and burnt nurseries lacked mycorrhizas; 44 indigenous tree species and 15 exotics had VA mycorrhizas.[100,101] A very limited survey in the Philippines found VA mycorrhizas on a dominant savanna tree species, but did not find mycorrhizas on *Trema orientalis,* a small gap-colonizing tree growing on the university campus at Los Baños.[96] Study of several vegetation types in subtropical south Florida showed a similar trend.[109] From a 17-year-old *Myrica* woodland, a 30-year-old *Schinus* forest, and a mature hardwood hammock forest, larger proportions of the species examined had VA mycorrhizas than species from 3-month-old and 2½-year-old burned and unburned seral communities. A comparison of 12 common tree species from 2 forest types with 10 of the most abundant early successional species from a 2-year-old abandoned farm plot at San Carlos de Rio Negro, Venezuela, however, found the ranges of VA mycorrhizal infection percentages of both groups to broadly overlap.[107]

Two studies somewhat contradict the nearly universal occurrence of VA mycorrhizas on tropical forest trees. In Puerto Rico, 9 of 32 forest tree species lacked mycorrhizas,[103] but that report was based upon examination of thin sections of roots in which it is difficult to detect sparse Va mycorrhizal infection. However, in a study conducted in forest near Manaus, Brazil, repeated samples of 25 species among 86 examined lacked mycorrhizas.[82] Members of the families Lecythidaceae and Sapotaceae especially lacked infection; this lack in the latter family agrees with observations in Sri Lanka. Nonmycorrhizal species in this Brazilian forest were estimated to account for almost one quarter of all species by importance value.

Among pioneer or colonizing species, members of the Phytolaccaceae,[11] Papaveraceae,[110] and Cyperaceae[111] that can dominate disturbed sites do not form mycorrhizas. Presumably, other weedy members of the Papaverales, Caryophyllales, Polygonales, Urticales, and Capparales among the dicots, and Commelinaceae, Juncaceae, and Cyperaceae among the monocots[10,11,96,110,111] will be demonstrated to always lack mycorrhizas, although by no means will all species included in these groups fail to form mycorrhizas.[10,11] Some weedy species such as *Eupatorium odoratum* do depend on VA mycorrhizas.[80] Invasive species of Casuarinaceae can form VA mycorrhizas or ectomycorrhizas, as can some forest trees in the Leguminosae and Dipterocarpaceae.[10]

In lowland dry tropical areas, as in humid forests, VA mycorrhizas appear to be common. Of 44 species in dry, lowland forest in Tanzania, 37 had VA mycorrhizas; the remainder had ectomycorrhizas.[5] All 23 species examined from a dry forest in Puerto Rico including trees, shrubs, and herbs had VA mycorrhizas.[6] I know of only a single report concerning humid tropical forests at high elevations which lists VA mycorrhizas in all three tree species that were examined.[5] For the humid tropical lowlands, the infection status of plants growing in forest understory at low light intensities, of epiphytes, and of canopy roots[112] has not been reported.

Butler[93] was the first to comment on the regularity of occurrence of VA mycorrhizas in cultivated perennial plants, and on their incidental presence in some annual crops in India. He noted that infection is common in tea, coffee, rubber, and citrus, and luxuriantly developed in bamboo and some shade trees, but infrequent in the common cereals and lacking in flooded rice. VA mycorrhizal infection has been reported for many perennial humid-tropical crops: sugar cane,[113] citrus,[114-116] cacao,[117] Brazilian rubber tree,[118] avocado,[119] litchi,[120] papaya,[121] peach palm,[122] cassava,[123] black pepper,[124] guava,[11] ramtil, and sweet mace.[67,68] Among lowland tropical perennial crops, only pineapple,[94] durian,[97] and betel nut[99] have been examined and not found to have VA mycorrhizas. Annual crops had low average intensity of infection in Trinidad,[102] and were more often uninfected than citrus in New Zealand.[116] Although upland, dry rice can form VA mycorrhizas,[45] and some tropical aquatic plants can sustain VA mycorrhizas,[98] lack of infection in flooded rice is similar to absence of infection from all of the 20 species examined from a seasonally flooded everglade marsh in Florida.[109]

B. Host/Fungus Interactions

Infrequent infection and low average infection intensities of seral species and annual crops in the lowland humid tropics may reflect intolerance of the soil environment by species of VA mycorrhizal fungi after land clearing, lack of VA mycorrhizal inoculum,[12] or independence of mycorrhizas by these plant species. Little is known concerning the tolerances of tropical VA mycorrhizal fungus species for soil environments altered by clearing. In general, phosphorus availability probably influences mycorrhiza formation more than any other factor. Profit from mycorrhizas is primarily limited by phosphorus availability and the capacity of uninfected individuals to take up phosphorus, and it is the ability of hosts to profit from mycorrhizas that constrains the frequency and intensity of infection that they sustain.

Many tropical plants have been shown to benefit from VA mycorrhizas when little phosphorus is available. Most studies, however, have involved fruit trees or herbaceous crops; few investigations have considered the responses of forest trees or seral species to VA mycorrhizas. Early research sought effects of VA mycorrhizas on the growth of coffee,[7] citrus,[116] cacao,[117] and Brazilian rubber tree,[118] but failed to detect responses to infection, probably because of the use of heat-sterilized soil of relatively high fertility, or because experiments were not long enough for large-seeded species to exhaust their seed mineral reserves[11] and become totally dependent on mineral nutrient uptake from the soil. Subsequent research has repeatedly demonstrated beneficial effects of VA mycorrhizas on citrus[63,72,125-133] and other perennial tropical crops,[11,71,121,122,124,134] especially cassava.[42,50,60,123,135,136] VA mycorrhizas have improved the growth of annual crops and pasture grasses in tropical soils,[44,45,61,67,68,73,75,102,111,123,137-141] as well as nodulation and growth of tropical crop and forage legumes.[44,46-49,123,138,142-145] VA mycorrhizas also have stimulated the growth of seedlings of lowland humid tropical forest trees[11,78,146,147] and seral woody species[11,73,74,80,148] in pot experiments. Because of their superficial, magnolioid roots, most tree species similarly are likely to benefit from VA mycorrhizas under the ambient conditions of low phosphorus availability in tropical soils.

1. Dependence on Mycorrhizas

Plant species differ greatly in the ability of individuals without mycorrhizas to grow, which reflects differences in their mineral nutrient requirements, growth rates, and capacities of uninfected root systems to take up required minerals. Root/shoot ratio, root distribution, geometry, morphology, and investiture with root hairs all may influence mineral uptake by uninfected roots.[8,11,149] Some species, such as members of the Phytolaccaceae,[11] Cyperaceae,[111] and Brassicaceae,[123] are totally independent of mycorrhizas for mineral nutrient uptake; they are nonmycorrhizal, depending on fine, highly branched root systems, root hairs, or secretion of organic acids (which may release occluded organic phosphorus from clay aggregates in acid tropical soils) for mineral nutrient uptake, or tolerating limited mineral supply by having slow growth rates or low tissue mineral requirements. Mycotrophic species — those of which growth can be stimulated by mycorrhizas — range from those facultatively dependent on mycorrhizas, to those unable to grow without mycorrhizas in the most fertile of their natural soils;[11] the latter are ecologically obligately mycotrophic.[18] Seral species tend to be nonmycorrhizal or facultatively mycotrophic; tropical forest trees and fruit trees with extensive, superficial roots that have low orders (3 to 4) and rates (branches per length) of branching, coarse ultimate rootlets lacking root hairs, high lignin, tannin, or other antiherbivore compound contents in their roots,[150] and low root turnover rates tend to be obligately mycotrophic.[11,12,122]

Species and varieties have been ranked with respect to dependence on VA mycorrhizas by the level of phosphorus availability below which they could not grow without mycorrhizas,[123] and according to their ability to grow without mycorrhizas relative to growth when infected.[11,129] Minerals other than phosphorus can affect assessment of dependence on mycorrhizas, however,[129] hence, ranking of species according to dependence on mycorrhizas may best be accomplished by comparing growth with and without mycorrhizas on all soils of different fertilities on which the species co-occur.[18] In practice, the latter approach is laborious if not impossible, so easily assessed correlates of dependence have been sought.

Baylis suggested that dependence on mycorrhizas is correlated with the presence of large-diameter ultimate rootlets which lack root hairs — the magnolioid root morphology.[151] Correlation of dependence on mycorrhizas with ultimate rootlet diameter is strong for tropical plants, but correlation of independence with presence of root hairs is poor. Citrus varieties with measurably different degrees of dependence on VA mycorrhizas did not possess root hairs, but root dry-masses of uninfected plants were inversely proportional to dependence.[129] Root/shoot ratios, however, were not correlated with dependence. Cassava, which is highly dependent on mycorrhizas, has a coarse, sparsely branched root system.[42] *Phytolacca rivinoides,* a nonmycorrhizal species,[11] lacks root hairs, but its root system is highly branched and its ultimate rootlets are very fine. Other studies that have not measured dependence directly, but have assumed that infection is correlated with dependence, have found similar relationships. VA mycorrhizal infection was significantly correlated with ultimate root diameter of 89 Brazilian forest trees of which few had root hairs.[152] Infection of 105 seedlings of an African mahogany was not correlated with root hair development.[100]

Unfortunately, root infection percentages may not accurately indicate dependence on mycorrhizas. Infection of a perennial species at one time need not reflect its dependence on mycorrhizas during its entire life. Mature trees in aseasonal environments that do not have coordinated leaf loss may have low mineral nutrient requirements except when fruiting, and may be much less dependent on mycorrhizas than rapidly growing seedlings and saplings. Infection can vary with depth in soil; superficial roots of tropical trees have greater infection than roots deeper in the soil.[81,82,100] In addition, conditions of very low phosphorus availability under which most mycotrophic species are highly dependent on mycorrhizas for mineral uptake and growth can sometimes result in low infection percentages.[42,46,111]

2. Infection Regulation

Little VA mycorrhizal infection at very low levels of phosphorus availability may lend credence to a recently proposed mechanism by which infection is controlled: phosphorus-nutrition-dependent changes in root cell plasma membrane permeability might influence exudation from roots of materials (including soluble carbon compounds) that stimulate fungus spore germination, germ-tube growth,[153] and the development of infection.[41] Phosphorus-deficient roots are thought to be leaky, hence, to favor infection provided that photosynthate is available for leakage. At very low levels of phosphorus availability there may be insufficient photosynthate to foster much infection of some hosts even though root membrane leakiness is maximal. Shading to 2% of full sun, however, had no effect on infection of an African mahogany.[78] Decreased light intensity similarly had no effect on infection of sudangrass when phosphorus availability was very low, but infection of plants grown under low light intensity was inhibited more strongly by added phosphorus than that of plants under high light intensity.[87] In contrast, exudation of reducing sugars from mycorrhizas of citrus was enhanced by lengthened photoperiod.[130] It is likely that the effects of light and phosphorus supply to uninfected roots on VA mycorrhizal infection will differ quantitatively among plant species with different shade tolerances and membrane biochemistries. Among 13 wheat cultivars, those with the least infection that were among the most independent of mycorrhizas for growth, had the lowest sugar contents in their root exudates in spite of having high quantities of sugars in their roots.[154]

Infection regulation under host control strongly implies that there can be situation-dependent disadvantages to hosts of mycorrhizal infection; disadvantage from mycorrhizas is most likely a reflection of carbon cost to hosts of growth and respiration of their mycorrhizal fungus associates. Free-living ability of VA mycorrhizal fungi has not been demonstrated, and it is probable that they depend on hosts for energy.[41,155] Carbon is translocated from hosts to fungi, and direct measurement of the carbon costs of mycorrhizas by supplying radioactively labeled carbon dioxide to hosts suggests that approximately 10 to 12% less net assimilate from photosynthesis is retained by mycorrhizal hosts than by uninfected plants.

Indirect estimates of the carbon costs of VA mycorrhizas have been made by supplying uninfected plants with sufficient phosphorus that they attain the same tissue phosphorus concentration or total phosphorus content as mycorrhizal individuals which are usually smaller; the difference in size representing the cost of mycorrhizas in terms of diminished host growth. Such cost estimates can be as high as 60%,[41] but may be misleading. These estimates assume equal rates of gross photosynthesis by infected and uninfected plants over the entire growth period so that differences in net plant assimilation are wholly attributable to the carbon drain imposed by mycorrhizal fungus associates; this assumption probably is untenable. Well-infected mycorrhizal individuals of some species have been demonstrated to have higher photosynthetic rates than uninfected individuals of similar phosphorus nutritional status[41,72] which would cause the growth difference method to underestimate the cost of mycorrhizas. It seems more likely, however, that during the initial phases of growth before mycorrhizal infection is well developed, the photosynthetic rate of uninfected plants receiving supplemental phosphorus would exceed that of inoculated plants given little or no phosphorus; if inoculated individuals do not produce the same total gross photosynthate as uninfected individuals over the course of an experiment, then the growth difference method overestimates the cost of mycorrhizas.

Although obligately mycotrophic species by definition always depend on mycorrhizas and must sustain their cost, mycorrhizas are not always advantageous to facultatively mycotrophic species. If infection occurred, it likely would be a detriment to nonmycorrhizal species.[11] When mycorrhizal individuals are smaller than uninfected plants of the same phosphorus content or concentration (as has sometimes been observed[41]), mycorrhizal individuals of the same size as uninfected individuals must have higher phosphorus contents and concentrations. This implies that mycorrhizal plants accumulate "luxury" phosphorus, i.e., are less efficient at utilizing phosphorus than are uninfected plants. If mycorrhizal individuals of a species are inefficient in phosphorus utilization when compared to uninfected plants over the range of soil fertilities that species naturally encounters, then natural selection should cause the species to reject mycorrhizal infection; that is, become nonmycorrhizal. Higher phosphorus contents of mycorrhizal than uninfected individuals, however, may be correlated with more numerous photosynthetic units, improved stomatal control and water use, and energetically less costly pH regulation during ion uptake.[41] Compensatory increases in rate of photosynthesis or leaf hydration of mycorrhizal plants[155] could reduce or eliminate apparent inefficiency in phosphorus utilization. Nevertheless, the reduced growth of mycorrhizal individuals compared to uninfected plants given supplemental phosphorus attests that VA mycorrhizas can be a stress to facultatively mycotrophic species under nonlimiting conditions of mineral nutrient availability, especially of phosphorus.

The consequence of the potential cost of mycorrhizas to facultative mycotrophs in excess of benefits at high levels of nutrient availability is infection regulation, perhaps by changes in membrane permeability and root exudation. This hypothesized mechanism is attractive because it incorporates simultaneous influences of both benefits (i.e., adequacy of phosphorus supply for membrane integrity) and costs relative to benefits (i.e., availability of photosynthate for exudation which is partly a function of plant nutritional status). Provided that the spread of VA mycorrhizal infection in a root system requires repeated reinfection from the soil, decreases in membrane permeability, root exudation, spore germination, and germ tube growth after initial infection have improved root phosphorus status could establish an equilibrium level of infection. To the extent that repeated reinfection from the soil is replaced by fungus spread within roots, the benefits of mycorrhizas which depend upon hyphae external to the root will decline and infection will not be effective. Ineffective infection would increase membrane permeability, however, with the possibility that more infection from soil-borne propagules would be stimulated; increased leakage of carbon compounds even might reduce the carbon supply to ineffective infection, thereby limiting its continued spread.

How a host plant partitions photosynthate between its own growth centers and mycorrhizal fungi, and how it concomitantly utilizes phosphorus influence tissue phosphorus concentration and probably determine host species-specific characteristic levels of mycorrhizal infection. The more a species depends on mycorrhizas for mineral nutrient uptake and growth, the less root membrane permeability should change with phosphorus uptake and the greater the proportional effect of photosynthate partitioning will be on root exudation and mycorrhizal infection. Rapidly growing, facultatively mycotrophic weedy species may divert photosynthate to the strong carbon sinks of their own growth centers away from mycorrhizal fungi when mineral nutrients are moderately available. In contrast, dependent species probably maintain infection by continuing root exudation. Among wheat cultivars, the most highly infected tended to show the greatest increases caused by mycorrhizal infection in the content of reducing sugars in root exudates.[154]

Continued root exudation requires that the root membranes of highly mycorrhiza-dependent plants retain their permeability characteristics as phosphorus is taken up. This could be accomplished by maintenance of relatively invariant tissue phosphorus concentrations; that is, dependent plant species should be less likely to accumulate what I have called "luxury" phosphorus than species that are more independent of mycorrhizas. There was no correlation between dependence on mycorrhizas and shoot phosphorus content of wheat cultivars,[154] but average percentage of phosphorus in the leaves of uninfected plants of five citrus cultivars when given moderate fertilization minus phosphorus was inversely correlated with dependence on mycorrhizas.[129] In general, wheat with its graminoid root system is less dependent on mycorrhizas than citrus, which has more nearly magnolioid roots. Uninfected citrus given 100 mg of phosphorus weekly (in 1200-mℓ containers) were similar in size to unfertilized mycorrhizal plants after 26 weeks, and did not differ significantly from mycorrhizal plants in leaf phosphorus concentration.[133] Cassava, which can depend greatly on VA mycorrhizas,[123] tended to have slightly higher phosphorus concentrations in the tops of uninfected plants than in the tops of mycorrhizal plants of similar size when eight different cultivars were grown in flowing solution culture at four levels of phosphorus availability.[42]

C. Host-Plant Interconnections

In consequence of the improbability of mycorrhiza-dependent species rejecting infection by nonspecific VA mycorrhizal fungi in impoverished tropical soils,[11] the predominance of obligately mycotrophic species in lowland tropical forests may have profound effects on nutrient cycling and plant competition. Several temperate-zone studies have demonstrated that the root systems of mycorrhizal individuals of the same and different species can be interconnected by VA mycorrhizal fungus hyphal bridges,[156-159] and that phosphorus[157,158,160] and carbon[159] can be transferred between host root systems by VA mycorrhizas. Although host-plant interconnections undoubtedly can occur, their persistence in nature will be influenced by rate of root turnover and predation on hyphae by soil organisms. In a California serpentine annual grassland, mycorrhizal interconnections were thought to fluctuate greatly such that transfer of radioactive phosphorus from donor plants to recipients could not be predicted by taxonomic affinity, distance from donor, or size of recipient.[158] In a lowland tropical forest, root turnover likely is less than in an annual grassland, but predation on hyphae might be substantial because of the lack of an adverse season that limits populations of fungus-feeding organisms. Nevertheless, the prevalence in some lowland humid tropical forests of small achlorophyllous herbs[17] that probably receive carbon via hyphal connections to autotrophic donors[161] suggests functional persistence of hyphal bridges.

VA mycorrhizal hyphal bridges among individuals of many different species in typically highly diverse lowland humid tropical forest could influence the distribution of mineral nutrients and carbon among hosts. In a series of short-term labeling experiments, radioactive phosphorus applied to the shoots of donors was transferred inter- and intraspecifically to the shoots of recipients in a soil of low phosphorus availability, but transfer could not be attributed uniquely to VA mycorrhizas.[160] Mineral nutrients sufficient to increase the relative growth rate of a recipient plant fourfold when compared to uninfected controls probably did move through interconnecting hyphae from the roots of a fertilized donor plant to a conspecific recipient plant in washed sand.[157] Similarly, radioactively labeled photosynthate moved from the leaves of a donor through interconnecting VA mycorrhizal hyphae to the roots of a heterospecific recipient in dune sand during a short-term experiment.[159] Recipient plants in the dark received six times more label than those in full light. Source-sink relationships most likely govern the movements of both mineral nutrients and carbon among interconnected plants.

If interspecific hyphal interconnections among hosts in lowland humid tropical ecosystems are persistent, then those hosts in effect share a common root system. Should shaded shoots be strong sinks for photosynthate from illuminated individuals as suggested by Francis and Read,[159] the cost of mycorrhizas to illuminated plants will be increased. The overall effect of such carbon transfer would be to homogenize plant growth rates and mineral utilization among species within the constraints of intrinsic differences in growth rate and tissue mineral requirements. I have suggested that the high within-habitat species diversity of lowland humid tropical forests may result in part because obligately VA mycotrophic host species simultaneously associated with the same few VA mycorrhizal fungus species differ little from one another in ability to compete for limited mineral nutrients.[10] If host species share photosynthate, then their ability to outcompete one another is additionally constrained.

The need of obligately mycotrophic hosts for consistent mycorrhizal infection without which they cannot take up required mineral nutrients may make them susceptible to sharing photosynthate. Over evolutionary time, however, natural selection would favor adaptations for mineral uptake alternative to mycorrhizas should costs be excessive relative to benefits. Stated another way, there should be strong constraints against "cheaters" — shoots that fail to foster mycorrhizal fungus growth sufficient to recompense the benefits derived from mycorrhizas — on shared mycorrhizal root systems. Tightly linked reciprocal inorganic phosphate and sugar transport across host or mycorrhizal fungus plasma membranes could prevent cheating, but such a mechanism is unknown in roots.[41] Nevertheless, if cheating is prevented in some way, sharing of photosynthate between donor and recipient shoots must somehow enhance phosphorus supply to the donor, perhaps by increasing overall phosphorus uptake.

If overall performance of VA mycorrhizas can be enhanced by simultaneous infection of more than one host species, then tropical polycultural-agricultural systems might be designed to profit from such an effect. Puga[111] has recently shown that VA mycorrhizas can cause overyielding in pots by mixtures of corn and a solanaceous weed that is common in the fields of farmers practicing shifting cultivation in Panama, although host interconnections were not demonstrated in his experiments. In both the absence and presence of mycorrhizas produced by inoculation of sterilized soil with field-collected roots of a forest palm species, mixture reduced shoot dry-mass of the solanaceous weed species by about two thirds of its growth in monoculture; mixture similarly halved corn shoot dry-mass in the absence of mycorrhizas, but mixture more than doubled corn shoot yield when mycorrhizas were formed. When compared with the most heavily infected monoculture, mixture markedly increased the average percent infection of combined corn and solanaceous roots which could not be separated, and probably increased the total number of mycorrhizas formed, as well. Increased infection may have resulted from increased photosynthate available to mycorrhizal fungi from large plants, or because of complementary stimulation by both host species of several species of mycorrhizal fungi that were potentially present. Single VA mycorrhizal fungus species preferentially enhanced the growth of kudzu or cassava in mixture,[136] suggesting that host species might correspondingly differentially stimulate mycorrhizal fungi. In species-rich, lowland humid tropical forests such facilitative interactions may especially contribute to high productivity of both obligately mycotrophic host plants and VA mycorrhizal fungi.

IV. CONCLUSIONS

VA mycorrhizas in lowland humid tropical ecosystems function similarly to temperate-zone VA mycorrhizas, but tropical soil and root environments are different from those of the temperate zones in several respects; the ecophysiology of VA mycorrhizas of lowland humid tropical forest trees and some perennial crops may represent an extreme of the spectrum of adaptations of VA mycorrhizas. Although indigenous tropical VA mycorrhizal fungi and host-fungus combinations adapted to prevailing conditions in tropical soils likely have different phosphorus, pH, moisture, oxygen, and temperature optima for infection and function than temperate-zone associates, tropical seral species and annual crops broadly overlap with temperate-zone species in interaction with VA mycorrhizal fungi.

Tropical seral species and herbaceous crops tend to be facultatively mycotrophic, as are many temperate-zone species that are growing in more generally fertile soils. The independence of mycorrhizas of tropical facultative mycotrophs probably results because chemical soil characteristics after land clearing often facilitate growth without mycorrhizas, at least temporarily, and physical conditions are unfavorable for the persistence of mycorrhizal fungi. Typical tropical shifting agriculture involves the burning of slash which produces a transient increase in available phosphorus, potassium, calcium, and magnesium, and elevated pH; soil structural changes associated with loss of organic matter increase moisture and nitrate retention. These changes favor fine, fibrous root systems which reject mycorrhizas that would impose needless carbon drain. Mycorrhiza dependence, moreover, would be a liability if mycorrhizal fungus hyphal growth or propagule survival is limited by elevated phosphorus, high temperature, decreased aeration, and reduced root development. After clearing, soil temperatures near the surface where the majority of mycorrhizas are located can greatly exceed the full range of temperature normally experienced by mycorrhizal fungi in forest; hyphal growth might be completely inhibited by such temperatures even if spores can survive them. VA mycorrhizal fungi from soils of low phosphorus availability similarly may be more sensitive to elevated phosphorus availability than fungi from more fertile soils, and may be susceptible to direct effects of phosphorus in addition to effects of phosphorus mediated by hosts. Ironically, mycorrhiza independence of crops and seral species and inability of VA mycorrhizal fungi to tolerate altered soil conditions after vegetation is cut and burned may lead to diminution of mycorrhizal fungus populations upon which continued agricultural productivity and ecosystem recovery increasingly depend as fertility declines.

Low mineral nutrient availability, especially that of phosphorus, in most tropical soils likely has led to obligate mycotrophy of many forest tree species and perennial crops. The pivotal role of organic matter in both provision and retention of phosphorus, sulfur, and nitrogen in the soil probably contributes predominantly to the superficiality of tree root systems and utilization of VA mycorrhizas from which external hyphae can colonize organic particles. Provided that mycorrhizal inocula are reliably present in forest soils, extreme dependence on mycorrhizas makes root hairs redundant. Root hairs can be a liability, for they are frequently sites of injury and invasion by pathogens and parasites. Moreover, because tropical soils drain very freely and canopy trees experience high transpirational water demands, root hairs may be less effective than hyphae in maintaining unbroken pathways for water uptake. Tree root systems spread greatly to collect water and mineral nutrients from a large area, hence, roots may tend to be coarse, with large-diameter vascular elements that have low resistance to water transport. Thus, obligate mycotrophy of tropical trees is correlated with magnolioid root morphology.

Tropical soil and tree root characteristics demand highly effective mycorrhizas; in consequence, several postulated physiological attributes of VA mycorrhizas most likely may be found in the lowland humid tropics. Tropical VA mycorrhizal fungus hyphae should vigorously colonize organic material, and the fungi might produce organic acids and metallic-ion ligands, in addition to acid phosphatase. High root densities, high leaching rates, low cation exchange capacities, and resultant low soil solution concentrations of cations may make mycorrhizas effective in uptake of potassium and ammonium; free drainage may reduce diffusion and enhance this effect. Mycorrhizas can enhance soluble sulfate uptake, and root density and leaching might emphasize this ability and confer effectiveness in nitrate uptake as well. Nitrate uptake by hyphae could ameliorate the net cost of ion uptake to hosts by reducing host need to sequester carboxylic anions in vacuoles together with cations. Because obligate mycotrophs cannot reject infection without which they do not survive, the carbon demands of tropical VA mycorrhizas may be high; demand for photosynthate may increase hydroxyl extrusion from roots, causing compensatory increases in direct anion absorption by roots which could be especially important for sulfate uptake from throughfall. Hydroxyl extrusion by both roots and hyphae might locally increase pH in acid tropical soils, thereby solubilizing phosphorus, precipitating aluminum, and elevating cation exchange capacity. Although organic nitrogen and ammonium may be used preferentially, nitrate also may be taken up by tropical VA mycorrhizal fungi.

Obligate mycotrophs must encourage mycorrhizal infection, perhaps by maintaining relatively high rates of root exudation. Exudation might best be continued by not accumulating high concentrations of phosphorus in tissues, but instead having strong sinks for phosphorus in root and shoot growth centers. Strong phosphorus sinks in hosts might increase host resistance to aluminum toxicity by reducing the chance of precipitation of phosphorus in root cell walls and cytoplasmic membranes, and could contribute to high VA mycorrhizal fungus affinities for phosphorus by increasing fungus throughput of phosphorus. Strong host phosphorus sinks might encourage nonarbuscular transfer of phosphorus to hosts which could lengthen the functional life of mycorrhizas, further enhancing phosphorus uptake, or might overcome constraints imposed by common lack of arbuscular infection in long-lived rootlets that are well-protected by tannins and other antiroot-predator compounds. As evidenced by the high primary production of lowland humid tropical forests, well-insolated canopy trees are quite capable of sustaining the demands for photosynthate of their own growth sinks and mycorrhizal fungi. Canopy trees may share photosynthate with suppressed individuals via VA mycorrhizal fungus hyphal bridges, but only if this accrues to their ultimate benefit by increasing overall abundance of mycorrhizas and mineral nutrient uptake by interconnected roots. In lowland humid tropical forests, sharing of photosynthate, and similar mineral uptake abilities among individuals of different obligately mycotrophic species may minimize their ability to outcompete one another, thereby contributing to high within-habitat diversity. Similar mycorrhiza-mediated interactions may cause overyielding by tropical intercrops. These interactions should be sought in attempts to develop productive polycultural agricultural systems for the tropics through utilization of combinations of crop species which have high dependence on VA mycorrhizas.

ACKNOWLEDGMENTS

I thank Inn-Siang Ooi and Cecilio Puga for discussion of many of the ideas presented in this paper. I also am grateful to Maureen Donnelly, Steven Oberbauer, and Inn-Siang Ooi for criticizing the manuscript, and to Gene Safir for forbearance in awaiting it.

This is contribution No. 155 from the Program in Tropical Biology, Ecology, and Behavior of the Department of Biology, University of Miami.

REFERENCES

1. Lambert, D. H., Cole, H., Jr., and Baker, D. E., Adaptation of vesicular-arbuscular mycorrhizae to edaphic factors, *New Phytol.*, 85, 513, 1980.
2. Godse, D. B., Madhusudan, T., Bagyaraj, D. J., and Patil, R. B., Occurrence of vesicular-arbuscular mycorrhizas on crop plants in Karnataka, *Trans. Br. Mycol. Soc.*, 67, 169, 1976.
3. Redhead, J. F., Endotrophic mycorrhizas in Nigeria: species of the Endogonaceae and their distribution, *Trans. Br. Mycol. Soc.*, 69, 275, 1977.
4. Diem, H. G., Gueye, I., Gianinazzi-Pearson, V., Fortin, J. A., and Dommergues, Y. R., Ecology of VA mycorrhizae in the tropics: the semi-arid zone of Senegal, *Acta Oecol. Plant.*, 2, 53, 1981.
5. Hogberg, P., Mycorrhizal associations in some woodland and forest trees and shrubs in Tanzania, *New Phytol.*, 92, 407, 1982.
6. Dunevitz, V. L., Mulligan, M. F., Murphy, P. G., and Lugo, A. E., The occurrence of vesicular-arbuscular mycorrhizae in a subtropical dry forest, in *Abstracts*, American Phytopathology Society, St. Paul, Minn., 1983.
7. Janse, J. M., Les endophytes radicaux de quelques plantes Javanaises, *Ann. Jard. Bot. Buitenzorg*, 14, 53, 1896.
8. Bowen, G. D., Mycorrhizal roles in tropical plants and ecosystems, in *Tropical Mycorrhiza Research*, Mikola, P., Ed., Clarendon Press, Oxford, 1980, 165.
9. Mikola, P., Mycorrhizae across the frontiers, in *Tropical Mycorrhiza Research*, Mikola, P., Ed., Clarendon Press, Oxford, 1980, 3.
10. Janos, D. P., Tropical mycorrhizas, nutrient cycles and plant growth, in *Tropical Rain Forest: Ecology and Management*, Sutton, S. L., Whitmore, T. C., and Chadwick, A. C., Eds., Blackwell Scientific, Oxford, 1983, 327.
11. Janos, D. P., Vesicular-arbuscular mycorrhizae affect lowland tropical rain forest plant growth, *Ecology*, 61, 151, 1980.
12. Janos, D. P., Mycorrhizae influence tropical succession, *Biotropica*, 12(Suppl.), 56, 1980.
13. National Research Council, *Conversion of Tropical Moist Forests*, National Academy of Sciences, Washington, D.C., 1980.
14. National Research Council, *Ecological Aspects of Development in the Humid Tropics*, National Academy of Sciences, Washington, D.C., 1982.
15. Keeley, J. E., Endomycorrhizae influence growth of blackgum seedlings in flooded soils, *Am. J. Bot.*, 67, 6, 1980.
16. Ovrebo, R. G. T., personal communication, 1984.
17. Richards, P. W., *The Tropical Rain Forest*, University Press, Cambridge, 1966.
18. Janos, D. P., Methods for vesicular-arbuscular mycorrhiza research in the lowland wet tropics, in *Physiological Ecology of Plants of the Wet Tropics*, Tasks for Vegetation Science 12, Medina, E., Mooney, H. A., and Vazquez-Yanes, C., Eds., Junk, The Hague, 1984, 173.
19. Sanchez, P. A. and Buol, S. W., Soils of the tropics and the world food crisis, *Science*, 188, 598, 1975.

20. Sanchez, P. A. and Cochrane, T. T., Soil constraints in relation to major farming systems of tropical America, in *Priorities for Alleviating Soil-Related Constraints to Food Production in the Tropics,* International Rice Research Institute, Los Baños, 1980, 107.
21. Went, F. W. and Stark, N., The biological and mechanical role of soil fungi, *Proc. Natl. Acad. Sci. U.S.A.,* 60, 497, 1968.
22. Jordan, C. F., Nutrient regime in the wet tropics: physical factors, in *Physiological Ecology of Plants of the Wet Tropics,* Tasks for Vegetation Science 12, Medina, E., Mooney, H. A., and Vazquez-Yanes, C., Eds., Junk, The Hague, 1984, 3.
23. Herrera, R., Jordan, C. F., Klinge, H., and Medina, E., Amazon ecosystems. Their structure and functioning with particular emphasis on nutrients, *Interciencia,* 3, 223, 1978.
24. Sanchez, P. A., *Properties and Management of Soils in the Tropics,* John Wiley & Sons, New York, 1976.
25. Jordan, C. F. and Herrera, R., Tropical rain forests: are nutrients really critical?, *Am. Nat.,* 117, 167, 1981.
26. Anderson, J. M. and Swift, M. J., Decomposition in tropical forests, in *Tropical Rain Forest: Ecology and Management,* Sutton, S. L., Whitmore, T. C., and Chadwick, A. C., Eds., Blackwell Scientific, Oxford, 1983, 287.
27. Janzen, D. H., Tropical blackwater rivers, animals, and mast fruiting by the Dipterocarpaceae, *Biotropica,* 6, 69, 1974.
28. McKey, D., Waterman, P. G., Gartlan, J. S., and Struhsaker, T. T., Phenolic content of vegetation in two African rain forests: ecological implications, *Science,* 202, 61, 1978.
29. Rice, E. and Pancholy, S. K., Inhibition of nitrification by climax ecosystems. II. Additional evidence and possible role of tannins, *Am. J. Bot.,* 60, 691, 1973.
30. Parker, G. G., Throughfall and stemflow in the forest nutrient cycle, in *Advances in Ecological Research,* MacFadyen, A. and Ford, E. D., Eds., Academic Press, London, 1983, 57.
31. St. John, T. V., Response of tree roots to decomposing organic matter in two lowland Amazonian rain forests, *Can. J. For. Res.,* 13, 346, 1983.
32. St. John, T. V., Coleman, D. C., and Reid, C. P. P., Association of vesicular-arbuscular mycorrhizal hyphae with soil organic particles, *Ecology,* 64, 957, 1983.
33. St. John, T. V., Coleman, D. C., and Reid, C. P. P., Growth and spatial distribution of nutrient-absorbing organs: selective exploitation of soil heterogeneity, *Plant Soil,* 71, 487, 1983.
34. St. John, T. and Machado, A. D., Evidência da ação de microorganismos na ramificação de raízes, *Acta Amazonica,* 8, 9, 1978.
35. Tinker, P. B., The soil chemistry of phosphorus and mycorrhizal effects on plant growth, in *Endomycorrhizas,* Sanders, F. E., Mosse, B., and Tinker, P. B., Eds., Academic Press, London, 1975, 353.
36. Vitousek, P. M., Litterfall, nutrient cycling, and nutrient limitation in tropical forests, *Ecology,* 65, 285, 1984.
37. Owusu-Bennoah, E. and Wild, A., Effects of vesicular-arbuscular mycorrhiza on the size of the labile pool of soil phosphate, *Plant Soil,* 54, 233, 1980.
38. Graustein, W. C., Cromack, K., and Sollins, P., Calcium oxalate: its occurrence in soils and effect on nutrient and geochemical cycles, *Science,* 198, 1252, 1977.
39. Coleman, D. C., Reid, C. P. P., and Cole, C. V., Biological strategies of nutrient cycling in soil systems, in *Advances in Ecological Research,* MacFadyen, A. and Ford, E. D., Eds., Academic Press, London, 1983, 1.
40. Uehara, G., An overview of the soils of the arable tropics, in *Exploiting the Legume-Rhizobium Symbiosis in Tropical Agriculture,* College of Tropical Agriculture Miscellaneous Publ. 145, Vincent, J. M., Whitney, A. S., and Bose, J., Eds., University of Hawaii, Honolulu, 1977, 67.
41. Cooper, K. M., Physiology of VA mycorrhizal associations, in *VA Mycorrhiza,* Powell, C. Ll. and Bagyaraj, D. J., Eds., CRC Press, Boca Raton, Fla., 1984, 155.
42. Howeler, R. H., Asher, C. J., and Edwards, D. G., Establishment of an effective endomycorrhizal association on cassava in flowing solution culture and its effects on phosphorus nutrition, *New Phytol.,* 90, 229, 1982.
43. Bagyaraj, D. J., Biological interactions with VA mycorrhizal fungi, in *VA Mycorrhiza,* Powell, C. Ll. and Bagyaraj, D. J., Eds., CRC Press, Boca Raton, Fla., 1984, 131.
44. Mosse, B., Hayman, D. S., and Arnold, D. J., Plant growth responses to vesicular-arbuscular mycorrhiza. V. Phosphate uptake by three plant species from P-deficient soils labelled with ^{32}P, *New Phytol.,* 72, 809, 1973.
45. Sanni, S. O., Vesicular-arbuscular mycorrhiza in some Nigerian soils: the effect of *Gigaspora gigantea* on the growth of rice, *New Phytol.,* 77, 673, 1976.
46. Mosse, B., The role of mycorrhiza in legume nutrition on marginal soils, in *Exploiting the Legume-Rhizobium Symbiosis in Tropical Agriculture,* College of Tropical Agriculture Miscellaneous Publ. 145, Vincent, J. M., Whitney, A. S., and Bose, J., Eds., University of Hawaii, Honolulu, 1977, 275.

47. Waidyanatha, U. P. de S., Yogaratnam, N., and Ariyaratne, W. A., Mycorrhizal infection on growth and nitrogen fixation of *Pueraria* and *Stylosanthes* and uptake of phosphorus from two rock phosphates, *New Phytol.*, 82, 147, 1979.
48. Guzman, P. R. A., Ferrera-Cerrato, R., Etchevers, B. J., and Corona, S. T., Evaluacion de cepas de *Rhizobium* sp. y de endomicorrizas vesiculo-arbusculares en *Leucaena leucocephala* bajo dos fuentes de fosforo (roca fosforica y superfosfato simple de calcio) a diferentes dosis, in *Resumenes*, 16th Cong. Nacional de la Ciencia del Suelo, Instituto Tecnologico Agropecuario de Oaxaca No. 23, Sociedad Mexicana de la Ciencia del Suelo A. C., 1983, 113.
49. Islam, R., Ayanaba, A., and Sanders, F. E., Response of cowpea *(Vigna unguiculata)* to inoculation with VA-mycorrhizal fungi and to rock phosphate fertilization in some unsterilized Nigerian soils, *Plant Soil*, 54, 107, 1980.
50. Kang, B. T., Islam, R., Sanders, F. E., and Ayanaba, A., Effect of phosphate fertilization and inoculation with VA-mycorrhizal fungi on performance of cassava *(Manihot esculenta* Crantz) grown on an Alfisol, *Field Crops Res.*, 3, 83, 1980.
51. Meiklejohn, J., Microbiology of the nitrogen cycle in some Ghana soils, *Empire J. Exp. Agric.*, 30, 115, 1962.
52. Forman, R. T. T., Canopy lichens with blue-green algae: a nitrogen source in a Colombian rain forest, *Ecology*, 56, 1176, 1975.
53. App, A. A., Watanabe, I., Alexander, M., Ventura, W., Daez, C., Santiago, T., and De Datta, S. K., Nonsymbiotic nitrogen fixation associated with the rice plant in flooded soils, *Soil Sci.*, 130, 283, 1980.
54. Jordan, C. F., Todd, R. L., and Escalante, G., Nitrogen conservation in a tropical rain forest, *Oecologia*, 39, 123, 1979.
55. Grove, T. L., *Nitrogen Fertility in Oxisols and Ultisols of the Humid Tropics*, Cornell International Agriculture Bulletin 36, New York State College of Agriculture and Life Sciences, Ithaca, 1979.
56. Ames, R. N., Reid, C. P. P., Porter, L. K., and Cambardella, C., Hyphal uptake and transport of nitrogen from two ^{15}N-labelled sources by *Glomus mosseae*, a vesicular-arbuscular mycorrhizal fungus, *New Phytol.*, 95, 381, 1983.
57. Haines, B. L. and Best, G. R., *Glomus mosseae*, endomycorrhizal with *Liquidambar styraciflua* L. seedlings retards NO_3, NO_2 and NH_4 nitrogen loss from a temperate forest soil, *Plant Soil*, 45, 257, 1976.
58. Bowen, G. D. and Smith, S. E., The effects of mycorrhizas on nitrogen uptake by plants, *Ecol. Bull.*, 33, 237, 1981.
59. Powell, C. Ll., Potassium uptake by endotrophic mycorrhizas, in *Endomycorrhizas*, Sanders, F. E., Mosse, B., and Tinker, P. B., Eds., Academic Press, London, 1975, 461.
60. Vander Zaag, P., Fox, R. L., De La Pena, R. S., and Yost, R. S., P nutrition of Cassava including mycorrhizal effects on P, K, S, Zn and Ca uptake, *Field Crops Res.*, 2, 253, 1979.
61. Swaminathan, K. and Verma, B. C., Responses of three crop species to vesicular-arbuscular mycorrhizal infection on zinc-deficient Indian soils, *New Phytol.*, 82, 481, 1979.
62. Cooper, K. M. and Tinker, P. B., Translocation and transfer of nutrients in vesicular-arbuscular mycorrhizas. II. Uptake and translocation of phosphorus, zinc and sulphur, *New Phytol.*, 81, 43, 1978.
63. Timmer, L. W. and Leyden, R. E., The relationship of mycorrhizal infection to phosphorus-induced copper deficiency in Sour Orange seedlings, *New Phytol.*, 85, 15, 1980.
64. Buwalda, J. G., Stribley, D. P., and Tinker, P. B., Increased uptake of anions by plants with vesicular-arbuscular mycorrhizas, *Plant Soil*, 71, 463, 1983.
65. Medina, E., Adaptations of tropical trees to moisture stress, in *Tropical Rain Forest Ecosystems: Structure and Function*, Golley, F. B., Ed., Elsevier, New York, 1983, 225.
66. Siqueira, J. O., Hubbell, D. H., and Mahmud, A. W., Effect of liming on spore germination, germ tube growth and root colonization by vesicular-arbuscular mycorrhizal fungi, *Plant Soil*, 76, 115, 1984.
67. Moawad, M., Nutzung der vesikulär-arbuskulären Mykorrhiza im tropischen Pflanzenbau, *Angew. Botanik*, 53, 99, 1979.
68. Moawad, M., Ecophysiology of vesicular-arbuscular mycorrhiza, in *Tropical Mycorrhiza Research*, Mikola, P., Ed., Clarendon Press, Oxford, 1980, 203.
69. Landsberg, J. J., Physical aspects of the water regime of wet tropical vegetation, in *Physiological Ecology of Plants of the Wet Tropics*, Tasks for Vegetation Science 12, Medina, E., Mooney, H. A., and Vazquez-Yanes, C., Eds., Junk, The Hague, 1984, 13.
70. Van Wambeke, A., Properties and potentials of soils in the Amazon basin, *Interciencia*, 3, 233, 1978.
71. Menge, J. A., Davis, R. M., Johnson, E. L. V., and Zentmyer, G. A., Mycorrhizal fungi increase growth and reduce transplant injury in avocado, *Calif. Agric.*, 32, 6, 1978.
72. Levy, Y. and Krikun, J., Effect of vesicular-arbuscular mycorrhiza on *Citrus jambhiri* water relations, *New Phytol.*, 85, 25, 1980.

73. Sieverding, E., Influence of soil water regimes on VA mycorrhiza. I. Effect on plant growth, water utilization and development of mycorrhiza, *Z. Acker Pflanzenbau,* 150, 400, 1981.
74. Sieverding, E., Influence of soil water regimes on VA mycorrhiza. II. Effect of soil temperature and water regime on growth, nutrient uptake, and water utilization of *Eupatorium odoratum* L., *Z. Acker Pflanzenbau,* 152, 56, 1983.
75. Sieverding, E., Influence of soil water regimes on VA mycorrhiza. III. Comparison of three mycorrhizal fungi and their influence on transpiration, *Z. Acker Pflanzenbau,* 153, 52, 1984.
76. Reid, C. P. P. and Bowen, G. D., Effects of soil moisture on VA mycorrhizae formation and root development in *Medicago,* in *The Soil-Root Interface,* Harley, J. L. and Russel, R. S., Eds., Academic Press, London, 1979, 211.
77. Nelsen, C. E. and Safir, G. R., Improved drought tolerance of mycorrhizal onion plants caused by improved phosphorus nutrition, *Planta,* 154, 407, 1982.
78. Redhead, J. F., Endotrophic mycorrhizas in Nigeria: some aspects of the ecology of the endotrophic mycorrhizal association of *Khaya grandifoliola* C. DC., in *Endomycorrhizas,* Sanders, F. E., Mosse, B., and Tinker, P. B., Eds., Academic Press, London, 1975, 447.
79. Krishna, K. R., Suresh, H. M., Syamsunder, J., and Bagyaraj, D. J., Changes in the leaves of finger millet due to VA mycorrhizal infection, *New Phytol.,* 87, 717, 1981.
80. Saif, S. R., The influence of soil aeration on the efficiency of vesicular-arbuscular mycorrhizae. I. Effect of soil oxygen on the growth and mineral uptake of *Eupatorium odoratum* L. inoculated with *Glomus macrocarpus, New Phytol.,* 88, 649, 1981.
81. St. John, T. and Machado, A. D., Efeitos da profundidade e do sistema de manejo de um solo de terra firme (Latossolo) em infestações por micorrizas, *Acta Amazonica,* 8, 139, 1978.
82. St. John, T. V., A survey of mycorrhizal infection in an Amazonian rain forest, *Acta Amazonica,* 10, 527, 1980.
83. Cooper, K. M. and Tinker, P. B., Translocation and transfer of nutrients in vesicular-arbuscular mycorrhizas. IV. Effect of environmental variables on movement of phosphorus, *New Phytol.,* 88, 327, 1981.
84. Parke, J. L., Linderman, R. G., and Trappe, J. M., Effect of root zone temperature on ectomycorrhiza and vesicular-arbuscular mycorrhiza formation in disturbed and undisturbed forest soils of southwest Oregon, *Can. J. For. Res.,* 13, 657, 1983.
85. Schenck, N. C. and Schroder, V. N., Temperature response of *Endogone* mycorrhiza on soybean roots, *Mycologia,* 66, 600, 1974.
86. Schenck, N. C. and Smith, G. S., Responses of six species of vesicular-arbuscular mycorrhizal fungi and their effects on soybean at four soil temperatures, *New Phytol.,* 92, 193, 1982.
87. Graham, J. H., Leonard, R. T., and Menge, J. A., Interaction of light intensity and soil temperature with phosphorus inhibition of vesicular-arbuscular mycorrhiza formation, *New Phytol.,* 91, 683, 1982.
88. Cunningham, R. K., The effect of clearing a tropical forest soil, *J. Soil Sci.,* 14, 334, 1963.
89. National Research Council, *Soils of the Humid Tropics,* National Academy of Sciences, Washington, D.C., 1972.
90. Ewel, J., Berish, C., Brown, B., Price, N., and Raich, J., Slash and burn impacts on a Costa Rican wet forest site, *Ecology,* 62, 816, 1981.
91. Uhl, C., Jordan, C., Clark, K., Clark, H., and Herrera, R., Ecosystem recovery in Amazon caatinga forest after cutting, cutting and burning, and bulldozer clearing treatments, *Oikos,* 38, 313, 1982.
92. Werner, P., Changes in soil properties during tropical wet forest succession in Costa Rica, *Biotropica,* 16, 43, 1984.
93. Butler, E. J., The occurrences and systematic position of the vesicular-arbuscular type of mycorrhizal fungi, *Trans. Br. Mycol. Soc.,* 22, 274, 1939.
94. Maeda, M., The meaning of mycorrhiza in regard to systematic botany, *Kumamoto J. Sci.,* 3, 57, 1954.
95. Thapar, H. S. and Khan, S. N., Studies on endomycorrhiza in some forest species, *Proc. Indian Natl. Sci. Acad.,* 39, 687, 1973.
96. Tupas, G. L. and Sajise, P. E., Mycorrhizal associations in some savanna and reforestation trees, *Kalikasan,* 5, 235, 1976.
97. Shamsuddin, M. N., Mycorrhizas of tropical forest trees, in *Abstracts, 5th Int. Symp. of Tropical Ecology,* Furtado, J., Ed., University of Malaya, Kuala Lumpur, 1979, 173.
98. Bagyaraj, D. J., Manjunath, A., and Patil, R. B., Occurrence of vesicular-arbuscular mycorrhizas in some tropical aquatic plants, *Trans. Br. Mycol. Soc.,* 72, 164, 1979.
99. de Alwis, D. P. and Abeynayake, K., A survey of mycorrhizae in some forest trees of Sri Lanka, in *Tropical Mycorrhiza Research,* Mikola, P., Ed., Clarendon Press, Oxford, 1980, 146.
100. Redhead, J. F., Mycorrhizal associations in some Nigerian forest trees, *Trans. Br. Mycol. Soc.,* 51, 377, 1968.

101. Redhead, J. F., Mycorrhiza in natural tropical forests, in *Tropical Mycorrhiza Research,* Mikola, P., Ed., Clarendon Press, Oxford, 1980, 127.
102. Johnston, A., Vesicular-arbuscular mycorrhiza in Sea Island cotton and other tropical plants, *Trop. Agric.,* 26, 118, 1949.
103. Edmisten, J., Survey of mycorrhiza and root nodules in the El Verde forest, in *A Tropical Rain Forest,* Odum, H. T. and Pigeon, R. F., Eds., U.S. Atomic Energy Commission, Oak Ridge, 1970, F15.
104. Thomazini, L. I., Mycorrhiza in plants of the 'Cerrado', *Plant Soil,* 41, 707, 1974.
105. St. John, T., Uma lista de espécies de plantas tropicais brasileiras naturalmente infectadas com micorriza vesicular-arbuscular, *Acta Amazonica,* 10, 229, 1980.
106. Herrera, R. A. and Ferrer, R. L., Vesicular-arbuscular mycorrhiza in Cuba, in *Tropical Mycorrhiza Research,* Mikola, P., Ed., Clarendon Press, Oxford, 1980, 156.
107. St. John, T. V. and Uhl, C., Mycorrhizae in the rain forest at San Carlos de Rio Negro, Venezuela, *Acta Cient. Venez.,* 34, 233, 1983.
108. Bononi, V. L. R. and Trufem, S. F. B., Endomicorrizas vesiculo-arbusculares do cerrado da reserva biologica de Moji-guaçu, SP, Brazil, *Rickia,* 10, 55, 1983.
109. Meador, R. E., The Role of Mycorrhizae in Influencing Succession on Abandoned Everglades Farmland, M. S. thesis, University of Florida, Gainesville, 1977.
110. Williams-Linera, G. and Ewel, J. J., Effect of autoclave sterilization of a tropical andept on seed germination and seedling growth, *Plant Soil,* 82, 263, 1984.
111. Puga, C., Influence of Vesicular-Arbuscular Mycorrhizas on Competition between Corn and Weeds in Panama, M. S. thesis, University of Miami, Coral Gables, 1985.
112. Nadkarni, N., Canopy roots: convergent evolution in rainforest nutrient cycles, *Science,* 214, 1023, 1981.
113. Ciferri, R., Preliminary observations on sugar cane mycorrhizae and their relationship to root diseases, *Phytopathology,* 18, 249, 1928.
114. Rayner, M. C., Mycorrhiza in the genus, *Citrus, Nature (London),* 131, 399, 1933.
115. Reed, H. S. and Frémont, T., Factors that influence the formation and development of mycorrhizal associations in citrus roots, *Phytopathology,* 25, 645, 1935.
116. Neill, J. C., *Rhizophagus* in citrus, N.Z. J. Sci. Technol., 25, 191, 1944.
117. Laycock, D. H., Preliminary investigations into the function of the endotrophic mycorrhiza of *Theobroma cacao* L., *Trop. Agric.,* 22, 77, 1945.
118. Wastie, R. L., The occurrence of an *Endogone* type of endotrophic mycorrhiza in *Hevea brasiliensis, Trans. Br. Mycol. Soc.,* 48, 167, 1965.
119. Ginsburg, D. and Avizohar-Hershenson, Z., Observations on vesicular-arbuscular mycorrhiza associated with avocado roots in Israel, *Trans. Br. Mycol. Soc.,* 48, 101, 1965.
120. Pandey, S. and Misra, A. P., *Rhizophagus* in mycorrhizal association with *Litchi chinensis* Sonn., *Mycopathol. Mycol. Appl.,* 45, 337, 1971.
121. Ramirez, B. N., Mitchell, D. J., and Schenck, N. C., Establishment and growth effects of three vesicular-arbuscular mycorrhizal fungi on papaya, *Mycologia,* 67, 1039, 1975.
122. Janos, D. P., Vesicular-arbuscular mycorrhizae affect the growth of *Bactris gasipaes, Principes,* 21, 12, 1977.
123. Yost, R. S. and Fox, R. L., Contribution of mycorrhizae to P nutrition of crops growing on an Oxisol, *Agron. J.,* 71, 903, 1979.
124. Krikun, J. and Bar Joseph, B., Pepper and mycorrhizae: a field study, in *Program and Abstr. 4th North American Conf. on Mycorrhiza,* Colorado State University, Fort Collins, 1979.
125. Marx, D. H., Bryan, W. C., and Campbell, W. A., Effect of endomycorrhizae formed by *Endogone mosseae* on the growth of citrus, *Mycologia,* 63, 1222, 1971.
126. Kleinschmidt, G. D. and Gerdemann, J. W., Stunting of citrus seedlings in fumigated nursery soils related to the absence of endomycorrhizae, *Phytopathology,* 62, 1447, 1972.
127. Schenck, N. C. and Tucker, D. P. H., Endomycorrhizal fungi and the development of citrus seedlings in Florida fumigated soils, *J. Am. Soc. Hortic. Sci.,* 99, 284, 1974.
128. Hattingh, M. J. and Gerdemann, J. W., Inoculation of Brazilian sour orange seed with an endomycorrhizal fungus, *Phytopathology,* 65, 1013, 1975.
129. Menge, J. A., Johnson, E. L. V., and Platt, R. G., Mycorrhizal dependency of several citrus cultivars under three nutrient regimes, *New Phytol.,* 81, 553, 1978.
130. Johnson, C. R., Menge, J. A., Schwab, S., and Ting, I. P., Interaction of photoperiod and vesicular-arbuscular mycorrhizae on growth and metabolism of sweet orange, *New Phytol.,* 90, 665, 1982.
131. Botello, G. J. J., Ferrera-Cerrato, R., and García, S. H., Respuesta de *Citrus aurantium* a la inoculacion con hongos endomicorricicos vesicular-arbuscular utilizando diferentes gradientes de inoculo, in *Resumenes, 16th Congr. Nacional de la Ciencia del Suielo,* Instituto Technologico Agropecuario de Oaxaca No. 23, Sociedad Mexicana de la Ciencia del Suelo A. C., 1983, 125.

132. Graham, J. H., Linderman, R. G., and Menge, J. A., Development of external hyphae by different isolates of mycorrhizal *Glomus* spp. in relation to root colonization and growth of Troyer citrange, *New Phytol.*, 91, 183, 1982.
133. Johnson, C. R., Phosphorus nutrition on mycorrhizal colonization, photosynthesis, growth and nutrient composition of *Citrus aurantium*, *Plant Soil*, 80, 35, 1984.
134. Pandey, S. P. and Misra, A. P., Mycorrhizas in relation to growth and fruiting of *Litchi chinensis* Sonn., *J. Indian Bot. Soc.*, 54, 280, 1975.
135. Howeler, R. H. and Sieverding, E., Potentials and limitations of mycorrhizal inoculation illustrated by experiments with field-grown cassava, *Plant Soil*, 75, 245, 1983.
136. Sieverding, E. and Leihner, D. E., Influence of crop rotation and intercropping of cassava with legumes on VA mycorrhizal symbiosis of cassava, *Plant Soil*, 80, 143, 1984.
137. Mosse, B., Effects of different *Endogone* strains on the growth of *Paspalum notatum*, *Nature (London)*, 239, 221, 1972.
138. Sanni, S. D., Vesicular-arbuscular mycorrhiza in some Nigerian soils and their effect on the growth of cowpea *(Vigna unguiculata)*, tomato *(Lycopersicon esculentum)* and maize *(Zea mays)*, *New Phytol.*, 77, 667, 1976.
139. Mosse, B., Specificity of vesicular-arbuscular mycorrhizas, in *Endomycorrhizas*, Sanders, F. E., Mosse, B., and Tinker, P. B., Eds., Academic Press, London, 1975, 469.
140. Pichot, J. and Truong, B., Effect of endomycorrhizae on growth and phosphorus uptake of *Agrostis* in a pot experiment, in *Tropical Mycorrhiza Research*, Mikola, P., Ed., Clarendon Press, Oxford, 1980, 206.
141. Ferrera-Cerrato, R. and Macedo, A. S., Susceptibilidad de dos variedades de cebolla *(Allium cepa)* a 7 especies de hongos endomicorricicos vesicular-arbuscular (V-A), in *Memorias, Tomo 1, 16th Congr. Nacional de la Ciencia del Suelo*, Sociedad Mexicana de la Ciencia del Suelo A. C., San Luis Potosi, 1981, 477.
142. Crush, J. R., Plant growth responses to vesicular-arbuscular mycorrhiza. VII. Growth and nodulation of some herbage legumes, *New Phytol.*, 73, 743, 1974.
143. Daft, M. J. and El-Giahmi, A. A., Studies on nodulated and mycorrhizal peanuts, *Ann. Appl. Biol.*, 83, 273, 1976.
144. La Torraca, S. M., Efeitos da Inoculaçao de Micorriza VA no Crescimento e Nodulaçao de *Vigna unguiculata* (L) Walp., em Três Solos de Terra Firme, M. S. thesis, Instituto Nacional de Pesquisas da Amazonia, Manaus, 1979.
145. Waidyanatha, U. P. de S., Mycorrhizae of *Hevea* and leguminous ground covers in rubber plantations, in *Tropical Mycorrhiza Research*, Mikola, P., Ed., Clarendon Press, Oxford, 1980, 238.
146. Janos, D. P., Effects of vesicular-arbuscular mycorrhizae on lowland tropical rainforest trees, in *Endomycorrhizas*, Sanders, F. E., Mosse, B., and Tinker, P. B., Eds., Academic Press, London, 1975, 437.
147. McHargue, L. A., V-A mycorrhizae improve growth and nodulation of two tropical leguminous trees, in *Program and Abstr. 5th North American Conf. on Mycorrhizae*, Universite Laval, Quebec, 1981, 52.
148. Johnson, C. R. and Michelini, S., Effect of mycorrhizae on container grown acacia, *Proc. Fla. State Hortic. Soc.*, 87, 520, 1975.
149. Mosse, B. and Hayman, D. S., Mycorrhiza in agricultural plants, in *Tropical Mycorrhiza Research*, Mikola, P., Ed., Clarendon Press, Oxford, 1980, 213.
150. McKey, D., The distribution of secondary compounds within plants, in *Herbivores: Their Interaction with Secondary Plant Metabolites*, Rosenthal, G. A. and Janzen, D. H., Ed., Academic Press, New York, 1979, 55.
151. Baylis, G. T. S., The magnolioid mycorrhiza and mycotrophy in root systems derived from it, in *Endomycorrhizas*, Sanders, F. E., Mosse, B., and Tinker, P. B., Eds., Academic Press, London, 1975, 373.
152. St. John, T. V., Root size, root hairs and mycorrhizal infection: a re-examination of Baylis's hypothesis with tropical trees, *New Phytol.*, 84, 483, 1980.
153. Graham, J. H., Effect of citrus root exhudates on germination of chlamydospores of the vesicular-arbuscular mycorrhizal fungus, *Glomus epigaeum*, *Mycologia*, 74, 831, 1982.
154. Azcon, R. and Ocampo, J. A., Factors affecting the vesicular-arbuscular infection and mycorrhizal dependency of thirteen wheat cultivars, *New Phytol.*, 87, 677, 1981.
155. Snellgrove, R. C., Splittstoesser, W. E., Stribley, D. P., and Tinker, P. B., The distribution of carbon and the demand of the fungal symbiont in leek plants with vesicular-arbuscular mycorrhizas, *New Phytol.*, 92, 75, 1982.
156. Heap, A. J. and Newman, E. I., Links between roots by hyphae of vesicular-arbuscular mycorrhizas, *New Phytol.*, 85, 169, 1980.
157. Whittingham, J. and Read, D. J., Vesicular-arbuscular mycorrhiza in natural vegetation systems. III. Nutrient transfer between plants with mycorrhizal interconnections, *New Phytol.*, 90, 277, 1982.

158. Chiariello, N., Hickman, J. C., and Mooney, H. A., Endomycorrhizal role for interspecific transfer of phosphorus in a community of annual plants, *Science,* 217, 941, 1982.
159. Francis, R. and Read, D. J., Direct transfer of carbon between plants connected by vesicular-arbuscular mycorrhizal mycelium, *Nature (London),* 307, 1, 1984.
160. Heap, A. J. and Newman, E. I., The influence of vesicular-arbuscular mycorrhizas on phosphorus transfer between plants, *New Phytol.,* 85, 173, 1980.
161. Furman, T. E. and Trappe, J. M., Phylogeny and ecology of mycotrophic achlorophyllous angiosperms, *Q. Rev. Biol.,* 46, 219, 1971.

Chapter 8

THE ECOLOGY OF VESICULAR-ARBUSCULAR MYCORRHIZAE IN GRASS- AND SHRUBLANDS

R. Michael Miller

TABLE OF CONTENTS

I. INTRODUCTION

Mycorrhizal fungi have the greatest influence on vegetation in natural or seminatural vegetation — most importantly in grasslands and shrublands where the soils, plants, and landscapes are typically in a state of precarious balance. These ecosystems are frequently characterized by climatic extremes; precipitation patterns, wind, and temperature fluctuate greatly, influencing water loss through evapotranspiration. The intent of this chapter is to review what is known about the ecology of vesicular-arbuscular (VA) mycorrhizae in grasslands and shrublands and, whenever possible, to emphasize the physiology and population ecololgy of the relationship.

Grasslands and shrublands develop in areas with less than 100 cm of annual precipitation and usually are located on plains in the interiors of great landmasses. Shrublands represent the xeric end of a caternary moisture sequence with grasslands. Shrubs are more prevalent than grasses in arid environments, where precipitation is erratic and unpredictable. Under conditions of extreme aridity, the role of grasses is more ephemeral than under more mesic conditions. Permanent grasslands are composed principally of natural grasses, grass-like plants, forbs, and shrubs. Generally, the proportion of grasses and shrubs varies according to grazing pressure, fire frequency, and climate; the drier the site and the greater the grazing, the more shrubs present. In many regions of the world, shrublands formerly were grasslands, but with abuse by man became dominated by shrubs. Another type of grassland, i.e., pasture, is characterized by perennial grasses and legumes in humid and subhumid regions which are grazed year after year, and typically are established on lands that have been plowed. Pastures are grasslands that normally would support forest vegetation and are not a result of past climate regimes. Generally, the continued existence of pastures depends on intervention by man. These are typically the kinds of grasslands associated with the eastern U.S. and most of Great Britain and Western Europe. The upland or montane grasslands of Great Britain differ from the typical lowland pasture-type grasslands, however, in that they can be maintained by climate rather than by man and grazing (even though this region was also previously occupied by forest). Thus, these grasslands have characteristics resembling those of the continental steppes, prairies, and savannas that are the result of climates too severe for forest development.[1] A good review of the different kinds of prairies in North America is given by Risser et al.[2]

In recent years, the study of the relationship of mycorrhizae to the ecophysiology of plants in grasslands and shrublands has become more important; these ecosystems occupy a major part of the earth's landmass, and a large part of the human population depends on them for food and clothing. In many parts of the world, grasslands and shrublands also contain large energy and mineral reserves and will be subjected to the perturbations associated with extraction of these minerals. Understanding how such ecosystems function may be the key to their best use. An experimental approach to this end is one that integrates aboveground and belowground components at the soil-root interface. A prominent feature of this interface is the VA mycorrhizal fungus.

Despite increased awareness of the widespread occurrence of mycorrhizae in natural environments, only during the last 10 years has emphasis been placed on putting into reasonable perspective how mycorrhizae influence population and ecosystem dynamics. For mycorrhizae to have any synecological significance, the association must have an effect at the individual or population level. Mycorrhizal effects on plant growth may influence survival in a competitive environment, such as in heavily grazed land, or where moisture and nutrients are limiting to plant growth.[3] To understand the functional role of mycorrhizae in a natural setting, a better understanding of this symbiotic relationship at the individual and population levels is necessary.

Very little research has been concerned with mycorrhizae in a nonagricultural setting. Past research has been concerned mainly with mycorrhizae in relation to crop species, i.e., plants that have evolved in highly productive environments and in many cases possessing a ruderal or weedy growth habitat. Our knowledge of the physiologies of natural vegetation often is quite limited, with only a few plants having been studied in detail. However, since the physiology of crop plants is quite different from those of natural or wild plants,[4] most likely the influence of mycorrhizal fungi on plants also will be different. Research on the influence of mycorrhizae on grassland and shrubland plants is scanty; very little information exists on the life cycle of mycorrhizal fungi in relation to plant growth over a complete growing season. A better definition is needed of resource or nutrient pools of grassland and shrubland sites in relation to availability to the plant via mycorrhizal fungi. In calcareous soils, the nutrient pools are quite different than in those associated with highly weathered, acidic soils. The question of whether nutrient pools are available to the mycorrhizae that may not be readily available to the root alone needs to be resolved. Research also needs to address the question of whether a VA mycorrhizal fungus has a capacity for selective exploitation of nutrients in the soil. Studies addressing these factors will be necessary if we are to understand the influences of mycorrhizae on grassland and shrubland vegetation.

A. Mycorrhizae and Plant Community Structure

If an understanding of the place of mycorrhizae in a natural setting is to be achieved, in addition to studies of how plants grow, research on why plants grow where they do will also be necessary. This means that in addition to studies on how mycorrhizae influence the physiology of plants, the occurrence of a plant must also be placed within a conceptual framework where there is a better definition of how plants and associated mycorrhizal fungi are affected by soils and their resources within a climatic regime in which they grow.

One approach is to study the influence of mycorrhizae on the behavior of plants based on physiological response to mineral nutrition. Most studies on nutrient aquisition have been of plants growing in soils of relatively high nutrient availability; a few studies have been directed at plants obtained from soils of low nutrient availability. The studies indicate that plants of poor soils do not "grow" by the same rules as plants of rich soils.[4,5] A cultivated plant in rich soil grows as fast as it can, and with increased nutrients the plant grows faster. However, when a nutrient is withheld, the plant continues to grow and may develop symptoms of nutrient deficiency. Alternatively, a wild plant from a nutrient-poor soil has a growth rate either genetically or physiologically determined that is slower under low-nutrient conditions. As nutrients increase, tissue concentrations of that nutrient increase instead of growth rate increasing and the plant readily exhibits toxicity symptoms. When a nutrient is limiting for these plants, the growth rate of the plant is reduced, allowing the plant to wait until nutrient supply is restored. Under this regime, tissue nutrient levels remain normal. These plants do not typically show nutrient deficiency symptoms as do plants adapted to high-nutrient soils.[4,5]

A similar approach has been put forth by Grime[6,7] based on an argument for the existence of three primary strategies for growth forms in higher plants that is based on a combination of interdependent physiological characteristics necessary for successful exploitation of a particular environment or habitat. Grime points out that the external factors limiting plant growth in any habitat are stress and disturbance. Stress is defined as those external constraints that restrict the rate of dry matter production, e.g., shortages of water, light, and mineral nutrients, as well as suboptimal temperatures. Disturbance is any factor associated with partial or complete destruction of plant biomass and can arise from the activities of herbivores, pathogens, man (trampling, mowing, plowing, or even surface mining), and from phenomena such as wind, frosts, soil erosion, fire, and desiccation. The three strategies are based on a relationship between intensities of stress and disturbance, thus, a matrix is formed to the type of strategy that will predominate in an area, i.e., the type of strategy that will produce the highest productivity for that area. If the intensities of disturbance and stress are both low, plants with a competitive strategy predominate in the vegetation. If disturbance is low and stress high, the community is dominated by stress-tolerant plant species. In areas of high disturbance and low stress, plants with a ruderal growth strategy will predominate. However, if the intensity of both stress and disturbance is high, no viable strategy is possible since no vascular plants can survive under such harsh conditions.

A theoretical approach developed by Tilman[8] using resource requirements of species to predict the outcome of multispecies competition for several limiting resources also appears to lend itself to studies with mycorrhizal fungi. This theory is an extension of equilibrium theory to multispecies competition in spatially heterogeneous habitats and is a concept in which equilibrium occurs when resource-dependent reproduction of each species balances total resource consumption for each resource. Since mycorrhizae can affect resource consumption via several mechanisms, this association can affect the outcome of competitive pairs for nutrient(s) and uses an approach incorporating graphical representations of both single and multiple species and resource pairs to aid in understanding mycorrhizal associations in a competitive environment.

B. Adaptations in Root Morphology

Morphological adaptations by which roots bring adequate nutrients to the root surface include altered root radius and root length, and spacing and positioning of root hairs. Particularly good reviews of root hairs, root clusters, and mycorrhizae as nutrient stress adaptations in plants are given by Lamont[9] and Coleman et al.[10] Since root hairs increase the effective radius of the root, it would appear that this structure would be very important for nutrient uptake. However, in many plants root hairs are rather short lived, being destroyed by the sloughing off of the epidermis and cortex during secondary root growth.[11] Under field conditions, the importance of root hairs to the plant is not thoroughly understood. It has been suggested that one of the functions of root hairs may be related to maintaining a continuity between root and soil, which is accomplished by the physical presence of the root hair and the secretion of mucilage.[12] Nye[13] has calculated that except for phosphorus uptake, the long term effects of root hairs would not appear to be very important. Even so, Baylis[14] has suggested that root hairs may be advantageous to roots with ion uptake since they develop with the recommencement of growth at the start of the growing season prior to the growth of the mycorrhizal fungus. In calcium-dominated soils, the value of root hair growth at this time is that nitrogen (N) in the form of NH_4^+ is released during mineralization at the root surface, allowing for a reduction in rhizosphere pH and enhanced phosphorus (P) solubilization with increased acidity.[15,16]

Root hairs are highly developed in plants which are typically nonmycorrhizal, like the chenopod *Salsola kali,*[17] in rushes and sedges,[18] and many grasses, e.g., *Lolium perenne.*[19] It has been suggested that plants with few or short root hairs appear to depend more on mycorrhizae than plants with well-developed root hairs for nutrient aquisition.[20] Until recently, sedges and rushes were thought to be nonmycorrhizal.[18] However, it appears now that even these grass-like plants can respond to infection by mycorrhiza-like fungi.[21]

Research by Bolan et al.[22] suggests that the type of root system a plant has may affect its response to applied phosphorus in the presence of iron hydroxide. Their research shows that in *Trifolium subterraneum,* a clearly marked threshold level of P application is evident below which plants take up little P and grow poorly. This response occurred only when iron hydroxide was added to the soil and when the plants were nonmycorrhizal. Also, mycorrhizal plants did not exhibit a threshold response with P application. For example, with *L. rigidum* a threshold response was not observed with either mycorrhizal or nonmycorrhizal plants with the addition of iron hydroxide to the growth medium. This indicates that in coarse-rooted plants such as clovers, a need or response to mycorrhizal fungi depends on the form of P. Plants possessing fine roots, as with *Lolium* spp., do not respond as readily to changes in P form unless they are under extremely deficient P levels.

Much of the interpretation of ion uptake by extramatrical hyphae of VA mycorrhizal fungi has been based on a mechanism by which greater exploitation of the soil occurs by hyphae through their extending the depletion zone found around nonmycorrhizal roots.[23] The distance an ion has to diffuse to the root is decreased by mycorrhizal hyphae,[24] and that selective rather than random exploitation of the soil by hyphae to nutrient-rich microsites can occur.[25] There also is research to suggest that mycorrhizal and nonmycorrhizal roots may differ in their ability to absorb P from solution,[26-28] that mycorrhizal roots may release exudates that increase P availability,[29] and that differential absorption of cations and anions by mycorrhizal plants may lead to differences in rhizosphere pH, resulting in a change of availability of absorbed P.[30,31] All of this suggests a direct mechanism where by mycorrhizae are better absorbers of mineral ions than roots.

II. PHYSIOLOGICAL RELATIONSHIPS

A. The Soil Resource

The mycorrhizal association is best characterized by its hyphal connections with the soil. These connections allow for extensive yet selective soil exploitation in which hyphae can bridge regions of nutrient deficiency of relatively immobile nutrients near the root to more nutrient-rich regions not otherwise available to the plant. This suggests that mycorrhiza increase the volume of soil normally available to the plant[32,33] and enhance the translocation of nutrients across diffusion zones.[34] It is well established that mycorrhizal fungus infection typically increases the rate of P uptake by a host from soil. Also, increase in the rate of uptake of zinc, copper, sulfur, calcium, and potassium by mycorrhizal hyphae may be occurring.[23,35-39]

How mycorrhizae influence vegetation may depend on the soils in which they occur, for it is the soil that provides the major mineral resources required by the plant and fungus. Nutrient uptake is highly dependent on ion solubility, especially for soil P where essentially all the P is derived directly or indirectly from the parent material. By comparison, very little of the soil N originates from the parent material, but rather is derived from precipitation and biological means.

The state or form of an ion, as well as its concentration and distribution, varies among soils depending on parent material, climate, topography, vegetation, and time.[40] However, all P is taken up by the plant in the oxidized PO_4^{+3} state. In relatively unweathered soils, such as Entisols associated with the shrublands of some of the semiarid regions of the western U.S., soil solution P is supplied to the root surface from hydrolysis of calcium-bound phosphates within the parent material. The solution inorganic phosphorus (P_i) may be precipitated by the presence of various cations (mainly calcium and magnesium) adsorbed to mineral surfaces, or may be incorporated into microbial and plant biomass and soil organic matter.[41] In weathered soils more characteristic of humid or subhumid grasslands, the bases and silica are lost, and Al- and Fe-oxy-hydroxides are generated, forming secondary Al- or Fe-phosphate.[41] In these soils the P_i forms shift from primary to secondary. Along with the soil P transformations, levels of organic carbon, nitrogen, and sulfur pools rise to a maximum, then decline with the disappearance of primary mineral phosphorus.[42] That portion of primary P_i that is taken up by the plant and microbial biomass essentially becomes part of the soil organic phosphorus (P_o) pool. This pool can be a major source of available phosphorus; through its mineralization by soil phosphatases it can be converted readily to P_i and taken up by the root. In most grassland soils ranging from shortgrass prairie to tallgrass prairie (aridisols and mollisols), the P_o pool represents a major source of P to soil solution P via mineralization.[43] These trends associated with soil development suggest a sequence in which unweathered soils are composed predominantly of primary P_i; secondary P_i and P_o increase in importance with weathering and vegetation development; in highly weathered soils, nearly all the soil P is in the occluded Al- and Fe-phosphate form.[42] Even more important to nutrient availability than the total amount of an ion is the mineralization rate of that ion. Another major controlling factor of P_o mineralization, besides the form of the soil P_o, is climate; there is positive association between annual precipitation and orthophosphate ions.[44]

It has been hypothesized that P released from microbial biomass during periodic population crashes, either environmentally induced or via invertebrate grazing, accounts for most of the P return to the available P pool and that the rate of return to this pool determines the rate of P uptake by plants.[45] In grassland soils, microbial biomass can account for from 5 to 24% of the total P_o pool, with an annual flux of P through microbial biomass as large as 23 kg P ha^{-1} $year^{-1}$, which suggests that biomass P could make a significant contribution to plant P nutrition in grasslands.[46]

The concentrations of nutrients in solution in natural grassland soils typically are very low. A typical soil may contain from 0.05 to 1.0 g of P per kilogram of soil, but the concentration in soil solution is likely to be between 0.1 and 10 μM, or roughly 60 μg P per kilogram of moist soil.[47] The rate of uptake of a nutrient in these soils is limited by its diffusion rate through the soil to the root surface or by the ability of the root to absorb the ion from low concentrations in the soil solution.[14] The levels of N, P, and K in soil solution in field soils are normally in the range of growth response, where uptake of ions by the root is remarkably insensitive to external ortho-P concentration unless it is a magnitude below 10^{-6} and 10^{-5} M for K^+ and NO_3^-.[48]

Mass flow (a result of transpiration of the leaf) and diffusion (a result of absorption of the root surface) at times do not bring adequate nutrients to the surface of roots to meet the needs of the plant, especially for P.[49] To compensate for these lower nutrient levels, a plant can increase nutrient absorption only by increasing root biomass, altering root morphology, or by associating with mycorrhizal fungi, but apparently not by increasing physiological capacity to absorb nutrients per unit mass of root.[50] Also, plants from habitats of differing fertility typically differ more in growth rate than they do in nutrient absorption capacity,[4] with the strongest determinant of nutrient absorption capacity being nutrient demand during shoot growth.[51]

The potential uptake rate of P does not appear to limit the growth of the coexisting British upland grassland species *Deschamsia flexuosa, Festuca ovina, Juncus squarrosa, and Nardus stricta*.[52] When grown in solution culture, these species show no preference for the P_i or P_o source, and the growth response does not depend on nutrient concentration. However, when P_i and P_o were presented in equal concentrations, the uptake of P_i far exceeded the hydrolytic capacity of the plants' acid phosphatases, indicating that P_i would be depleted preferentially from soil solution. It was only in *D. flexuosa* that the P_i level reduced acid phosphatase activity when plants were grown with higher P_i levels. Also, the rush *J. squarrosa* responds to lower P_i by increasing root growth. The root morphology is probably more important for this species than for the other species. The considerable number of long root hairs produced by this rush can be an alternative method for obtaining sufficient P.[39,52] *Juncus* is also nonmycotrophic, although some mycorrhizal infection can be encountered in the field.[18,39]

This trend of plants typically differing more in growth rate than in nutrient adsorption capacity holds even when grasses of the same genus which are obtained from habitats of differing fertility are studied.[50] In similar environments in New Zealand, the alpine tussuck grass *Chionochloa pallens* occurs on a wider range of soils, having a higher P availability than *C. crassiuscula*. When these grasses were grown in solution culture with P_i ranging from 1 to 100 μmol ℓ^{-1}, the high-P-adapted *C. pallens* produced more biomass at all P levels tested than the low-P-adapted *C. crassiuscula*.[50] However, there appears to be no major difference in either short term or long term phosphorus metabolism in that P_i adsorption rates per unit mass of root were similar for both *Chionochloa* species. This indicates that high nutrient absorption capacity of the root has not been an important adaptation to infertile soils.[4]

B. Growth Responses

To date only a few studies have been concerned with growth responses of natural vegetation to mycorrhizal fungi. Of the studies conducted with grassland plant species, most investigations have focused on pasture grasses (Table 1). Much of the early research indicated that tropical grasses often responded to infection by mycorrhizal fungi by increased biomass production,[53-55] but that the responses of temperate and alpine grasses to mycorrhizal fungi have not been as dramatic.[56-59] These studies indicate that growth responses to infection by mycorrhizal fungi varies with climate, soil type, and habitat from which both host and endophyte originated.[57,59,60] In the tussock grasslands of New Zealand and the upland grasslands of Great Britain, mycorrhizal fungi are widespread.[57,61,62] Although the response to mycorrhizal infection for a particular grass species has not always been as predictable as its distribution, it appears that mycorrhizal fungi can be efficient symbionts with many native grasses under low P fertility.[57,58,63] Interestingly, the response to infection by mycorrhizal fungi for many of the grasses investigated is one of growth suppression, especially under moderate to high P fertility.[19,59] For example, *Lolium perenne* has been shown repeatedly to exhibit a parasitic response to infection by mycorrhizal fungi.[19,64,65] However, research with the tussuck grasses *C. rigida, Poa colensoi,*[58] *Dactylis glomerata,*[63] and *F. ovina,*[58] and the dune grasses *Agropyron junceiforme* and *Ammophila arenaria*[66] indicates that mycorrhizae confer an advantage to plants under nutrient-limiting conditions.

One of the main characteristics of those grasses that does not appear to respond to mycorrhizal fungus infection is a well-developed, fine root system with an abundance of long root hairs.[20] Also, many of these grasses have evolved in habitats along the forest edge[67] and require soils with a high nutrient status and/or a continued disturbance regime for sustained productivity. The grasses that appear to benefit from mycotrophy are those plants that have a rather coarse root morphology, produce few root hairs, and occur in nutritionally sparse habitats.

It appears that an increase in biomass or productivity to mycorrhizal infection may not always be the best measure of how the mycorrhizal association influences a host. Research with *Bouteloua gracilis*, a co-dominant grass of the shortgrass prairie, has suggested a high degree of physiological dependency on mycorrhizae as measured by stomatal resistance and photosynthetic rates.[68] Even though higher photosynthetic rates were measured for mycorrhizal plants, no increase in biomass was evident; however, P may not have been limiting. Also, the fate of a substantial portion of the fixed carbon may have been incorporated in the biomass of the mycorrhizal fungus. What is unusual about this host is the high degree of infection encountered in its roots, even though the soils in which it grows are moderately high in bicarbonate-extractable P (15 to 20 $\mu g \cdot g^{-1}$).[68-70] It appears now that an interaction between P and N may explain this apparent anomaly in the field where low N levels are limiting root growth.[70] When *B. gracilis* was grown over a range of P and N, enhanced growth occurred with increased

Table 1
GRASS AND GRASS-LIKE PLANTS FOR WHICH MYCORRHIZAL GROWTH RESPONSES HAVE BEEN DETERMINED

Host	Growth response to VA mycorrhizae	Habitat	Ref.
Gramineae			
Agropyron junceiforme	+	Dunes	66
A. smithii	+	Prairie/steppe	73, 227
Agrostis tenuis	+/−	Upland grassland	59
Ammophilia arenaria	+	Dunes	66
Anthoxanthum odoratum	+/−	Upland grassland	19, 59
Bouteloua curtipendula	+	Prairie	227
B. gracilis	+/0/−	Prairie	68, 70, 227
Bromus tectorum	0	Steppe/disturbed	229
Chionochloa macra	−	Montane grassland	57
C. rigida	+	Montane grassland	57
Cynosurus cristatus	+/−	Upland grassland	59
Dactylis glomerata	+/−	Upland grassland	19, 63
Distichlis spicata	0	Saline soils	228
Festuca arundinaceae	−	Pasture/meadow	215
F. novae-zelandiae	−	Montane grassland	57
F. ovina	+	Upland grassland	58
F. rubra	−	Upland grassland	59
Hilaria jamesii	+	Deserts	226
Holcus lanatus	+	Pasture/meadow	174
Lolium perenne	+/−	Pasture/meadow	19, 64, 65, 230
L. rigida	−	Pasture/meadow	60
Nardus stricta	+	Upland grassland	59
Oryzopsis hymenoides	+	Steppe	227
Panicum coloratum	+	Savanna	74
Poa colensoi	+	Montane grassland	18, 57
P. laevis	−	Montane grassland	57
Sporobolus airoides	+	Prairie/steppe	227
S. cryptandrus	0	Prairie/steppe	227
Cyperaceae			
Carex cariacea	−[a]	Montane grassland	18
C. firma	+[b]	Alpine meadow	21
C. sempervirens	+[b]	Alpine meadow	21
Uncinia divaricata	−[a]	Montane grassland	18
Juncaceae			
Juncus articulatus	−[a]	Montane grassland	18
J. planifolius	−[a]	Montane grassland	18
J. novae-zelandiae	−[a]	Montane grassland	18

Note: + = Positive growth response; 0 = no response; − = decrease in growth as measured in shoot or total biomass produced.

[a] Inoculated plants did not become infected.
[b] Endophyte not a typical mycorrhizal fungus, but reported as a "dark-sterile-form".

nutrient supply over the entire range of fertilities.[70] In addition, infected plants were consistently smaller than noninfected plants; the researchers suggest that this carbon drain most likely was related to the parasitic nature of the early phase of infection, as has been reported for other plants.[71] It may be that during the early phases of infection, carbon allocation favors the mycorrhizal fungus.[72]

Few studies for which many of our conclusions about growth responses of grasses to infection by mycorrhizal fungi have been conducted using soils having a narrow fertility range, or have required the addition of P or other nutrients in order to achieve a range of soil fertility. It appears that for many soils the addition of fertilizer P can eliminate a mycorrhiza-associated positive plant growth response. Furthermore, micronutrients can be immobilized through such additions. A means to avoid the complications associated with fertilization can be achieved by using a soil dilution series for obtaining an appropriate growth medium. Providing the soil used has a high initial value of the ion in question and providing care is taken to supplement for other ions, a more natural ion pool can be achieved. *Agropyron smithii* showed a positive growth response after inoculation with mycorrhizal fungi at all soil P concentrations when a relatively high-P sandy-loam soil (30 $\mu g \cdot g^{-1}$ bicarbonate extractable P) was diluted with sand and supplemented with Hoaglands solution minus P to achieve a range of bicarbonate extractable P (2 to 20 $\mu g \cdot g^{-1}$) (Figure 1).[73] The mycorrhizal dependency index* reduced from 35% for plants at 2 $\mu g \cdot g^{-1}$ bicarbonate-extractable soil P to 0.3% for plants at 20 $\mu g \cdot g^{-1}$ P. Although mycorrhizal plants may produce more dry matter, they were shorter and produced more tillers; thus, mycorrhizal fungi can influence a host's biomass allocation without affecting total biomass produced. When tiller production is expressed in terms of carbon cost per unit P, a relationship where mycorrhizal grasses are more efficient than nonmycorrhizal ones is observed (Figure 2). It appears that this tillering response is more than just an increase in P concentration. When leaf surface area is corrected for differences in tissue P concentration and tiller number regressed against this value, a relationship becomes evident where the amount of surface area required to produce a tiller in a mycorrhizal grass is significantly less than that for a nonmycorrhizal one.[73] The increase in efficiency of carbon cost to phosphorus gain as expressed in leaf surface area per unit P can result in a redirecting of carbon into increased tiller production via stimulation of axillary meristems.

A positive association for tiller production and mycorrhizae also has been reported for *D. glomerata*[63] and *Panicum coloratum*,[74] while in *B. gracilis*[70] a reduction in tillerage was evident with infection. Dahl and Hyder[75] have pointed out that a plant that is best adapted to withstand grazing pressure does so via prostrate, spreading growth. The research with *A. smithii* and *P. coloratum* indicates that mycorrhizae may confer grazing tolerance on plants by maximizing plant biomass production below the grazing zone via promotion of tillerage and a prostrate growth habit.[73,74] These studies raise the question of whether mycorrhizae influence all grasses similarly, or do rhizominous forms, i.e., those grasses with highly developed axillary meristems, respond differently than those with more apically dominated meristems like the tussock grasses. Since the roots of *A. smithii* only occupy the soil directly beneath the shoots, tillering is the major means by which an increase in exploitation of the soil can occur by this host.[76]

In soils of low P status, mycorrhizae strongly stimulate nodulation and growth of legumes.[55,77-80] It appears that when legumes are inoculated with appropriate strains of *Rhizobium,* they nodulate in the most P-deficient soils only if mycorrhizal fungi are present.[77] Additionally, other studies have found that mycorrhizae enhance N_2 fixation.[79] However, this enhancement by mycorrhizal fungi may be indirect and probably is nothing more than the bacterial symbiont responding to the increased vigor of the mycorrhizal plant.[81] Very few investigations have evaluated the relationship of mycorrhizal fungi to legumes found in grassland and shrubland ecosystems, with most of the information available coming from studies conducted in an agricultural setting. The studies so far indicate that this group of plants is highly dependent on mycorrhizae in

* [(Mycorrhizal plant dry weight − nonmycorrhizal plant dry weight ÷ mycorrhizal plant dry weight) × 100].

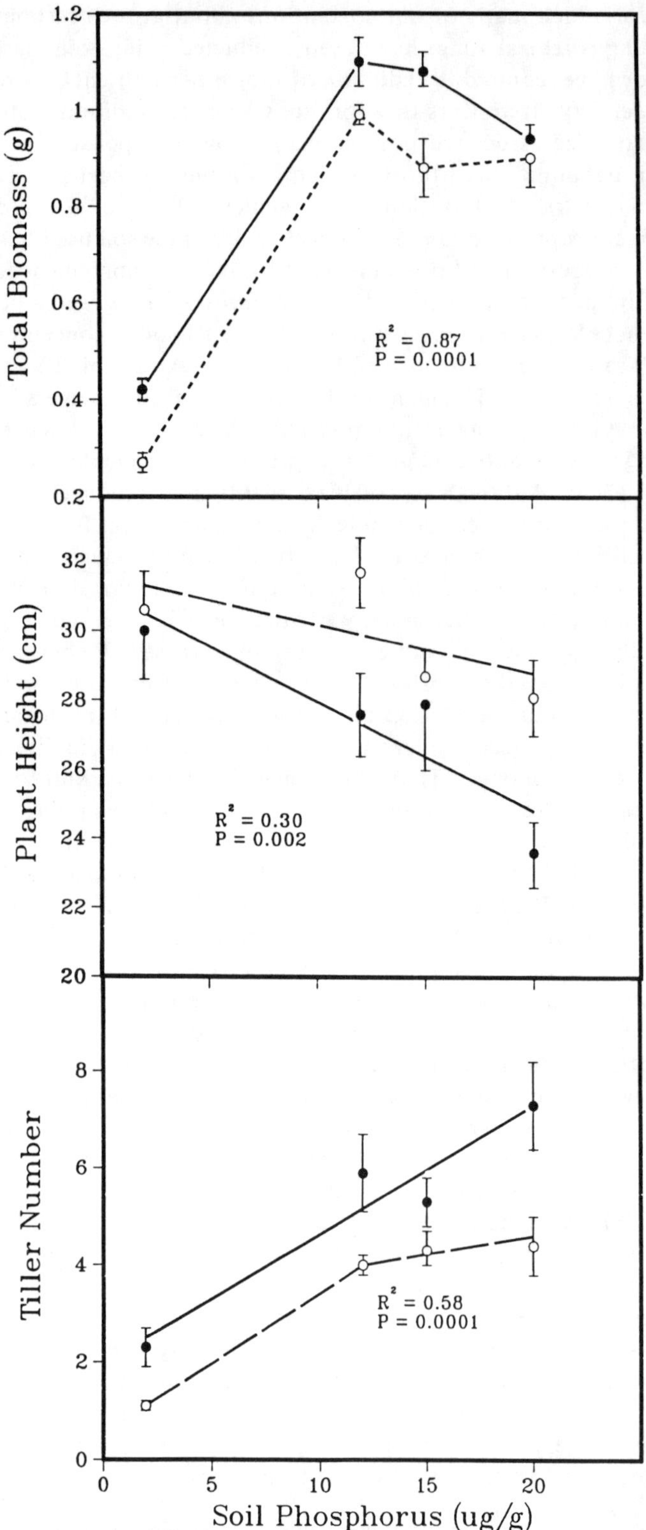

FIGURE 1. The influence of mycorrhizal fungi on biomass allocation in *Agropyron smithii* over a soil phosphorus gradient. Each data point represents a mean and standard error derived from a sample size of n = 9; —•— = inoculated with *Glomus mosseae/monosporum*, —○— = uninoculated control plants. The R^2 and P are for the ANOVA consisting of the dependent variables of soil phosphorus level and inoculation.[73]

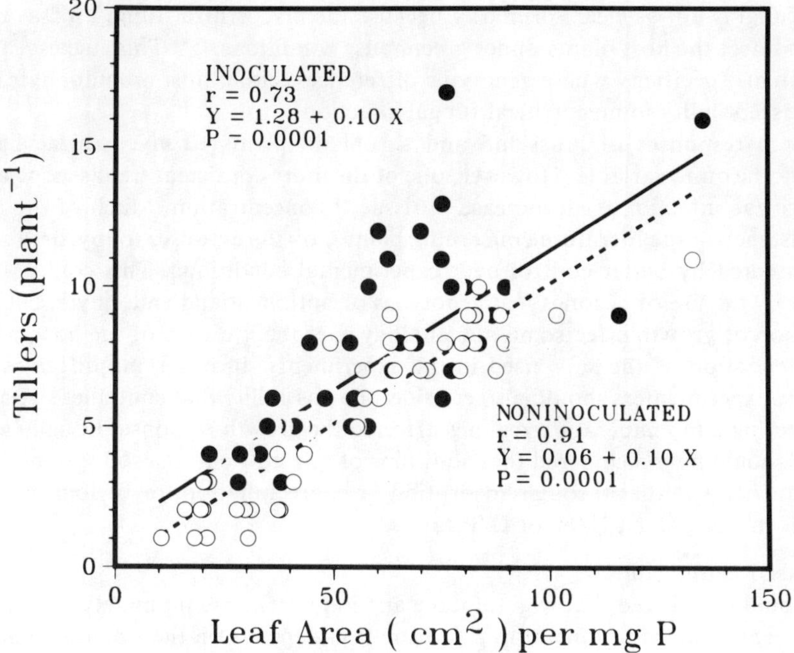

FIGURE 2. The relationship of leaf surface area corrected for shoot P concentration to tillering. The intercepts of inoculated vs. noninoculated plants are significantly different (P = 0.001) and the slopes are parallel (P = 0.001).[73]

natural soils. In *Hedysarum boreale*, a legume native to the western U.S., mycorrhizal fungi stimulated both growth and N_2 fixation.[80] For *Cassia fasiculata*, a legume that can grow in a relatively broad range of environments but that is more characteristic of disturbed areas, one would expect that a need for mycorrhizae would be rather facultative (*sensu* Janos[82]), however, what was found was that those plants inoculated with mycorrhizal fungi grew quite vigorously and flowered, while plants inoculated with Rhizobia alone or uninoculated controls were stunted,[83] indicating the obligate nature of the mycorrhizal association even in legumes associated with disturbance.

Only a few studies have been done on the effects of mycorrhizae on the growth of shrub species. Lindsey[84] has put together a thorough review of these investigations along with a survey of shrub species in which mycorrhizae have been reported. Excluding the studies with chenopod species, generally a significant growth response occurs in shrubs following mycorrhizal inoculation. Depending on the species of endophyte used for inoculum, the composite shrubs *Artemisia tridentata* and *Chrysothamnus nauseosus* showed increases in biomass production up to sevenfold.[84] These shrubs are a major component of much of the shrublands of western North America. A similar response to infection has been reported for the latex-producing shrub *Parthenium argentatum* found in northern Mexico and West Texas, where plants infected with mycorrhizal fungi have increased latex production and produced fivefold increases in biomass over noninoculated controls.[85,86] Growth response to mycorrhizae is not as straightforward for the chenopod shrub *Atriplex canescens*. Early research by Williams et al.[87] suggested that a significant increase in plant growth occurred following inoculation with mycorrhizal fungi. Subsequent studies by Aldon[88] found that mycorrhizae increased survival and growth of *A. canescens* when grown on spoil material in northwestern New Mexico. Even though subsequent research has been able to repeat these earlier growth increases, the apparent stimulation of growth to the plants by mycorrhizal fungi is not as clear, primarily because the mycorrhizal fungi used as inoculum did not infect the host plants under greenhouse conditions.[84,89] This suggests a possible host-fungus specificity where genotypic differences, either host or endophyte, may exist in susceptibility to mycorrhizal fungal infection.

Growth responses of grassland and shrubland plants to mycorrhizae have been shown to be quite variable. However, one of the more consistent trends associated with mycorrhizae infection is an increase in tissue P concentration. Much of the apparent inconsistency in the literature concerning plant growth responses to mycorrhizae could be eliminated by better control over experimental conditions. This could be accomplished by the use of ''clones'' or genotypes of both host and endophyte, better documentation of growth effects and morphology over the life span of the host, and better characterization of the soils used in the experiments. In observing differences in responses, experimenters should also consider the possibility that container size and varying watering and temperature regimes affected the growth response. Equally as important, it should be remembered that additions of one nutrient to a soil system can affect other nutrients, either through adsorption or coprecipitation, or biologically through changes in, e.g., C/N, N/P, or C/P ratios.

C. Water Relationships

It should be realized that water uptake and ion uptake are intimately related because the concentration of a mineral in a soil solution depends on the water content of that soil. In a drying soil, absorption at the root surface is reduced because ion movement is slowed. Also, because ion uptake occurs in a water milieu, the sites of water uptake are also where solute transfer occurs.

The integral nature of water in the photosynthetic process is just as important as that with ion uptake, since the short-term productivity of a plant depends on maintaining its photosynthetic tissue at a high-water status (i.e., while the stomata remain open). To maintain the high-water status, maintaining positive turgor is necessary. In mesophytes, the symplast osmotic pressure under natural conditions ranges from 1.0 to 2.5 MPa; in xerophytes osmotic pressure can be twice these values.[90] The loss of water by transpiration from the leaves of plants inevitably accompanies photosynthesis. To make up for this evaporative water loss, a plant must absorb water from the soil through its roots.

Mycorrhizae can alter the water relations of the host plant and appear to increase drought tolerance.[91-93] Safir and co-workers[92,95-97] have put forth a nutritional hypothesis to explain these altered water relations. Using Fiscus's model root system[98] in which P plays a large role in membrane permeability, Safir et al. propose that the changes in hydraulic conductivity associated with mycorrhizae may actually be P induced, since it is under low-P conditions that mycorrhizae alter the water relations of host plants. Their data suggest that the water relations of plants are affected similarly by amending the growth medium with P and by inoculating with mycorrhizal fungi. The primary mechanism of these mycorrhizal-mediated water relation responses is nutritional. Also, under conditions of high-water and P availability, infection with mycorrhizal fungi does not appear to affect plant-water relationships. This observation can lead to the oversimplified conclusion that with P fertilization or in soil with high-P status, mycorrhizae may not be necessary. However, when plants are growing where soil moisture either limits growth or varies cyclically, plant growth stimulation by mycorrhizal fungi is not duplicated by P addition.[92,95]

Several studies have been reported on the influence of mycorrhizae on water relations of grassland or shrubland plant species. Three studies have been conducted with the steppe grasses *B. gracilis*[68,99] and *Agropyron smithii*,[100] and one with the pasture legume *Trifolium pratense*.[91] Leaf diffusive resistances of mycorrhizal *T. pratense* were higher and transpiration fluxes were lower compared with nonmycorrhizal plants. Also, mycorrhizal plants were able to extract moisture at lower water potentials than nonmycorrhizal plants (difference of 1 MPa). The hydraulic conductance of mycorrhizal roots on a per unit root length basis was two to three times higher than nonmycorrhizal roots, suggesting that the difference in water use is due to hyphal growth in the soil.[91] In *B. gracilis,* the total plant weight, leaf area, and root length of mycorrhizal plants had fewer and shorter root hairs.[99] Mycorrhizal plants had 50% lower leaf resistance without a change in leaf or root water potentials,[68,99] while transpiration rates increased by more than 100% with infection.[68] *A. smithii* responds similarly to infection; mycorrhizal plants had 47% lower leaf resistance at low soil and plant water tensions than their noninfected counterparts.[100] These investigations all suggest that the mycorrhizal state is an advantage to a host in times of limiting moisture. Mycorrhizal plants have higher transpiration rates that appear to be related to increased conductivity of roots, which appear to be related to increased P status of mycorrhizal roots,[92,94] possibly via an increase in vessel size and length.[101] Alternatively, altered hormonal levels of mycorrhizal roots affecting their hydraulic conductivities have been proposed as a possible mechanism.[99]

The interface between the soil and the root surface may present resistance to water flow as a result of a limited number of contact points between soil particles and the root surface for areas where water transfer can occur.[102] This interfacial resistance would increase with increasing water stress as contact points were reduced by the contraction of root tissue forming gaps between the surrounding soil and the root surface.[103] Conceivably, a fundamental ecological role for mycorrhizal fungus hyphae is the bridging of the annular space or voids created by a root's expansion and contraction, resulting in the continuing physical continuity between the root surface and soil particles. In doing so, the hyphae provide a means of increasing the effective surface area of the root while also achieving a decrease in resistance to water flow to the root surface by allowing closer contact with the soil. The mycorrhizal hyphae also act as contact points with soil particles and, unlike root hairs, may provide a longer-term mechanical link across the soil-root interface. This relationship between the root surface, fungus, and soil matrix would be especially important to plants occurring in soils of high-water conductive resistance and where a dry down cycle, i.e., a drought cycle, is commonplace. The controlling factors of this phenomena are soil texture and fertility, the number and spacing of the hyphal links, root hairs, and mucilage production, and the swelling and shrinking tendency of host roots.

Mycorrhizal fungi appear to have the ability to facilitate a change to its own environment through the influence of its hyphae on soil aggregate stability. This is supported by research in which plants, roots, and mycorrhizal fungal hyphae have been found to play an intricate role in the formation of stable soil aggregates in association with other soil organisms.[104] The importance of these aggregates is their influence on soil structure; soils with good structure by definition have a high degree of aggregation. Soils having good soil structure hold sufficient water to prevent moisture deficits around plant roots during dry periods, but allow sufficient drainage to prevent waterlogging during wet periods. It appears than that mycorrhizal fungi, through their association with aggregation, may be intimately associated with large pore space formation resulting in increased hydraulic conductance of the soil they inhabit. Any means by which an increase in large pore space can occur would be advantageous to survival of a host, especially in ecosystems where climatic patterns are such that a major portion of a year's precipitation can occur in just a few events. Under these circumstances, infiltration rates would be very important; i.e., the ability of a soil (mediated by mycorrhizal fungi) to slow down runoff of moisture by increasing infiltration can be important to plant survival in semiarid and arid environments. This attribute of mycorrhizal roots may be one of the more overlooked areas in mycorrhiza research.

When the responses of plants to infection by mycorrhizal fungi are put into the context of their growth requirements, several trends become evident. Plants needing relatively fertile soil and low stress growing conditions do not appear to require mycorrhizae for survival or reproduction unless they have a coarse root morphology. Similarly, a positive growth response to mycorrhizae may occur in plants possessing a fine root morphology when grown in extremely low-P soils. The typical response to infection for these plants under adequate fertility, i.e., their normal range of fertility, may be no effect or even growth suppression. Plants that best fit these descriptions typically are found growing under humid pasture conditions. Plants that show a positive growth response to mycorrhizae typically occur in soils of low fertility or under moderate- to high-stress growing conditions or where climate limits the growing season. These plants are likely to have a coarse root morphology, but also may develop fine roots and have a slow growth rate with luxury consumption of mineral ions. These plants may show stunting in the absence of mycorrhizae. Thus, an approach based on mycorrhiza influence to the total biomass of natural vegetation may be misleading. As shown with *B. gracilis,* the effects of infection by mycorrhizal fungi on plant growth are on physiology and morphology,[68] and in *A. smithii* the effects are on biomass allocation (Figure 1) and possibly its survival rather than total production or yield.[73]

III. OCCURRENCE OF MYCORRHIZAE

A. Distribution of Mycorrhizae

Upon reviewing the literature on the occurrence of mycorrhizae in rangeland plants, Trappe[105] has noted that except for the Cistaceae, Fagaceae, and Pinaceae (all of which possess ectomycorrhizae), all mycorrhizal plant taxa probably form associations with VA mycorrhizal fungi. Nonmycorrhizal taxa include the Cruciferae and Zygophyllaceae. The Cactaceae, Chenopodiaceae, Cyperaceae, and Juncaceae typically are thought to be nonmycorrhizal, but some species appear to become infected by mycorrhizal fungi under certain conditions.

Infection by mycorrhizal fungi occurs in most plant species in temperate seminatural grasslands,[106-109] tallgrass prairie,[110] shortgrass steppe,[69] alpine and tundra grasslands,[58,111,112] and in the savanna of the Serengeti.[74] Similarly, the occurrence of mycorrhizal fungi is extensive in shrubland plant communities,[10,84,105] high-elevation sagebrush communities,[113-115] a high-elevation cold desert,[116] the Chihuahuan Desert,[117] the Sonoran Desert,[118,119] and the deserts of Asia.[120]

The literature on the occurrence of mycorrhizae in the Chenopodiaceae, an important family in shrubland ecosystems, varies, particularly with regard to shrubs of the genus *Atriplex,* in which both presence and absence of vesicular fungal infection[113,114,116,121-124] and related growth effects[84,87] have been reported. Hirrel et al.,[125] working with annual chenopods, suggested that infection by mycorrhizal fungi in the Chenopodiaceae may occur in dead root material or in the stele as only vesicles are encountered; they believed the lack of arbuscules indicates that the mycotrophic relationship may be nonfunctional. Miller,[116] studying plants in the Red Desert of Wyoming, found that mycotrophy may be associated with the growth habit of a host, where woody chenopods (e.g., *A. confertifolia*) having a stress-tolerant growth strategy may be mycorrhizal, while annual chenopods with a weedy or ruderal habit, such as *Halogeton glomeratus* and *Salsola kali,* are nonmycorrhizal. It has also been shown that the occurrence of infection by mycorrhizal fungi may be related to the type of interspecific plant associations occurring.[124] The solitary occurrence of *A. confertifolia* or occurrence in the presence of nonmycorrhizal *A. gardneri* lead to a nonmycorrhizal state;

however, when occurring near grasses or *Artemisia spinescens, Atriplex confertifolia* roots were colonized by mycorrhizal fungi.

The presence of arbuscules in the Chenopodiaceae has been reported only occasionally in the literature. Allen[123] found arbuscules in *A. gardneri* to be associated primarily with the fine roots of this half-shrub. He also suggests that the occurrence of mycorrhizae, based on the detection of fungal structures in the roots, may be seasonal and related to the phenology of the plant. Arbuscules also have been induced in the normally nonmycorrhizal plant (Chenopodium quinona with the herbicide simazine.[126]

The occurrence of infection in the Cyperaceae also has been reported.[18,61,106,109,110,127] In a mesic tallgrass prairie, *Carex* spp. and *Scirpus fluviatilis* appeared not to be infected by mycorrhizal fungi, while infection was found sporadically in *Eleocharis.*[110] Powell,[18] surveying sedges and rushes from a variety of habitats in New Zealand, found no infection. High levels of infections by mycorrhizal fungi rarely are reported for these plants. It appears that in some alpine grasslands the predominant type of infection for *Carex* spp. was not the mycorrhizal type, but rather a "dark-sterile form".[111] This particular type of mycorrhiza has been reported to increase shoot P uptake and dry-matter production in *Carex* spp.[128]

B. Propagule Population Dynamics and Species Diversity

The assessment of propagule population densities of mycorrhizal fungi is a significant undertaking. Previously, only chlamydospores and azygospores of mycorrhizal fungi were counted as propagules, with the spread of the fungi through the roots considered a measure of fungal activity. The primary types of measurements used to assess the occurrence of propagules are based on the percentage of root system occupied by the endophyte — typically the percentage is based on root length[129] — and by spore counts usually expressed per unit of soil.[130] Neither measure is a good indicator of the infectivity of a particular endophyte or soil because nonsporing forms of mycorrhizal fungi may constitute a significant portion of the population of propagules occurring at a site.[131,132] The primary means of infection for a seedling or young plant may most likely be initiated through root contact with mycorrhizal hyphae than via spores,[61,133] except very early during the growing season. Organic matter and fragmented root pieces containing mycorrhizal hyphae can act as propagules and may be a major source of inoculum for infection under a range of conditions, especially for those soils with a low spore density and actively growing roots. On a dry-weight basis, root fragments containing mycorrhizal fungi can be a more effective inoculum than spores, since mycorrhizal root fragments can initiate infection faster than spores.[134-136] However, the longer the inoculum is kept from a host the more likely is infection from a spore, since spores can survive longer in the absence of a host than other propagules.[136] When correlation between spore numbers and infection rate is poor, it is an indication of competition between the indigenous mycorrhizal fungi and/or it is often due to differential infection rates of the different fungi.[137] It also appears that in grassland and shrubland soils a residual spore pool exists.[138] This pool results from the carryover of spores from previous growing seasons. In soils of harsher sites a residual spore population is minimal.[139]

The early work of Nicolson[140,141] characterizing VA mycorrhizae in the Gramineae initiated an era of ecological studies to many researchers interested in the ecological aspects of this association. These studies, along with the life history studies of Mosse,[142] are still quite relevant today. However, it was not until Gerdemann and Trappe[143] revised the taxonomy of the Endogonaceae that a systematic approach could be taken to field studies, with the distributions of species of mycorrhizal fungi determined along with host species associations. In recent years there has been an increase in the number of field studies addressing those factors associated with the occurrence of mycorrhizal fungi. Nicolson,[66,144] studying grasses associated with sand dune communities, suggested that species of mycorrhiza fungi may be associated more with community type than with a particular plant-endophyte association. More recent research in dune environments suggests that there is an association between endophyte type and dune developmental stage or position.[66,145] Koske[146] has suggested that the most important determinants affecting the propagule population of mycorrhizal fungi are site characteristics, species of host plants, and species of mycorrhizal fungi.

The overriding factors associated with propagule population density in a grassland or shrubland setting are climate related. On a macroscale this would include climatic influences on both plant distribution and productivity, and on a microscale, precipitation influences on soil moisture. For example, in a mesic tallgrass prairie in Illinois, other than vascular plant cover, soil moisture is the dominant factor associated with spore population density for mycorrhizal fungi.[147,148] In one study over a soil moisture gradient ranging from xeric to hydric, a negative association between gravimetric soil moisture and spore population density existed.[147] However, on a second study on a microsite level, the spore population density of soils surrounding roots of *Schizachyrum scoparium* had a positive association with soil moisture.[148] The discrepancy in findings between these two studies is most likely explained by the distribution of *S. scoparium,* which occurs mainly in the drier parts of the soil moisture gradient. In addition, one study consists of spore populations from a specific plant's rhizosphere, while the other study obtained spores through random sampling. The factors in a high-elevation cold desert shrubland in southwestern Wyoming that best predict the propagule population density of mycorrhizal fungi are percentage cover occupied by mycorrhizal plant species and soil water-retaining capacity.[149] In this ecosystem, a gradient exists for mycorrhizal plant cover; there is a transition from an essentially nonmycorrhizal plant community composed primarily of nonmycorrhizal chenopods to a community dominated solely by mycorrhizal plant species. Another characteristic of these communities is that total plant cover typically occupies only 30 to 40% of the soil surface leaving gaps between shrubs. Using a soil bioassay to determine the potential for infection by mycorrhizal fungi along this gradient, mycorrhizal plant cover was the best predictor of infection potential of a soil when a positive association between infection level and mycorrhizal plant cover occurred (Figure 3A). A negative relationship between mycorrhizal fungi infection potential and water-retaining capacity also was present. This suggests that the more droughty a soil the higher the propagule density (Figure 3C); however, water-retaining capacity may also be an indirect measure of soil aeration and temperature.

It has been demonstrated that nonspore propagules can occur for *Glomus fasciculatum, G. monosporum,* and *Gigaspora calospora,* where hyphae can extend radically up to 5 cm from the broken ends of cortex of dead roots.[150] It also appears that those mycorrhizal fungi capable of forming intraradical or extraradical vesicules produce nonspore propagules.[151] The survival of nonspore propagules depends on dried soil; hyphae of infected roots kept in soils at a water potential of −50 MPa were capable of initiating infection, whereas hyphae of infected roots kept under moist conditions could not grow.[150] This may explain why under field conditions in shrubland soils, a high mycorrhizal propagule density which has a large nonspore component is related to low water-retaining capacity (Figures 3B and 3C) — i.e., the more xeric the site the more likely that a higher mycorrhizal infection potential will exist for that site.

The occurrence of species of mycorrhizal fungi and spore population density have been reported for grassland and shrubland soils of Australia,[146,152-154] New Zealand,[58,132,155,156] Spain,[157] and Pakistan.[122,158] The distribution of species in North

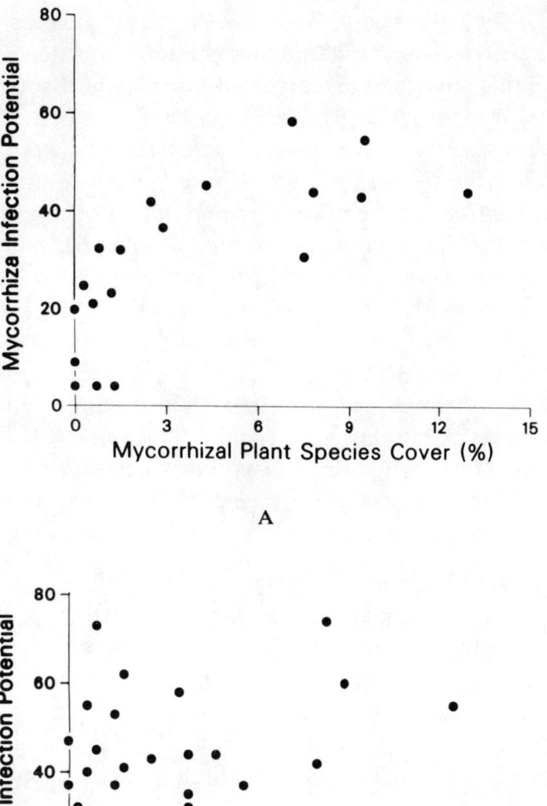

FIGURE 3. (A) The relationship of mycorrhizal plant species cover to mycorrhizal infection potential for soils obtained from a high elevation cold desert community in Wyoming; (B) the association between mycorrhizal fungus spore numbers and mycorrhizal infection potential for soils obtained from a depth of 0 to 10 cm; (C) the relationship between water retaining capacity (WRC) of a soil and mycorrhizal infection potential, where WRC is the amount of water held at field capacity minus the amount of water held at wilting point corrected for soil bulk density.

American grassland and shrublands has been reported for tallgrass species,[110,147,148,159] in the shortgrass prairie and shrubland steppe,[160] and in the Sonoran Desert.[118,119] Also, the species of mycorrhizal fungi occurring with various North American grasses have been reported.[148,156,161]

In the grasslands and shrublands of North America, the most common species of mycorrhizal fungi encountered are *G. fasciculatum, G. macrocarpum, G. microcarpum,* and *G. mosseae.*[110,118,119,159,160,162] In the hill country soils of New Zealand, the most-encountered endophytes typically are *Acaulospora laevis* and *G. fasciculatum,*

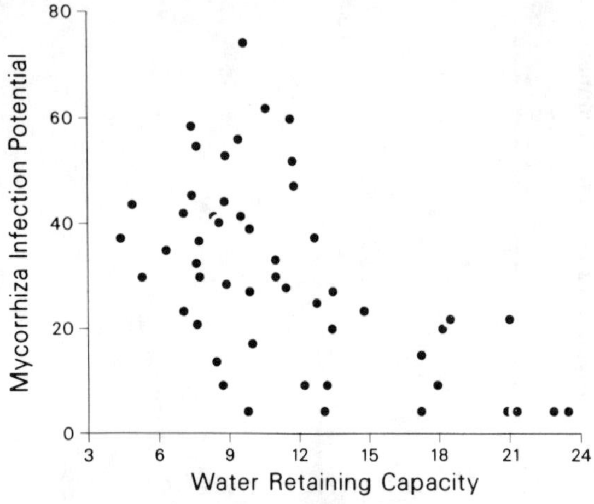

FIGURE 3C.

with the fine-endophyte *G. tenue* also being common.[19,132,155,157,163] Mycorrhizal fungi spore density is lowest in tussuck grassland soils;[156] however, a predominance of nons-pore form strains appears common since colonization was not associated with spore population densities in these studies.[132] In Australian soils *A. laevis* and *G. mosseae,* along with *G. fasciculatum* and *G. monosporum,* are isolated typically; also *G. tenue* is abundant in roots from western Australian soils.[137,153] In the soils of northern Paki-stan, *G. mosseae* and *G. fasciculatum* typically are encountered in association with sandy soils.[122]

IV. ASSOCIATION EFFECTS OF MYCORRHIZAE

The biology of plants when grown alone does not compare to their biology when grown in mixtures.[164] When plants grow in close proximity to each other, both their morphology and physiology may be changed. Typically, the increased proximity of neighbors results in a plant showing a lower growth rate or a reduction in dry matter yield which may be due to competition for environmental resources such as light, water, and mineral nutrients; or as the result of allelopathic effects, or to the presence of neighbors promoting disease incidence.[165] Harper[166] has suggested the term interfer-ence for hardships caused by the proximity of a neighboring plant. The effects of neighbors to a species may also be positive, via providing protection from wind, graz-ing, or against lodging; or they may also be neutral (see Trenbath and Harper[165]). However, in the absence of grazing or disease, a plant's response to neighboring species is more likely a measure of differences in the microenvironment. A plant's response to differences in the microenvironment can be influenced by the effectiveness of the my-corrhizal symbiosis. To complicate matters, the success of an individual is dependent not only on plant-to-plant relationships, but also plant-to-fungus and fungus-to-fun-gus interactions (Figure 4).

A. Plant-Fungus Association Effects

When comparing infection levels for roots obtained from the field, a factor that must be taken into account is the total weight of the infected root, since data consisting of percentage of root infected by the mycorrhizal fungus itself can be misleading. For example, in upland grasslands of Great Britain, the greatest weight of infected grass roots occurs during the summer, although the highest percentage of roots infected by

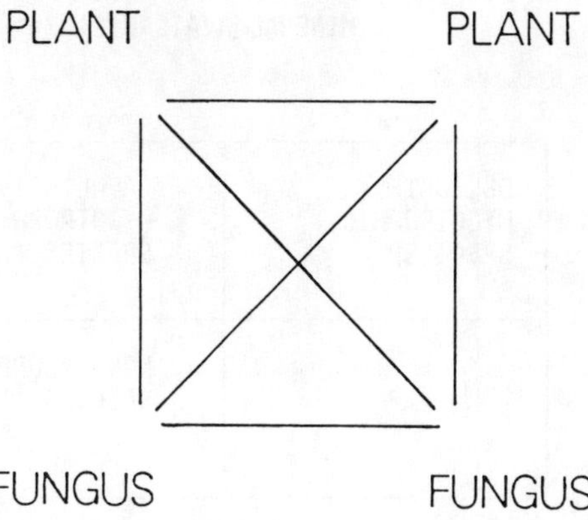

FIGURE 4. The interactions of organisms can be best represented by a diagram where not only plant-to-plant interactions can occur (both positive and negative), but also plant-to-fungus and fungus-to-fungus interactions.

mycorrhizal fungi occurs in winter.[62] It may be that as a root system responds to infection, an increase in growth rate mediated by increased internal P concentration due to infection by mycorrhizal fungi occurs. This can result in an apparent decrease in percentage of root colonized by mycorrhizal fungi even though a greater total fungus biomass is present.

The distribution and density of mycorrhizal fungi within roots can vary with soil depth, host species, and season of collection. For example, the mycorrhizal fungi entry points in roots of *Molinia coerulea* and *Galium verum* growing in meadows of Central Europe decrease with soil depth — significantly more entry points occur in the fall than in the spring.[105] Also, there does not appear to be any consistent trend for the occurrence of the highest levels of infection; Gay et al.[109] found that mycorrhizae density varied not only with sampling season but also year. Other studies suggest that soil moisture and nutritional factors are more important.[61,108,148] Many of the seasonal responses reported in the literature could be related to a host or endophyte responding to changes in soil moisture and nutrient pulses. Studies with *Glomus tenue* in grasses found in pastures in the eastern U.S. indicate that seasonal differences in percentage length of roots infected were generally highest in spring in P-deficient soils low in moisture.[108] Mycorrhizal infection levels can also be related to both total plant density and its composition. Mycorrhizal infection in *Plantago lanceolata* is negatively associated with vascular plant cover,[167] whereas in *Schizachyrum scoparium*, mycorrhizal colonization was higher in sites where it was the dominant species.[148]

The most important factors associated with spread of infection within a root are root morphology and root density. In grasses *Festuca rubra* and *Agrostis tenuis*, the rate of infection spread within the roots normally is less than 1 cm $week^{-1}$,[168] while in *Trifolium* spread rates are greater than 1 cm $week^{-1}$.[168,169] Linear spread of infection is favored by a high root density in plants with a coarse root morphology; plants with fine root morphology are positively correlated with root density up to a maximum level at which root growth outstrips linear spread of fungal colonization.[148,168] The distribution of infection rarely is uniform throughout a root system unless the inoculum is dispersed evenly through the soil; however, the absolute rate of infection is more rapid when inoculum placement is more localized.[170] Infection of a root can be initiated by growth of mycorrhizal hyphae from an adjacent infected root provided an interroot

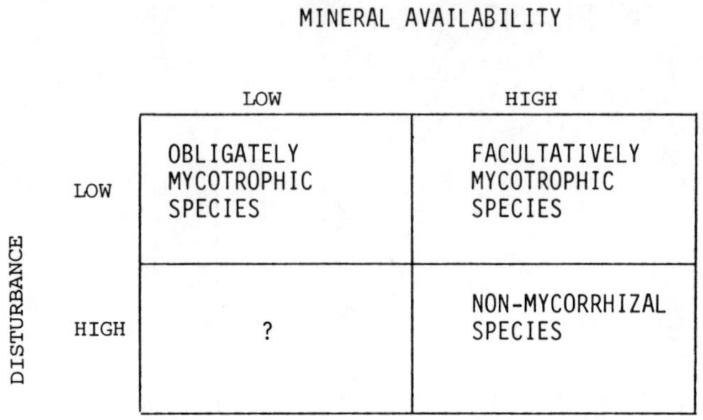

MINERAL AVAILABILITY

DISTURBANCE	LOW	HIGH
LOW	OBLIGATELY MYCOTROPHIC SPECIES	FACULTATIVELY MYCOTROPHIC SPECIES
HIGH	?	NON-MYCORRHIZAL SPECIES

FIGURE 5. The kinds of plants that can occupy a site as described by their need for mycorrhizae as determined by the degrees of disturbance and mineral ion availability.

distance of less than 30 mm exists;[171] also, the ability to bridge a root gap does not appear to differ among mycorrhizal fungal species.

The types of microbial associations a plant has can influence its growth and survival. Discussion of single host-endophyte associations represents an oversimplification of what occurs in nature. To understand the potential role of the mycorrhizal symbiosis in plant community structure, we have to be able to understand what happens when another plant is added to the above combination.

Competition between plants, whenever they occur in close proximity to each other, results in a tendency for the plants "to utilize the same quantum of light, ion of mineral nutrient, molecule of water or volume of space" whether they are of the same or different species, and no matter how dissimilar in vegetative growth or seed production method.[172] Using this definition, it can easily be seen how mycorrhizae could influence the outcome of an interaction, especially if there is a differential response to the indigenous mycorrhizal fungi by each plant host. Hypotheses need to be tested about how plant A may influence plant B's growth, and if the addition of mycorrhizal fungi alters the outcome (Figure 5). Also important is whether one plant is preferentially stimulated to grow when mycorrhizal fungi are present, and whether one plant affects the other by influencing the fungal partner. Similar questions can be asked about fungus-to-fungus interactions. Harley and Smith[173] have suggested that the outcome of competition between mycorrhizal hosts resides in their ability to act as sinks for soil-derived nutrients absorbed by the fungus, while at the same time releasing less carbohydrate to the fungus than their competitors. However, what is probably more important is the amount of carbohydrate utilized per unit gain. To date, the studies necessary to address accurately the influence of mycorrhiza on competition between plants have yet to be done. A first attempt at quantifying competition and infection by mycorrhizal fungi was with the pasture grass *Holcus lanatus* against *Lolium perenne;* each root competition and infection by mycorrhizal fungi slightly favored *H. lanatus,* but in combination, allowed for considerable suppression of *L. perenne.* This resulted partly from the reduction in the length of root of *Lolium* caused by infection and an increase in its extent of infection in the presence of *Holcus.*[174]

The competitive aspects of a grass-legume association confer an advantage to the legume associate.[55,78] Mycorrhizal *T. repens* had an advantage over the nonmycorrhizal counterparts in competition with *L. perenne* by increasing the volume of soil explored by the legume root system. This was especially evident when mycorrhizal *T. repens* was grown alone in a low-P-status soil — a fivefold increase in total dry weight was measured; however, when grown in combination with *Lolium*, a fourfold increase was observed.[78] The clover appears to be infected in preference to the grass, resulting in a mycorrhiza-induced depression in grass yield. The poor response of the grass was attributed to N deficiency.

The influences of neighboring plants appear complex and are both habitat and species specific. Research in lowland grasslands in Great Britain indicate that infection levels of mycorrhizal fungi are not significantly influenced by the species of neighboring plants; however, the shoot N concentration and levels of root-zone bacteria can be affected.[175,176] In the acid soils of upland grasslands, the particular combination of neighboring plants does appear to influence mycorrhizal abundance.[177] When *A. tenuis, Dechampsia flexuosa,* or *Festuca ovina* were grown alone or as partners, the effect of each partner on the infection level of the other was partner specific. For example, when *Agrostis* was grown in association with *D. flexuosa* or *F. ovina,* infection was not affected by a partner compared with plants grown solitarily; but in *Deschampsia,* infection levels were raised when *F. ovina* was a partner, and in *Festuca* infection was reduced when the plant was grown in association with *A. tenuis.*[177] The abundance of root-zone bacteria was not influenced by which partner was present; however, fungal abundance was consistently lowest when the partner species was *A. tenuis* and highest when the partner species was *F. ovina.* The mechanisms for these association effects are not known, but may be influenced by the kinds and levels of root exudates released by plant pairs.

Evidence for the existence of mycorrhizal fungus hyphal connections between roots of separate plants that allow for interplant transfer of nutrients, such as P, has been presented.[178-182] It is not surprising for these connections to occur, since more than 80 cm fungus hyphae per centimeter root infected have been measured surrounding the roots they infect.[34] Davidson and Christensen[69] record hyphal bridges between roots; however, infection points are not clearly visible. Using buried slides, connections between roots of *L. perenne* and *Lolium* and *P. lanceolata* have been shown clearly by Heap and Newman,[178] and between *Clarkia rubicunda* and *P. erecta* by Chiariello and co-workers.[180] Connections also have been demonstrated between *F. ovina* and *P. lanceolata* using light microscope and autoradiography.[182] Interplant transfer of carbon and phosphorus has been reported.[179-181,183] None of these studies have ruled out indirect means of nutrient uptake, i.e., uptake by the fungus of the receiver plant occurring only after leakage of nutrients from the donor root. Francis and Read,[182] using autoradiography, showed that transfer of carbon by mycorrhizal fungus hyphae occurs by a direct pathway via hyphal connections. Their research also suggests a source-sink relationship where the magnitude of transfer can be governed by shading of a receiver plant, indicating that carbon will move between plants along a gradient. The connections of plants via mycorrhizas at both the intraspecific and interspecific levels suggest that an understanding of belowground-neighbor effects may be necessary to explain aboveground interactions. This indicates that research addressing the factors controlling nutrient transfer via these links will be necessary if the significance of this nutrient transfer to community processes is to be understood.

B. Fungus-Fungus Association Effects

Fungus-to-fungus interactions are very important and should be included in considerations of plant interactions. Abbott and Robson[184] have shown that a major factor contributing to the effectiveness of a mycorrhizal fungus to a host is the ability of that fungus to produce rapid and extensive infection. However, in the soil, many of the indigenous mycorrhizal fungi are not equally effective endophytes; the effectiveness of a particular endophyte can vary with host species. The ability of a particular fungus to successfully compete with another fungus for an infection site depends partly on the aggressiveness of the fungus.[185-187] An aggressive isolate or fungus is one that is capable of maintaining infectivity in a mixed inoculum relative to its infectivity from a single inoculum, better than the fungus with which it is competing.[185,186] Aggressiveness does not take into account differences in absolute infectivity, but rather it is a quantitative measure of the ability of an isolate to exclude another fungus from a root, whereas infectivity is a measure of the frequency of infection of an isolate at a point in time.

The interactions between vesicular arbuscular mycorrhizal fungi are affected by the spatial distributions of their interacting propagules, as well as any host, soil, and climatic factors that could result in a lag time in growth between the interacting fungi.[187] The intensity of infection depends on inoculum density, along with the strategies used for spread of infection, and is sensitive to all of the host, soil, and climatic factors affecting fungal interactions. Some fungi, i.e., *G. fasciculatum,* spread from within the root, while species sich as *G. tenue* and *Gigaspora decipiens* depend more on external spread by formation of secondary entry points.[183] Prior occupation of a root by one fungus also can reduce infection by a second colonizer, but the ability to exclude a competitor is fungus dependent.[185-187] The host also can influence the success of a fungus by inhibiting an increase in secondary entry points if they would be in close proximity to a primary colonized site; thus, the host can prevent potentially viable propagules from forming infection points. It has been suggested that this may be a type of induced resistance, but the mechanisms of the phenomenon have not been researched.[187]

C. Host-Mycorrhizae-Grazing Effects

Mycorrhizae appear to confer a tolerance to plants in a grazing environment by increasing biomass allocation to growth below the grazing zone via promotion of tillerage and prostrate growth habit, while still maintaining root growth[73,74] (Figure 1). Under field conditions in a savanna dominated by *Panicum coloratum,* there was a positive correlation between grazing intensity and mycorrhizal infectivity.[74] However, a negative association between grazing and grasses was found for the occurrence of mycorrhizal infection in a Great Basin shrubland.[188] In the laboratory, when *P. coloratum* was nonmycorrhizal and subjected to severe clipping and N-limiting growth conditions, a reduction in photosynthesis occurred as measured by gas exchange. When these plants were treated similarly and inoculated with mycorrhizal fungi, no reduction in photosynthesis was evident.[74] The mechanisms by which mycorrhizae reduce the effects of clipping and overcome low N levels are not known, but most likely are related to improved plant nutritional status since mycorrhizae appear to increase leaf and sheath N content.[189] Grazing of mycorrhizal *Bouteloua gracilis* did not affect the overall infection level, but did alter the density of vesicles in roots.[190] Clipping of *L. perenne* appeared to stimulate hyphal growth, as measured by an increase in stable aggregates.[191] As long as adequate moisture was supplied to *L. perenne,* a significant increase was evident in stable aggregates associated with an increase in mycorrhizal fungi hyphae growth which developed around clipped grass roots. Wilting plants resulted in a decrease in the degree of stable aggregate formation. Even after plants died, mycorrhizal hyphae persisted for several months, continuing to bind soil particles.

V. MYCORRHIZAE, DISTURBANCE, AND SUCCESSION

At the conceptual level, the ecological principle of succession can be defined as an assemblage of species that succeeds one another until a "steady state" is reached, i.e., constant in species composition, though varying in time and space.[192] In the process of succession, plant species in an area modify each other's environment in such a way that they progressively replace one another. The species involved in the process are referred to as seral species. When the steady state is reached, the vegetation is in equilibrium with its environment and is known as climax vegetation.

The intensity of disturbance and size of area disturbed appear to be important determinants in the course of ecological succession, whereas the rate of disturbance may be associated more with maximizing species diversity.[193] Large disturbed areas such as those resulting from surface mining, cultivation, or forest fires, where disturbance is extreme, often require recruitment of plant and microbial propagules from outside. These areas are colonized by early successional plants that are able to go through successive generations until late seral species become established by spreading from the edge. Return to the original state may require a long period of time.[194]

In large areas where disturbance is not severe and the resident population is not eliminated, such as is caused by overgrazing or brush and grass fires, regrowth of survivors and recruitment from seeds will occur. Establishment can be by several means — by opportunistic species able to establish themselves through rapid seed germination or by suppression of seedlings of later sere species. Adults may survive through sprouting and may have an initial advantage over pioneer species, if their root systems are not damaged. The delay to climax under such conditions would not be as long as in large areas that were severely disturbed.[194]

Severe disturbance of even small areas requires recruitment from outside by seed or through vegetative growth of neighbors. Under such conditions climax species may be able to fill these small gaps even during the earlier stages.[192,194] Small gaps may be produced by lightning, or small rockslides, and by activities of man, i.e., damage resulting from herbicides.

In the case where disturbance is slight and the area is small (as with the loss of a single adult) the space is filled by later successional species either by seed or vegetatively from the surrounding adults. Since the opening is small and resources are limiting, few invading species become established.[194]

A. A Mycotrophy Continuum

The dependency of a host on a mycorrhizal fungus can be defined as either faculative or obligate, depending on the advantage or necessity of obtaining symbiont-supplied minerals.[82] Faculatively mycotrophic species can attain reproductive maturity without mycorrhizae in the more fertile of their natural habitats and have fewer mycorrhizal associations in fertile soil than in nutritionally poor soils. Pasture grasses are an example of plants that are generally considered faculative mycotrophs. It has been suggested by Baylis[20] that these grasses can persist without mycorrhizae at low P levels because of a well-developed fine-root system. Obligate mycotrophs are defined as those plant species that cannot survive to reproductive maturity without mycorrhizae at the fertility levels encountered in their natural habitat.[82] There is some indication that after maturity, obligate mycorrhizal plants could become independent, providing nutrient demands were lowered.[20] The concept of mycorrhizal dependency as defined by Janos[82] is compelling. It is relatively easy to categorize plants as either facultative or obligate as to their need for symbiont-supplied nutrients, however, what is really needed now are the experiments necessary to test this model.

If the mycotrophy continuum and the degrees of mycotrophy are superimposed over a matrix formed by mineral ion availability and degree of disturbance, the need of mycorrhizae can be put into context of the site where it occurs (Figure 5). This approach suggests that obligate mycotrophic species would predictably dominate plant communities found where mineral ion availability was low and disturbance minimal; facultative mycotrophic species would be found on sites of relatively higher mineral ion availability and where disturbance is not severe. Alternatively, when mineral ion availability is high, as is typical with high-disturbance sites, plant communities dominated by species are nonmycorrhizal species; however, it appears that when disturbance is high and mineral ion availability is low, no plants, or relatively few, can occupy a site.[7]

The suggestion that plant dependency on mycorrhizae may change with the successional stage of the ecosystem is of particular relevance to studies of ecosystem recovery from disturbance. In shrublands of the western U.S., many herbaceous or semishrub pioneer species belong to families that are nonmycorrhizal, whereas plants that belong to mycorrhizal families are more prevalent in older seral-stage seres.[113,114,116,195] It appears that for late seral communities, a host plant may be highly dependent on mycorrhiza,[113,117] but some facultative mycotrophs that occur in early- through midseral communities may be mycorrhizal only in the presence of other mycorrhizal plants.[124] These facultative mycotrophs (and other midseral plant species) appear to facilitate succession by acting as windbreaks and water catchments, and by accumulating organic matter (preventing nutrient loss) and building-up inoculum through root turnover, finally allowing obligate mycotrophs to become established.

The type of plant community found on a site could influence the next generation of plants through its effects on the mycorrhizal fungus population. This has been suggested to occur in semiarid and arid shrublands for soils that are highly disturbed.[113,116,195] These soils normally support plant communities that are mycorrhizal, but when disturbed, nonmycorrhizal annual chenopods and crucifers are the initial colonizers. The extended presence of nonmycorrhizal plants in these soils which contain mycorrhiza propagules results in a lowering of the mycorrhizal propagule reserve. The means by which this decline occurs is senescence of propagules and a lack of new propagules being produced, due in turn to the absence of host plants. This could result in either a continued presence of nonhost plants or a delay in the establishment of host plant species. There is some evidence that the presence of nonmycorrhizal species inhibits the development of mycorrhizal infection in plants whose roots normally are highly mycorrhizal.[196] However, Schmidt and Reeves[197] have found that the presence of the nonmycorrhizal species *Salsola kali* may prolong the survival of mycorrhizal fungus propagules compared with soils having no vegetative cover; the mechanism of survival may be through inhibition of propagule germination by either chemical or physical means. Disturbances that favor uninfected plant establishment might lead to a plant community differing from that occupying a site before disturbance.[113,116] Since the competitive ability of plants may be influenced by mycorrhizal fungi, any disruption or decoupling of this symbiotic association may be a determinant of either the direction of or time of succession. The dispersal of a host within a site could be limited by the growth rate of hyphae and thus through the occurrence of hyphae-to-root contact. This suggests the possibility of an "island effect", in which recolonization of an area is through the outward expansion of host-endophyte islands.

Disturbance also affects the species composition of the mycorrhizal fungus community. The disturbance of prairie soils by agricultural practices or by the practice of topsoil stockpiling can result in an altered species composition compared with relatively undisturbed soils.[93,159,198] The importance of a shift in dominance from one endophyte species to another is illustrated by host response. Wheat infected by the dominant mycorrhizal fungus from a virgin prairie soil elicited more negative leaf osmotic potentials which allowed positive turgor pressure to be maintained, even at low (more negative) soil water potentials. However, when wheat was infected by the dominant mycorrhizal fungus from a tilled soil, there was no similar response to drought stress, resulting in reduced vegetative growth.[93]

Severe disturbance can influence the ability of mycorrhizal fungi within the soil to initiate infection.[114,117,199-201] As previously mentioned, disturbance results in an extended annual nonmycorrhizal weed stage in most semiarid grassland and shrubland

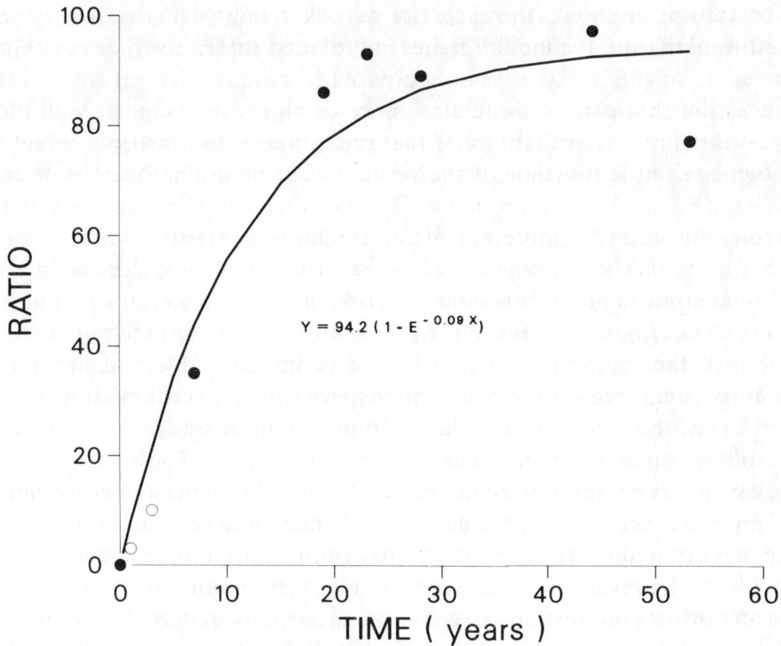

FIGURE 6. The time sequence for mycorrhizal plant species establishment in a primary sere (surface mine spoil) in North Dakota.

ecosystems. The occurrence of such a stage could result in a reduction of the mycorrhizal infection potential through either the exclusion of potential host species or possibly by allelochemical means. More likely, disturbance affects distribution of mycorrhizal propagules such that the number of viable mycorrhizal fungus propagules necessary to initiate successful infection may be reduced to a critical level. Undoubtedly, disturbance also influences turnover rates of mycorrhizal fungus propagules. This balance between propagule production and senescence, as well as to host productivity, is related to climatic and edpahic factors.

The time sequence involved for the establishment of the mycorrhizal component of a community where primary succession is occurring can be shown through the data from chronosequence studies of spoils in North Dakota.[149,202] Using plant species richness indexes from mine spoils ranging from 0 to 53 years, and grouping the plants into presumed mycorrhizal or nonmycorrhizal species (where the number of mycorrhizal to nonmycorrhizal plant species occurring is determined and corrected for mycorrhizal plant species richness from adjacent prairie areas), it is calculated that a time course of approximately 40 years is needed to achieve 90% of the undisturbed prairie level for mycorrhizal species (Figure 6). This suggests that for severely disturbed sites, succession may be delayed, while a protracted mycorrhizal stage exists until both propagule levels and microhabitat development are at a point conducive to host establishment.

Contrary to the finding in semiarid communities on the relationship between plant establishment and mycotrophy, the majority of plants colonizing disturbed sites in more mesic environments possesses mycorrhizae. The difference may be related more to nutritional factors that increased moisture. Studies of colonizing and intermediate successional species occurring on coal wastes and spoil material indicate that mycorrhizal fungi are associated with plant establishment and persistence.[203-207] Similarly, when perturbation is less severe than that associated with mining, such as with the disturbance of a 10-year *Solidago-Aster* old field by burning, herbicide application, disking, or topsoil removal, the majority of colonizing plants possess mycorrhizal fungi.[208] An explanation for the difference in the trend for mycorrhizae associated with plant species colonizing mesic sites as compared to nonmycorrhizal colonizers of arid and semiarid sites appears to be related more to moisture (evapotranspiration), and indirectly to the nutrient availability of these sites where both propagule and resource reserves, along with plant productivity, have a higher potential than for the more arid sites.

Pendleton and Smith,[115] surveying the mycorrhizae in weedy and colonizing plants of disturbed sites in Utah, suggest that the occurrence of mycorrhizae in colonizing species is related more along taxonomic lines rather than a weedy or ruderal growth habitat. They also found that soil moisture could account for much of the variability associated with the mycorrhizal status of the community. Flat, semiarid sites were predominantly composed of nonmycorrhizal species, whereas rocky, sloping sites were dominated by mycorrhizal species. The difference in these findings to those of Reeves et al.,[113] Miller,[116] and Allen and Allen[195] is one of degree. These studies all indicate that the disturbance severity of a site, especially as it influences stress components as measured by water relations and mineral ion availability, is most important in structuring the mycorrhizal component of an early seral community.[201] Also, the influence of mycorrhizae in later seral stages appears related to community structure; a strong negative correlation exists for nonmycorrhizal-species cover and the diversity of a community when measured over a variety of soils and habitats in the Wyoming shrub steppe.[209]

The recent research showing connections between roots of different plant species via the hyphae of mycorrhizal fungi suggests that in nutrient-limited ecosystems nutrients could be obtained from associated neighboring plants through an interconnecting mycorrhizal fungus network. This nutrient sharing may result in shunting of nutrients from less efficient hosts. If such links exist to any substantial degree, a disturbance to the soil system could result in a loss of these interconnections, potentially affecting the resiliency of that community to further perturbation. The most likely effect of a loss of these hyphal connections may be in the seedling establishment stage, in which, through connections, mature plants may act as nurse plants to seedlings.[182]

B. Mycorrhizae and Ecosystem Restoration

The vertical distribution of mycorrhizal propagules appears confined to the surface layer of soil in most semiarid ecosystems; at a depth of 0.5 m, spore number or infection potential is minimal[210,211] and is related to the rooting zone of the vegetation. This suggests that during the reclamation phase of surface mining, the salvaging of topsoil should occur in two lifts: the first for the shallow surface layer containing most of the mycorrhizal fungi propagule reserve, and a second lift containing that soil having the physical and chemical characteristics conducive to good plant establishment and growth. If direct application of the topsoil is not possible (e.g., during surface mining, the extraction of coal or minerals occurs at a rate faster than the reclamation phase), stockpiling becomes necessary.

The survival dynamics of mycorrhizal fungi propagules in stockpiled "topsoils" have been obtained for soils of grassland and shrubland ecosystems[133,139,212,213] and for stockpiled topsoils in other ecosystems.[214,215] The results of these studies indicated a decrease in mycorrhizal infection ability for topsoils which are stockpiled. At an arid site in the Red Desert of Wyoming, a senescence rate for mycorrhizal fungi propagules of 13% a year was predicted from a logistic model.[148] The data of Rives et al.[133] indicate that the change in propagule reserve may initially be quite subtle, and soil dilution is necessary to observe an effect because a propagule reserve for grassland soils can mask any decrease if present (due to saturation). However, in an arid shrubland site this reserve does not appear to be very large.[139] A comparison of bioassay infection levels with soil water potential suggests that the moisture content in topsoil stockpiles may influence propagule survival,[139] the effect does not appear to be uniform throughout a range of soil moistures. When water potential of stockpiled soils was less than −2 MPa, moisture was an important predictor of infection, while duration of storage had little predictive power. However, when water potential was greater than −2 MPa, the predictive importance of water potential and soil age was reversed. This suggests that survival of mycorrhizal propagules in stockpiled topsoils is related to the soil-water potential under which they were stored. This relationship should be predictive in most arid ecosystems, since, in such an environment, it is not likely that water recharge occurs during the lifespan of a stockpile. If soils are to be stored, stockpiling should be limited to time periods when soil water potentials are below −2 MPa to ensure adequate propagule survival.[139] An important factor that could influence propagule levels during topsoil storage is establishment of host plants on the stockpile surface. This would allow the production of new propagules to compensate for the loss of spores or sporocarps during storage. In arid ecosystems, host establishment will require supplemental moisture, if available, whereas in mesic environments, host establishment is less of a problem. Thus, in ecosystems where moisture is not limiting, the establishment of host and concomitantly associated VA mycorrhizal propagule production, through root colonization, can readily compensate for propagule losses associated with increased soil moisture during storage.

A delay in host seedling establishment is associated with reapplication of stored topsoil.[116,195,214] However, the rate at which mycorrhizae become established appears to be moisture dependent. In mesic environments, mycorrhizal fungi-propagule reserves recover from storage quite rapidly,[214] while in arid systems, even after 3 years of recovery, roots of *Agropyron intermedium* and *A. smithii* showed 50% less infection compared with field-collected roots.[195] Spore numbers were affected similarly. Studies in Wyoming indicate that even after 8 years of recovery, mycorrhizal infection potential for 2-year stored topsoil was significantly lower than direct-applied or adjacent native soils.[148] In a greenhouse study using a native prairie soil and 3-year stored prairie soil, *A. trachycaulum* mycorrhizal infection rate was slower than when the soil was stored, although after 3 weeks of growth no difference in infection of roots was found.[198] A shift in endophyte species composition was evident after 8 weeks of growth; *Glomus fasciculatum* dominated in prairie soils, while in stored soils *G. mosseae* was dominated.

Minesoils, i.e., spoil or mining wastes, initially lack a viable mycorrhizal fungus population which may result in a delay of vegetation establishment.[216] Probably more important initially to vegetation success in most mine wastes is the establishment of a nitrogen reserve necessary to support a self-sustaining ecosystem. For china clay waste in Great Britain, a minimum of 1000 kg N ha^{-1} appears to be required.[217] Bradshaw et al.[217] have found that legumes can contribute to soil N reserves at a rate of 50 to 150 kg N ha^{-1} $year^{-1}$, with approximately 40 years required before sufficient N is fixed to allow for a self-sustaining ecosystem, providing P was not limiting. Nitrogen accumulations of the magnitudes reported are necessary because the bulk of N is held in soil only in organic matter and is released through mineralization. Since organic matter decomposition rates are relatively slow in temperate climates (approximately 2% of the soil N capital $year^{-1}$), a large N reserve is necessary so that a sufficient supply of N is released by mineralization to meet the annual requirements of the vegetation. Harley[218] has suggested that plants with dual symbiotic associations, i.e., N_2 fixation and mycotrophy, possess nutritional and ecological advantages to compensate for nutrient-limiting situations, especially where P is limiting. He goes on to indicate that these associates are ideal for colonizing disturbed areas. The key then to successful N input in these developing systems is an adequate P supply to the legumes. Legumes have repeatedly been shown to require high levels of P for growth and effective nodulation,[219] and for the processes of N_2 fixation,[220] suggesting mycorrhizae play an important role in fixed N input.

The majority of studies concerning mycorrhizae and mine spoils has been conducted either to determine whether mycorrhizae occur in the vegetation on these sites or to monitor its development over a time sequence, and implies a benefit from the mycorrhizae association.[221] The research of Lambert and Cole[216] suggests that mycorrhizae may be a low-cost alternative to fertilization. Research with surface mine spoil[207] and oil sand tailings[222] indicates fertilization may be counterproductive, since reduced numbers of mycorrhizal spores are encountered under high fertility as compared to unamended spoil where infection by mycorrhizal fungi may be quite high.[223] In some cases plants can remain uninfected in mine spoil even 10 years after disturbance.[216] The addition of amendments containing inoculum can decrease the delay in infection and may be necessary under certain circumstances.[225] However, once infection is established, good growth responses to indigenous mycorrhizal fungi by legumes have been reported.[216,226]

VI. CONCLUSIONS

An understanding of the role played by mycorrhizal fungi in grassland and shrubland communities is just beginning to be realized. As with any developing area of research, we are now leaving the descriptive phase. Much of what we have learned from agricultural systems about mycorrhizae may not be relevant to these more stressed environments where maximum biomass production may not be advantageous. The relationship of mycorrhizal hyphae to aggregate stability and the importance of the hyphae to the physical continuity of the soil system is now being recognized. Also, the potential of both intra- and interspecific links occurring between plants via mycorrhizal hyphae indicates a high degree of interaction, suggesting the possibility of chemical communication between individuals. If future research supports this contention, a reevaluation of the concept of the individual and its relationship to the community may be in order.

Another area in which an increase in research emphasis is needed is with soil resource pools, nutrient availability, and the relationship of mycorrhizae to this availability. Questions concerning the relationship between modes of growth, or with nutrient acquisition strategies of plants and the requirement for mycorrhizae, need to be addressed. For example, do grasses occurring in fertile soils differ in their need for mycorrhizae from those grasses found in nutritionally poor soils (*sensu* Chapin or Grime)? Does Janos' continuum of mycotrophic responses actually exist? Experiments addressing these questions need to be carried out.

ACKNOWLEDGMENTS

I should like to thank A.-C. McGraw, Ray Franson, Vince Gutschick, Roger Anderson, and Sue Rabatin for their discussions and comments on an early draft of this manuscript. The preparation of this manuscript was supported in part by the Office of Health and Environmental Research of the U.S. Department of Energy.

REFERENCES

1. Duffey, E., Morris, M. G., Shenil, J., Ward, L. K., Wells, D. A., and Wells, T. C. E., *Grassland Ecology and Wildlife Management,* Halsted Press, New York, 1974, 281.
2. Risser, P. G., Birney, E. C., Blocker, H. D., May, S. W., Parton, W. J., and Wiens, J. A., *The True Prairie Ecosystem,* Hutchinson Ross Publishing, Stroudsburg, 1981, 557.
3. Janos, D. P., Vesicular-arbuscular mycorrhizae affect lowland tropical rain forest plant growth, *Ecology,* 63, 151, 1980.
4. Chapin, F. S., III, The mineral nutrition of wild plants, *Annu. Rev. Ecol. Syst.,* 11, 233, 1980.
5. Bieleski, R. L. and Laüchli, A., Synthesis and outlook, in *Inorganic Plant Nutrition, Encyclopedia of Plant Physiology 15B,* Laüchli, A. and Bieleski, R. L., Eds., Springer-Verlag, New York, 1983, 746.
6. Grime, J. P., Evidence for the existence of three primary strategies in plants and its relevance to ecological and evolutionary theory, *Am. Nat.,* 111, 1169, 1977.
7. Grime, J. P., *Plant Strategies and Vegetation Processes,* John Wiley & Sons, New York, 1979, 222.
8. Tilman, D., *Resource Competition and Community Structure,* Monogr. Population Biol. No. 17, Princeton University Press, Princeton, N.J., 1982, 296.
9. Lamont, B., Mechanisms for enhancing nutrient uptake in plant, with particular reference to mediterranean South Africa and Western Australia, *Bot. Rev.,* 48, 597, 1982.
10. Coleman, D. C., Reid, C. P. P., and Cole, C. V., Biological strategies of nutrient cycling in soils systems, *Adv. Ecol. Res.,* 13, 1, 1983.
11. Kramer, P. J., *Water Relations of Plants,* Academic Press, New York, 1983, 489.
12. Barley, K. P., The configuration of the root system in relation to nutrient uptake, *Adv. Agron.,* 22, 159, 1970.
13. Nye, P. H., The rate-limiting step in plant nutrient absorption from soil, *Soil Sci.,* 123, 292, 1977.
14. Baylis, G. T. S., Fungi, phosphorus, and the evolution of root systems, *Search,* 3, 257, 1972.
15. Riley, D. and Barber, S. A., Effect of ammonium and nitrate fertilization on phosphorus uptake as related to root-induced ph changes at the root-soil interface, *Soil Sci. Soc. Am. Proc.,* 35, 301, 1971.
16. Bagshaw, R., Vaidyanathan, L. V., and Nye, P. H., The supply of nutrient ions to plant roots in soil. V, *Plant Soil,* 37, 617, 1972.
17. Itoh, S. and Barber, S. A., Phosphorus uptake by six plant species as related to root hairs, *Agron. J.,* 75, 457, 1983.
18. Powell, C. L., Rushes and sedges are non-mycotrophic, *Plant Soil,* 42, 481, 1975.
19. Crush, J. R., The effects of *Rhizophagus tenuis* mycorrhizas on ryegrass, cocksfoot and sweet vernal, *New Phytol.,* 72, 965, 1973.
20. Baylis, G. T. S., The magnoloid mycorrhiza and mycotrophy in root systems derived from it, in *Endomycorrhizas,* Sanders, E. F., Moose, B., and Turker, P. B., Eds., Academic Press, New York, 1975, 373.
21. Haselwandter, K. and Read, D. J., Fungal associations of roots of dominant and sub-dominant plants in high-alpine vegetation systems with special reference to mycorrhiza, *Oecologia,* 45, 57, 1980.
22. Bolan, N. S., Robson, A. D., Barrow, N. J., and Arlmore, L. A. G., Specific activity of phosphorus in mycorrhizal and non-mycorrhizal plants in relation to the availability of phosphorus to plants, *Soil Biol. Biochem.,* 16, 299, 1984.
23. Tinker, P. B., Effects of vescular-arbuscular mycorrhizas in plant nutrition and plant growth, *Physiol. Veg.,* 16, 743, 1978.
24. Rhodes, L. H. and Gerdemann, J. W., Phosphate uptake zones of mycorrhizal and non-mycorrhizal onions, *New Phytol.,* 75, 555, 1975.
25. St. John, T. V., Coleman, D. C., and Reid, C. P. P., Association of vesicular-arbuscular mycorrhizal hyphae with soil organic particles, *Ecology,* 64, 957, 1983.
26. Bowen, G. D., Bevege, D. I., and Mosse, B., Phosphorus physiology of vescular-arbuscular mycorrhizas, in *Endomycorrhizas,* Sanders, F. E., Mosse, B., and Tinker, P. B., Eds., Academic Press, London, 1975, 241.
27. Cress, W. A., Throneberry, G. O., and Lindsey, D. L., Kinetics of phosphorus absorption by mycorrhizal and nonmycorrhizal tomato roots, *Plant Physiol.,* 64, 484, 1979.
28. Howeler, R. H., Asher, C. J., and Edwards, D. G., Establishment of an effective mycorrhizal association on cassava in flowing solution culture and its effects on phosphorus nutrition, *New Phytol.,* 90, 229, 1982.
29. Tinker, P. B., The soil chemistry of phosphorus and mycorrhizal effects on plant growth, in *Endomycorrhizas,* Sanders, F. E., Mosse, B., and Tinker, P. B., Eds., Academic Press, New York, 1975, 253.
30. Buwalda, J. G., Stribley, D. P., and Tinker, P. B., Increased uptake of bromide and chloride by plants infected with vesicular-arbuscular mycorrhizas, *New Phytol.,* 93, 217, 1983.

31. Hedley, M. J., White, R. E., and Nye, P. H., Plant-induced changes in the rhyzosphere of rape seedlings. III. Changes in L-value, soil phosphate fractions and phosphatase activity, *New Phytol.,* 91, 45, 1982.
32. Sanders, F. E. and Tinker, P. B., Mechanism of absorption of phosphate from soil by *Endogone* mycorrhizas, *Nature,* 233, 278, 1971.
33. Hayman, D. S. and Mosse, B., Plant growth responses to vesicular-arbuscular mycorrhiza. III. Increased uptake of labile P from soil, *New Phytol.,* 71, 41, 1972.
34. Sanders, F. E. and Tinker, P. B., Phosphate flow into mycorrhizal roots, *Pestic. Sci.,* 4, 385, 1973.
35. Cooper, K. M. and Tinker, P. B., Translocation and transfer of nutrients in vesicular-arbuscular mycorrhizas. IV. Effect of environmental variables on movement of phosphorus, *New Phytol.,* 83, 327, 1981.
36. Timmer, L. W. and Leyden, R. F., The relationship of mycorrhizal infection to phosphorus-induced copper deficiency in sour orange seedlings, *New Phytol.,* 85, 15, 1980.
37. Rhodes, L. H. and Gerdemann, J. W., Translocation of calcium and phosphate by external hyphae of vesicular-arbuscular mycorrhizae, *Soil Sci.,* 126, 125, 1978.
38. Rhodes, L. H. and Gerdemann, J. W., Hyphal translocation and uptake of sulfur by vesicular-arbuscular mycorrhizae of onion, *Soil Biol. Biochem.,* 10, 355, 1978.
39. Powell, C. L., Potassium uptake by endotrophic mycorrhizas, in *Endomycorrhizas,* Sanders, F. E., Mosse, B., and Tinker, P. B., Eds., Academic Press, New York, 1975, 461.
40. Stevens, P. R. and Walker, T. W., The chronosequence concept and soil formation, *Q. Rev. Biol.,* 45, 333, 1970.
41. Tiessen, H., Stewart, J. W. B., and Cole, C. V., Pathways of phosphorus transformations in soils of differing pedogenesis, *Soil Sci. Soc. Am. J.,* 48, 853, 1984.
42. Walker, T. W. and Adams, A. F. R., Studies on soil organic matter. I. Influence of phosphorus content of parent materials on accumulations of carbon, nitrogen, sulfur and organic phosphorus in grassland soils, *Soil Sci.,* 85, 307, 1958.
43. Bowman, R. A. and Cole, C. V., An exploratory method for fractionation of organic phosphorus from grassland soils, *Soil Sci.,* 125, 95, 1978.
44. Tate, K. R. and Newman, R. H., Phosphorus fractions of a climosequence of soils in New Zealand tussuck grassland, *Soil Biol. Biochem.,* 14, 191, 1982.
45. Chapin, F. S., III, Barsdale, R. J., and Barel, D., Phosphorus cycling in Alaskan coastal tundra: a hypothesis for the regulation of nutrient cycling, *Oikos,* 31, 189, 1978.
46. Brookes, P. C., Powlson, D. S., and Jenkinson, D. S., Phosphorus in the soil microbial biomass, *Soil Biol. Biochem.,* 16, 169, 1984.
47. Fitter, A. H. and Hay, R. K. M., *Environmental Physiology of Plants,* Academic Press, New York, 1981, 355.
48. Moorby, J. and Besford, R. T., Mineral nutrition and growth, in *Inorganic Plant Nutrition, Encyclopedia of Plant Physiology,* New Series Vol. 15B, Springer-Verlag, New York, 1983, 481.
49. Bieleski, R. L., Phosphate pools, phosphate transport, and phosphate availability, *Annu. Rev. Plant Physiol.,* 24, 255, 1973.
50. Chapin, F. S., III, Follett, J. M., and O'Connor, K. F., Growth, phosphate absorption, and phosphorus chemical fractions in two *Chionochloa* species, *J. Ecol.,* 70, 305, 1982.
51. Clarkson, D. T. and Hanson, J. B., The mineral nutrition of higher plants, *Annu. Rev. Plant Physiol.,* 31, 239, 1980.
52. Atkinson, C. J., Phosphorus aquisition in four co-existing species from montane grassland, *New Phytol.,* 95, 427, 1983.
53. Mosse, B., Effects of different Endogone strains on the growth of *Paspalum notatum, Nature,* 239, 221, 1972.
54. Mosse, B., Hayman, D. S., and Arnold, D. J., Plant growth responses to vesicular-arbuscular mycorrhiza. V. Phosphate uptake by three plant species from P-deficient soils labelled with ^{32}P, *New Phytol.,* 72, 809, 1973.
55. Crush, J. R., Plant growth responses to vesicular-arbuscular mycorrhiza. VII. Growth and nodulation of some herbage legumes, *New Phytol.,* 73, 743, 1974.
56. Ali, B., Influence of the endophyte of *Nardus* on plant growth, *Plant Soil,* 44, 329, 1976.
57. Crush, J. R., Significance of endomycorrhizas in tussock grassland in Otago, New Zealand, *N.Z. J. Bot.,* 11, 645, 1973.
58. Koucheki, H. K. and Read, D. J., Vesicular-arbuscular mycorrhiza in natural vegetation systems. II. The relationship between infection and growth in *Festuca ovina, New Phytol.,* 77, 655, 1976.
59. Sparling, G. P. and Tinker, P. B., Mycorrhizal infection in Pennine Grassland. II. Effects of mycorrhizal infection on the growth of some upland grasses on γ-irradiated soils, *J. Appl. Ecol.,* 15, 951, 1978.
60. Jasper, D. A., Robson, A. D., and Abbott, L. K., Phosphorus and the formation of vesicular-arbuscular mycorrhizas, *Soil Biol. Biochem.,* 11, 501, 1979.

61. Read, D. J., Koucheki, H. K., and Hodgson, J., Vesicular-arbuscular mycorrhizae in natural vegetation systems. I. The occurrence of infection, *New Phytol.*, 77, 641, 1976.
62. Sparling, G. P. and Tinker, P. B., Mycorrhizal infection in Pennine Grassland. I. Levels of infection in the field, *J. Appl. Ecol.*, 15, 943, 1978.
63. Powell, C. L., The need for mycorrhizas for cocksfoot growth at high altitude: a note, *N.Z. J. Agric. Res.*, 18, 95, 1975.
64. Powell, C. L., Mycorrhizas in hill country soils. V. Growth responses in ryegrass, *N.Z. J. Agric. Res.*, 20, 495, 1977.
65. Buwalda, J. G. and Goh, K. M., Host-fungus competition for carbon as a cause of growth depressions in vesicular-arbuscular mycorrhizal ryegrass, *Soil Biol. Biochem.*, 14, 103, 1982.
66. Nicolson, T. H. and Johnston, C., Mycorrhiza in the Gramineae. III. *Glomus fasiculatus* as the endophyte of pioneer grasses in a maritime sand dune, *Trans. Br. Mycol. Soc.*, 72, 261, 1979.
67. Haynes, R. J., Competitive aspects of the grass-legume association, *Adv. Agron.*, 33, 227, 1980.
68. Allen, M. F., Smith, W. K., Moore, T. S., Jr., and Christensen, M., Comparative water relations and photosynthesis and mycorrhizal and nonmycorrhizal *Bouteloua gracilis* H.B.K. Lag ex STEUD, *New Phytol.*, 88, 683, 1981.
69. Davidson, D. E. and Christensen, M., Root-microfungual and mycorrhizal associations in a shortgrass prairie, in *The Belowground Ecosystem: A Synthesis of Plant-Associated Processes*, Marshall, J. K., Ed., Range Science Department Science Series No. 26, Colorado State University, Fort Collins, 1977, 279.
70. Hays, R., Reid, C. P. P., St. John, T. V., and Coleman, D. C., Effects of nitrogen and phosphorus on blue grama growth and mycorrhizal infection, *Oecologia*, 54, 260, 1982.
71. Bethlenfalvay, G. J., Bayne, H. G., and Pacousky, R. S., Parasitic and mutualistic associations between a mycorrhizal fungus and soybean: the effect of phosphorus on host plant-endophyte interactions, *Physiol. Plant*, 57, 543, 1983.
72. Stribley, D. P., Tinker, P. B., and Rayner, J. H., Relation of internal phosphorus concentration and plant weight in plants infected by vesicular-arbuscular mycorrhizas, *New Phytol.*, 86, 261, 1980.
73. Miller, R. M., Jarstfer, A. G., and Pillai, J., Biomass allocation in an *Agropyron Smithii-Glomus* sympiosis, *Am. J. Bot.*, 74(1), 1987.
74. Wallace, L. L., Growth, morphology and gas exchange of mycorrhizal and nonmycorrhizal *Panicum coloratum* L., a C_4 grass species, under different clipping and fertilization regimes, *Oecologia*, 49, 272, 1981.
75. Dahl, B. E. and Hyder, D. N., Developmental morphology and management implications, in *Rangeland Plant Physiology*, Range Sciences Series No. 4, Sosebee, R. E., Ed., Society Range Management, Denver, 1977, 257.
76. Weaver, J. E., *Prairie Plants and Their Environment*, University of Nebraska Press, Lincoln, 1968, 276.
77. Mosse, B., Powell, C. L., and Hayman, D. S., Plant growth responses to vesicular-arbuscular mycorrhiza. IX. Interactions between VA mycorrhiza, rock phosphate and symbiotic nitrogen fixation, *New Phytol.*, 76, 331, 1976.
78. Hall, I. R., Effects of endomycorrhizas on the competitive ability of white clover, *N.Z. J. Agric. Res.*, 21, 509, 1978.
79. Azcon-G. de Aguilar, C., Azcon, R., and Barea, J. M., Endomycorrhizal fungi and *Rhizobium* as biological fertilisers for *Medicago sativa* in normal cultivation, *Nature*, 279, 325, 1979.
80. Redente, E. F. and Reeves, F. B., Interactions between vesicular-arbuscular mycorrhiza and *Rhizobium* and their effect on sweetvetch growth, *Soil Sci.*, 132, 410, 1981.
81. Cluett, H. C. and Boucher, D. H., Indirect mutualism in the legume *Rhizobium*-mycorrhizal fungus interaction, *Oecologia*, 59, 405, 1983.
82. Janos, D. P., Mycorrhizae influence tropical succession, *Biotropica (Trop. Succession)*, 12, 56, 1980.
83. McGraw, A.-C., Jarstfer, A., and Miller, R. M., unpublished data.
84. Lindsey, D. L., The role of vesicular-arbuscular mycorrhizae in shrub establishment, in *VA Mycorrhizae and Reclamation of Arid and Semi-Arid Lands*, Williams, S. E. and Allen, M. F., Eds., University of Wyoming Agriculture Experiment Stn., Laramie, 1984, 53.
85. Bloss, H. E. and Pfeiffer, C. M., Latex content and biomass increase in mycorrhizal guayule *(Parthenium argenlatum)* under field conditions, *Ann. Appl. Biol.*, 104, 175, 1984.
86. Bloss, H. E. and Pfeiffer, C. M., Growth and nutrition of mycorrhizal gayule plants, *Ann. Appl. Biol.*, 99, 267, 1981.
87. Williams, S. E., Wallum, A. G., III, and Aldon, E. F., Growth of *Atriplex cauescens* (Pursh) Nutt. improved by formation of vesicular-arbuscular mycorrhizae, *Soil Sci. Soc. Am. J.*, 38, 962, 1974.
88. Aldon, E. F., Endomycorrhizae enhance survival and growth of fourwing saltbush on coal mine spoil, U.S. Department of Agriculture Forest Service Research Note RM-294, Rocky Mt. For. Serv. and Range Exp. Stn., Fort Collins, Colo., 1975.

89. Lindsey, D. L., Cress, W. A., and Aldon, E. F., The effects of endomycorrhizae on growth of rabbitbrush, fourwing saltbush and corn in coal mine spoil material, Rocky Mountain Forest and Range Exp. Stn., U.S. Forest Serv. Res. Note RM-343, 1977.
90. Larcher, W., *Physiological Plant Ecology*, Springer-Verlag, Berlin, 1975, 252.
91. Hardie, K. and Leyton, L., The influence of vesicular-arbuscular mycorrhiza on growth and water relations of red clover, *New Phytol.*, 89, 599, 1981.
92. Nelsen, C. E. and Safir, G. R., Increased drought tolerance of mycorrhizal onion plants caused by improved phosphorus nutrition, *Planta*, 154, 407, 1982.
93. Allen, M. F. and Boosalis, M. G., Effects of two species of VA mycorrhizal fungi on drought tolerance of winter wheat, *New Phytol.*, 83, 7, 1983.
94. Graham, J. H. and Syvertsen, J. P., Influence of vesicular-arbuscular mycorrhiza on the hydraulic conductivity of roots of two citrus rootstocks, *New Phytol.*, 97, 277, 1984.
95. Safir, G. R., Boyer, J. S., and Gerdemann, J. W., Mycorrhizal enhancement of water transport in soybean, *Science*, 172, 581, 1971.
96. Safir, G. R., Boyer, J. S., and Gerdemann, J. W., Nutrient status and mycorrhizal enhancement of water transport in soybean, *Plant Physiol.*, 49, 700, 1972.
97. Nelsen, C. E. and Safir, G. R., The water relations of well-watered, mycorrhizal and non-mycorrhizal onion plants, *J. Am. Soc. Hortic. Sci.*, 107, 271, 1982.
98. Fiscus, E. L., The interaction between osmotic- and pressure-induced water flow in plant roots, *Plant Physiol.*, 55, 917, 1975.
99. Allen, M. F., Influence of vesicular-arbuscular mycorrhizae on water movement through *Bouteloua gracilis* (H.B.K.) Lag ex steud, *New Phytol.*, 91, 191, 1982.
100. Stahl, P. D. and Smith, W. K., Effects of different geographic isolates of *Glomus* on the water relations of *Agropyron smithii*, *Mycologia*, 76, 261, 1984.
101. Daft, M. J. and Okusanya, B. O., Effect of *Endogone* mycorrhiza on plant growth. VI. Influence of infection on the anatomy and reproductive development in four hosts, *New Phytol.*, 72, 1333, 1973.
102. Weatherley, P. E., Water uptake and flow in roots, in *Physiological Plant Ecology II, Water Relations and Carbon Assimilation, Encyclopedia of Plant Physiology*, New Series, Vol. 12B, Springer-Verlag Berlin, 1982, 79.
103. Tinker, P. B., Roots and waters — transport of water to plant roots in soil, *Philos. Trans. R. Soc. London Ser. B*, 273, 445, 1976.
104. Tisdall, J. M. and Oades, J. M., Organic matter and water-stable aggregates in soils, *J. Soil Sci.*, 33, 141, 1982.
105. Trappe, J. M., Mycorrhizae and productivity of arid and semiarid rangelands, in *Advances in Food Producing Systems for Arid and Semiarid Levels*, Academic Press, New York, 1981, 581.
106. Mejstrik, V. K., Vesicular-arbuscular mycorrhizas of the species of a *Molinietum coeruleae* L. I. Association: the ecology, *New Phytol.*, 71, 883, 1972.
107. Mejstrik, V., Ecology of endomycorrhizae of *Trisetum flavescens* (L.) P. Beaur. and *Alopecurus pratenis* L., and the intensity of soil cultivation, *Acta Mycol.*, 13, 179, 1977.
108. Rabatin, S. C., Seasonal and edaphic variation in vesicular-arbuscular mycorrhizal infection of grasses by *Glomus tenuis*, *New Phytol.*, 83, 95, 1979.
109. Gay, P. E., Grubb, P. J., and Hudson, H. J., Seasonal changes in the concentrations of nitrogen, phosphorus and potassium, and in the density of mycorrhiza, in biennial and matrix-forming perennial species of closed chalkland turf, *J. Ecol.*, 70, 571, 1982.
110. Anderson, R. C., Liberta, A. E., and Dickman, L. A., Interaction of vascular plants and vesicular-arbuscular mycorrhizal fungi across a soil moisture-nutrient gradient, *Oecologia*, 64, 111, 1984.
111. Read, D. J. and Haselwandter, K., Observations on the mycorrhizal status of some alpine plant communities, *New Phytol.*, 88, 341, 1981.
112. Christie, P. and Nicolson, T. H., Are mycorrhizas absent from the antartic?, *Trans. Br. Mycol. Soc.*, 80, 557, 1983.
113. Reeves, F. B., Wagner, D., Moorman, T., and Kiel, J., The role of endomycorrhizae in the revegetation practices in the semi-arid West. I. A comparison of incidence of mycorrhizae in severely disturbed vs. natural environments, *Am. J. Bot.*, 66, 6, 1979.
114. Call, C. A. and McKell, C. M., Vesicular-arbuscular mycorrhizae — a natural revegetation strategy for disposed processed oil shale, *Reclam. Reveg. Res.*, 1, 337, 1982.
115. Pendleton, R. L. and Smith, B. N., Vesicular-arbuscular mycorrhizae of weedy and colonizer plant species at disturbed sites in Utah, *Oecologia*, 59, 296, 1983.
116. Miller, R. M., Some occurrences of vesicular-arbuscular mycorrhiza in natural and disturbed ecosystems of the Red Desert, *Can. J. Bot.*, 57, 619, 1979.
117. Staffeldt, E. E. and Vogt, K. B., Mycorrhizae of desert plants, U.S./I.B.P. Desert Biome. Rep. 1974 Prog., 3, 63, 1975.
118. Rose, S. L., Vesicular-arbuscular endomycorrhizal associations of some desert plants of Baja California, *Can. J. Bot.*, 59, 1056, 1981.

119. Bethlenfalvay, G. J., Dakessian, S., and Pacovsky, R. S., Mycorrhizae in southern California desert: ecological implications, *Can. J. Bot.*, 62, 519, 1984.
120. Khan, A. G., The occurrence of mycorrhizas in halophytes, hydrophytes and xerophytes, and endogone spores in adjacent soils, *J. Gen. Microbiol.*, 81, 7, 1974.
121. Williams, S. E. and Aldon, E. F., Endomycorrhizal (vesicular-arbuscular) associations of some arid zone shrubs, *Southwest. Nat.*, 20, 437, 1976.
122. Saif, S. R. and Iffat, N., Vesicular-arbuscular mycorrhizae in plants and endogonaceous spores in the soil of northern areas of Pakistan. I. Hunza, Nagar and Gilgit, *Pak. J. Bot.*, 8, 163, 1976.
123. Allen, M. F., Formation of vesicular-arbuscular mycorrhizae in *Atriplex gardneri* (Chenopodiaceae): season response in a cold desert, *Mycologia*, 75, 773, 1983.
124. Miller, R. M., Moorman, T. B., and Schmidt, S. K., Interspecific plant association effects on vesicular-arbuscular mycorrhiza occurrence in *Atriplex confertifolia*, *New Phytol.*, 95, 241, 1983.
125. Hirrel, M. C., Mehravaran, H., and Gerdemann, J. W., Vesicular-arbuscular mycorrhizae in the Chenopodi-aceae and Cruciferae: do they occur?, *Can. J. Bot.*, 56, 2813, 1978.
126. Schwab, S. M., Johnson, E. L. V., and Menge, J. A., Influence of simazine on formation of vesicular-arbuscular mycorrhizae in *Chenopodium quinona* Willd, *Plant Soil*, 64, 283, 1982.
127. Liberta, A. E., Anderson, R. C., and Dickman, L. A., VA mycorrhiza fragments as a means of endophyte identification at hydrophytic sites, *Mycologia*, 75, 169, 1983.
128. Haselwandter, K. and Read, D. J., The significance of a root-fungus association in two *Carex* species of high-alpine plant communities, *Oecologia*, 53, 352, 1982.
129. Giovannetti, M. and Mosse, B., An evaluation of techniques for measuring vesicular arbuscular mycorrhizal infection in roots, *New Phytol.*, 84, 489, 1980.
130. Daniels, B. A. and Skipper, H. D., Methods for the recovery and quantitative estimation of propagules from soils, in *Methods and Principles of Mycorrhizal Research*, Schenke, N. C., Ed., American Phytopathology Society, St. Paul, 1982, 29.
131. Baylis, G. T. S., Host treatment and spore production by Endogone, *N.Z. J. Bot.*, 7, 173, 1969.
132. Powell, C. L., Mycorrhizas in hill-country soils. I. Spore-bearing mycorrhizal fungi in thirty-seven soils, *N.Z. J. Agric. Res.*, 20, 53, 1977.
133. Rives, C. S., Bajwa, M. I., Liberta, A. E., and Miller, R. M., Effects of topsoil storage during surface mining on the viability of VA mycorrhiza, *Soil Sci.*, 129, 253, 1980.
134. Powell, C. L., Development of mycorrhizal infections from *Endogone spores* and infected root segments, *Trans. Br. Mycol. Soc.*, 66, 439, 1976.
135. Hayman, D. S. and Stovold, G. E., Spore populations and infectivity of vesicular-arbuscular mycorrhizal fungi in New South Wales, *Aust. J. Bot.*, 27, 227, 1979.
136. Smith, G. W., Effects of inoculation level and sieving on the *Gigaspora gigantea*-soybean mycorrhizal symbiosis, *Soil Biol. Biochem.*, 13, 539, 1979.
137. Abbott, L. K. and Robson, A. D., Infectivity of vesicular arbuscular mycorrhizal fungi in agricultural soils, *Aust. J. Agric. Res.*, 33, 1049, 1981.
138. Hayman, D. S., *Endogone* spore numbers in soil and vesicular-arbuscular mycorrhiza in wheat as influenced by season and soil treatment, *Trans. Br. Mycol. Soc.*, 54, 53, 1970.
139. Miller, R. M., Carnes, B. A., and Moorman, T. B., Factors influencing survival of vesicular-arbuscular mycorrhiza propagules during topsoil storage, *J. Appl. Ecol.*, 22, 259, 1985.
140. Nicolson, T. H., Mycorrhiza in the Gramineae. I. Vesicular-arbuscular edophytes, with special reference to the external phase, *Trans. Br. Mycol. Soc.*, 42, 421, 1959.
141. Nicolson, T. H., Mycorrhiza in the Gramineae. II. Development in different habitats, particularly sand dunes, *Trans. Br. Mycol. Soc.*, 43, 132, 1960.
142. Mosse, B., Advances in the study of vesicular-arbuscular mycorrhiza, *Ann. Rev. Phytopathol.*, 11, 171, 1973.
143. Gerdemann, J. W. and Trappe, J. M., The Endogonaceae in the Pacific Northwest, *Mycol. Mem.*, 5, 1, 1974.
144. Nicolson, T. H., Vesicular-arbuscular mycorrhiza — a universal plant symbiosis, *Sci. Prog. (Oxford)*, 55, 561, 1967.
145. Koske, R. E. and Halvorson, W. L., Ecological studies of vesicular-arbuscular mycorrhizae in a barrier sand dune, *Can. J. Bot.*, 59, 1413, 1981.
146. Koske, R. E., A preliminary study of interactions between species of vesicular-arbuscular fungi in a sand dune, *Trans. Br. Mycol. Soc.*, 76, 411, 1981.
147. Anderson, R. C., Liberta, A. E., Dickman, L. A., and Katz, A. J., Spatial variation in vesicular-arbuscular mycorrhiza spore density, *Bull. Torrey Bot. Club*, 110, 519, 1983.
148. Dickman, L. A., Liberta, A. E., and Anderson, R. C., Ecological interaction of little bluestem and vesicular-arbuscular mycorrhizal fungi, *Can. J. Bot.*, 62, 2272, 1984.
149. Miller, R. M., Microbial ecology and nutrient cycling in disturbed arid ecosystems, in *Ecological Studies of Disturbed Landscapes*, DOE/NBM-5009372, USDOE, 1984.

150. Tommerup, I. C. and Abbott, L. K., Prolonged survival and viability of VA mycorrhizal hyphae after root death, *Soil Biol. Biochem.*, 13, 431, 1981.
151. Biermann, B. and Linderman, R. C., Use of vesicular-arbuscular mycorrhizal roots, intraradical vesicles and extraradical vesicles as inoculum, *New Phytol.*, 95, 97, 1983.
152. Koske, R. E., *Endogone* spores in Australian sand dunes, *Can. J. Bot.*, 53, 668, 1975.
153. Abbott, L. K. and Robson, A. D., The distribution and abundance of vesicular arbuscular endophytes in some western Australian soils, *Aust. J. Bot.*, 77, 515, 1977.
154. Sward, R. J., Hallam, N. D., and Holland, A. A., *Endogone* spores in a heathland area of south-eastern Australia, *Aust. J. Bot.*, 26, 29, 1978.
155. Mosse, B. and Bowen, G. D., The distribution of *Endogone* spores in some Australian and New Zealand soils, and in an experimental field soil at Rottamsted, *Trans. Br. Mycol. Soc.*, 51, 485, 1968.
156. Hayman, D. S., Mycorrhizal populations of sown pastures and native vegetation in Otago, New Zealand, *N.Z. J. Agric. Res.*, 21, 271, 1978.
157. Barea, J. M., Escudero, J. L., and Azcon-G. de Aguilar, C., Effects of introduced and indigenous VA mycorrhizal fungi on nodulation, growth and nutrition of *Medicago sativa* in phosphate-fixing soils as affected by P fertilizers, *Plant Soil*, 54, 283, 1980.
158. Khan, A. G., Occurrence of *Endogone* spores in West Pakistan soil, *Trans. Br. Mycol. Soc.*, 56, 217, 1971.
159. Hetrick, B. A. D. and Bloom, J., Vesicular-arbuscular mycorrhizal fungi associated with native tall grass prairie and cultivated winter wheat, *Can. J. Bot.*, 61, 2140, 1983.
160. Stahl, P. D. and Christensen, M., Mycorrhizal fungi associated with *Bouteloua* and *Agropyron* in Wyoming sagebrush-grasslands, *Mycologia*, 74, 877, 1982.
161. Molina, R. J., Trappe, J. M., and Strickler, G. S., Mycorrhizal fungi associated with *Festuca* in the western United States and Canada, *Can. J. Bot.*, 56, 1691, 1978.
162. McGraw, A.-C. and Miller, R. M., unpublished data.
163. Crush, J. R., Occurrence of endomycorrhizas in soils of the Mackenzie Basin, Canterburg, New Zealand, *N.Z. J. Agric. Res.*, 18, 361, 1975.
164. Harper, J. L., The individual in the population, *J. Ecol.*, 52 (Suppl.), 149, 1964.
165. Trenbath, B. R. and Harper, J. L., Neighbour effects in the genus *Avena*. I. Comparison of crop species, *J. Appl. Ecol.*, 10, 379, 1973.
166. Harper, J. L., Approaches to the study of plant competition, *Symp. Soc. Exp. Biol.*, 15, 1, 1961.
167. Newman, E. I., Heap, A. J., and Lawley, R. A., Abundance of mycorrhizas and root-surface microorganisms of *Plantago lanceolata* in relation to soil and vegetation: a multi-variate approach, *New Phytol.*, 89, 95, 1982.
168. Warner, A. and Mosse, B., Factors affecting the spread of vesicular mycorrhizal fungi in soil. I. Root density, *New Phytol.*, 90, 529, 1982.
169. Powell, C. L., Spread of mycorrhizal fungi through soil, *N.Z. J. Agric. Res.*, 22, 335, 1979.
170. Abbott, L. K. and Robson, A. D., The effect of root density, inoculum placement and infectivity of inoculum on the development of vesicular-arbuscular mycorrhizas, *New Phytol.*, 97, 285, 1984.
171. Warner, A. and Mosse, B., Spread of vesicular-arbuscular mycorrhizal fungi between separate root systems, *Trans. Br. Mycol. Soc.*, 80, 353, 1983.
172. Grime, J. P., Competitive exclusion in herbaceous vegetation, *Nature*, 242, 344, 1973.
173. Harley, J. L. and Smith, S. E., *Mycorrhizal Symbiosis*, Academic Press, New York, 1983.
174. Fitter, A. H., Influence of mycorrhizal infection on competition for phosphorus and potassium by two grasses, *New Phytol.*, 79, 119, 1977.
175. Christie, P., Newman, E. I., and Campbell, R., The influence of neighbouring grassland plants on each others' endomycorrhizas and root-surface microorganisms, *Soil Biol. Biochem.*, 10, 521, 1978.
176. Newman, E. I., Campbell, R., Christie, P., Heap, A. J., and Lawley, R. A., Root microorganisms in mixtures and monocultures of grassland plants, in *The Soil-Root Interface*, Harley, J. L. and Russel, R. S., Eds., Academic Press, London, 1979, 161.
177. Lawley, R. A., Newman, E. I., and Campbell, R., Abundance of endomycorrhizas and root-surface microorganisms on three grasses grown separately and in mixtures, *Soil Biol. Biochem.*, 14, 237, 1982.
178. Heap, A. J. and Newman, E. I., Links between roots by hyphae of vesicular-arbuscular mycorrhizas, *New Phytol.*, 85, 169, 1980.
179. Heap, A. J. and Newman, E. I., The influence of vesicular-arbuscular mycorrhizas on phosphorus transfer between plants, *New Phytol.*, 85, 173, 1980.
180. Chiariello, N., Hickman, J. C., and Mooney, H. A., Endomycorrhizal role for interspecific transfer of phosphorus in a community of annual plants, *Science*, 217, 941, 1982.
181. Whittingham, J. and Read, D. J., Vesicular-arbuscular mycorrhiza in natural vegetation systems. III. Nutrient transfer between plants with mycorrhizal interconnections, *New Phytol.*, 90, 277, 1982.
182. Francis, R. and Read, D. J., Direct transfer of carbon between plants connected by vesicular-arbuscular mycorrhizal mycelium, *Nature*, 307, 53, 1984.

183. Hirrel, M. C. and Gerdemann, J. W., Enhanced carbon transfer between onions infected with a vesicular-arbuscular mycorrhizal fungus, *New Phytol.*, 83, 731, 1979.
184. Abbott, L. K. and Robson, A. D., Infectivity and effectiveness of five endomycorrhizal fungi: competition with indegenous fungi in field soils, *Aust. J. Agric. Res.*, 32, 61, 1981.
185. Wilson, J. M. and Trinick, M. J., Infection development and interactions between vesicular-arbuscular mycorrhizal fungi, *New Phytol.*, 93, 543, 1983.
186. Wilson, J. M., Competition for infection between vesicular-arbuscular mycorrhizal fungi, *New Phytol.*, 97, 427, 1984.
187. Wilson, J. M., Comparative development of infection by three vesicular-arbuscular mycorrhizal fungi, *New Phytol.*, 97, 413, 1984.
188. Bethlenfalvay, G. J. and Dakessian, S., Grazing effects on mycorrhizal colonization and foristic composition of the vegetation on a semiarid range in northern Nevada, *J. Range Manage.*, 37, 312, 1984.
189. Wallace, L. L., McNaughton, S. J., and Goughenour, M. B., The effects of clipping and fertilization on nitrogen nutrition and allocation by mycorrhizal and nonmycorrhizal *Panicum coloratum* L., a C4 grass, *Oecologia*, 54, 68, 1982.
190. Reece, P. E. and Bonham, C. D., Frequency of endomycorrhizal infection in grazed and ungrazed blue grama plants, *J. Range Manage.*, 31, 149, 1978.
191. Tisdall, J. M. and Oades, J. M., The management of ryegrass to stabilize aggregates of red-brown earth, *Aust. J. Soil Res.*, 18, 415, 1980.
192. Horn, H. S., The ecology of secondary succession, *Ann. Rev. Ecol. Syst.*, 5, 25, 1974.
193. Huston, M., A general hypothesis of species diversity, *Am. Nat.*, 113, 81, 1979.
194. Connell, H. J. and Slatyer, R., Mechanisms of succession in natural communities and their role in community stability and organization, *Am. Nat.*, 111, 1119, 1977.
195. Allen, E. B. and Allen. M. F., Natural re-establishment of vesicular-arbuscular mycorrhizae following strip mine reclamation in Wyoming, *J. Appl. Ecol.*, 17, 139, 1980.
196. Iqbal, S. H. and Qureshi, K. S., The influence of mixed sowing (cereals and crucifers) and crop rotation on the development of mycorrhiza and subsequent growth of corps under field conditions, *Biologia*, 22, 287, 1976.
197. Schmidt, S. K. and Reeves, F. B., Effects of the non-mycorrhizal pioneer plant *Salsola kali* L. (Chenopodiaceae) on vesicular-arbuscular mycorrhizal (VAM) fungi, *Am. J. Bot.*, 71, 1035, 1984.
198. Visser, S., Griffiths, C. L., and Parkinson, D., Topsoil storage effects on primary production and rates of vesicular-arbuscular mycorrhizal development in *Agropyron trachycaulum*, *Plant Soil*, 82, 51, 1984.
199. Moorman, T. and Reeves, F. B., The role of endomycorrhizae in revegetation practices in the semi-arid west. II. A bioassay to determine the effect of land disturbance on endomycorrhizal populations, *Am. J. Bot.*, 66, 14, 1979.
200. Powell, C. L., Mycorrhizal infectivity of eroded soils, *Soil Biol. Biochem.*, 12, 247, 1980.
201. Doerr, T. B., Redente, E. F,. and Reeves, F. B., Effects of soil disturbance on plant succession and levels of mycorrhizal fungi in a sagebrush-grassland community, *J. Range Manage.*, 37, 135, 1984.
202. Wali, M. K. and Freeman, P. G., Ecology of some mined areas in North Dakota, in *Some Environmental Aspects of Strip Mining in North Dakota*, ed. Ser. 5, Wali, M. K., Ed., North Dakota Geological Survey, Grand Forks, 1973, 24.
203. Daft, M. J. and Nicolson, T. H., Arbuscular mycorrhizas in plants colonizing coal wastes in Scotland, *New Phytol.*, 73, 1129, 1974.
204. Daft, M. J. and Hacskaylo, E., Arbuscular mycorrhizas in the anthracite and bituminous coal wastes of Pennsylvania, *J. Appl. Ecol.*, 13, 523, 1976.
205. Khan, A. G., Vesicular-arbuscular mycorrhizas in plants colonizing black wastes from bitumunous coal mining in the Illawana region of New South Wales, *New Phytol.*, 81, 53, 1978.
206. Rothwell, F. M. and Vogel, W. G., Mycorrhizae of planted and volunteer vegetation on surface-mined sites, U.S.D.A. For. Serv. Gen. Tech. Rep. NE-66, 1982.
207. Kiernan, J. M., Hendrix, J. W., and Maronek, D. M., Endomycorrhizal fungi occurring on orphan strip mines in Kentucky, *Can. J. Bot.*, 61, 1798, 1983.
208. Medve, R. J., The mycorrhizae of pioneer species in disturbed ecosystems in western Pennsylvania, *Am. J. Bot.*, 71, 787, 1984.
209. Rabatin, S. C. and Miller, R. M., unpublished data.
210. Smith, T. F., A note on the effect of soil tillage on the frequency and vertical distribution of spores of vesicular-arbuscular endophytes, *Aust. J. Soil Res.*, 16, 359, 1978.
211. Schwab, S. and Reeves, F. B., The role of endomycorrhizae in revegetation practices in the semi-arid west. III. Vertical distribution of vesicular-arbuscular (VA) mycorrhiza inoculum potential, *Am. J. Bot.*, 68, 1293, 1981.
212. Gould, A. B. and Liberta, A. E., Effects of topsoil storage during surface mining on viability of vesicular-arbuscular mycorrhiza, *Mycologia*, 73, 914, 1981.

213. Liberta, A. E., Effects of topsoil-storage duration on inoculum potential of vesicular-arbuscular mycorrhizae, in *Symp. on Surface Mining Hydrology, Sedimentology and Reclamation,* University of Kentucky, Lexington, 1981, 45.
214. Warner, A., Re-establishment of indigenous vesicular-arbuscular mycorrhizal fungi after topsoil storage, *Plant Soil,* 73, 387, 1983.
215. Abdul-Kareem, A. W. and McRae, S. G., The effects of topsoil of long term storage in stockpiles, *Plant Soil,* 76, 357, 1984.
216. Lambert, D. H. and Cole, H., Jr., Effects of mycorrhizae on establishment and performance of forage species in mine spoil, *Agron. J.,* 72, 257, 1980.
217. Bradshaw, A. D., Marro, R. H., Roberts, R. D., and Skeffington, R. A., The creation of nitrogen cycles in derelict land, *Philos. Trans. R. Soc. London,* 296, 557, 1982.
218. Harley, J. H., The importance of microorganisms to colonizing plants, *Trans. Proc. Not. Soc. Edinburg,* 41, 65, 1970.
219. LaRue, T. A. and Patterson, T. G., How much nitrogen do legumes fix?, *Adv. Agron.,* 34, 15, 1981.
220. Bergersen, F. J., Biochemistry of symbiotic nitrogen fixation in legumes, *Ann. Rev. Plant Physiol.,* 22, 121, 1971.
221. Danielson, R. M., Mycorrhizae and reclamation of stressed terrestrial environments, in *Soil Reclamation Processes, Microbial Analyses and Applications,* Tate, R. L. and Klein, D. A., Eds., Marcel Dekker, New York, 1984.
222. Zak, J. C., Danielson, R. M., and Parkinson, D., Mycorrhizal fungal spore numbers and species occurrence in two amended mine spoils in Alberta, Canada, *Mycologia,* 74, 785, 1982.
223. Mott, J. B. and Zuberer, D. A., Abundance of *Rhizobium trifolii* and endomycorrhizal fungi in surface mine spoils in Texas, *Agron. Abstr. Am. Soc. Agron. (Madison),* p. 20, 1982.
224. Zak, J. C. and Parkinson, D., Initial vesicular-arbuscular mycorrhizal development of slender wheatgrass on two amended mine spoils, *Can. J. Bot.,* 60, 2241, 1982.
225. Zak, J. C. and Parkinson, D., Effects of surface amendation of two mine spoils in Alberta, Canada, on vesicular-arbuscular mycorrhizal development of slender wheatgrass: a 4-year study, *Can. J. Bot.,* 61, 798, 1983.
226. Mott, J. B. and Zuberer, D. A., Effect of endomycorrhizae on growth and nitrogen fixation of clovers, in Proc. 40th Southern Pasture and Forage Crop Improvement Conf., Baton Rouge, 1984.
227. Aldon, E. F. and Green, N., Effects of endomycorrhizae on shrub and grass growth in the southwest, Mineral Waste Stabilization Liaison Committee, 1980, 98.
228. Allen, E. B. and Cunningham, G. L., Effects of vesicular-arbuscular mycorrhizae on *Distichlis spicata* under three salinity levels, *New Phytol.,* 93, 227, 1983.
229. Allen, E. B., VA mycorrhizae and colonizing animals: implications for growth, competition, and succession, in *VA Mycorrhizae and Reclamation of Arid and Semi-Arid Lands,* Williams, S. E. and Allen, M. F., Eds., University of Wyoming Agriculture Experimental Stn., Laramie 1984, 42.
230. Hall, I. R., Johnstone, P. D., and Dolby, R., Interactions between endomycorrhizae and soil nitrogen and phosphorous on the growth of ryegrass, *New Phytol.,* 97, 447, 1984.

Chapter 9

VA MYCORRHIZAS IN FIELD CROP SYSTEMS

David S. Hayman

TABLE OF CONTENTS

I. INTRODUCTION

The widespread occurrence of VA mycorrhizas in diverse soil-plant systems is now a well-established fact. Numerous studies since the 1890s[1] and especially in the past 20 years or so[2-6] have highlighted the ubiquity of these symbiotic associations on a global scale and their abundance within most ecosystems and throughout the plant kingdom. As VA mycorrhizas are found in most of the major crop families and can markedly influence plant growth, it is not surprising that there is currently much interest in their potential role in field crop production.

In this chapter, emphasis is placed on the ecology and physiology of vesicular-arbuscular mycorrhizal (VAM) associations in field crop systems. The effects on crops of indigenous VAM fungi and of VAM fungi introduced as inoculants are considered, together with their behavior in arable soils and methods of manipulating them. In addition, the identification of mycorrhiza-responsive situations and the development of new practical inoculation technologies are discussed.

II. ECOLOGY

A. The Fungal Endophytes

Nearly 100 species of VAM fungi are presently recognized,[7] but the number reported from arable soils is considerably less. Still fewer species have been isolated and tested with hosts in pot trials or as field inoculants. Consequently, our knowledge of endophyte behavior, adaptability to different conditions, and their potential benefits to crop plants is based on a small proportion of known species which is not necessarily fully representative of their genetic diversity and symbiotic effectiveness. Comprehensive studies are needed with as many species and strains as possible to see whether "better" ones exist in nature or can perhaps be created by novel techniques like genetic manipulation.

Some VAM species that have been studied in detail, e.g., *Glomus mosseae* and *Acaulospora laevis,* have been reported from many countries around the world. Moreover, some species occur across a broad spectrum of plant habitats and soils, although they can vary in their sensitivity to environmental factors. Generally, VAM fungi are not limited in their host range, as individual isolates are usually capable of infecting a broad assortment of plant species. However, different VAM fungi are not always equally infective to any one plant species, and they certainly vary in their physiological interactions with different plants and hence in their effects on plant growth. This will, in turn, affect their own reproduction and survival in the field.

B. Influence of the Plant

Most plant families containing crops of major economic importance are mycorrhizal, including the Gramineae, Leguminosae, Solanaceae, and Rosaceae. Exceptions are the Cruciferae and Chenopodiaceae. Some field crops are often heavily infected by VAM fungi, e.g., maize, cassava, cowpea, soybean, clover, beans, sorghum, barley, upland rice, cacao, coffee, tea, rubber, oil palm, tobacco, cotton, sugarcane, lettuce, onions, etc. Others have little or no infection, e.g., the brassicas and sugar beet. Some crops are intermediate, e.g., tomato and ryegrass may be only moderately infected. The picture is complicated by evidence of differences in mycotrophy between cultivars of the same crop species, e.g., wheat,[8] alfalfa (lucerne),[9] and Phaseolus beans.[10] Furthermore, in mixed cropping systems, VAM infection in a host plant such as wheat can be decreased by the presence of a nonhost such as mustard.[11]

The fact that VAM fungi are obligate symbionts means that the distribution of VA mycorrhizas in cultivated soils is greatly affected by the species of crop grown. This has important implications in crop rotations where a heavily infected VAM-susceptible crop would leave a larger residual mycorrhizal inoculum in the soil to infect subsequent crops than would a lightly infected or VAM-resistant crop. It also implies that mycorrhizal populations should drop drastically during fallow periods, and this has been reported for arable soils.[12-15] However, field observations are not always consistent in practice. For example, wheat is usually more mycorrhizal than oats, but Kruckelmann,[16] despite finding most VAM spores in wheat monoculture, counted fewer spores associated with wheat after oats than in oat monoculture. Possibly endophytes with some host specificity built up during oat monoculture, making them more infective to oats than to wheat. A similar possibility was envisaged by Ocampo et al.[17] to account for the greater infectivity of indigenous endophytes to maize than to onions in soil collected from a maize field, in contrast to the greater infectivity of *Gigaspora margarita* to onion than to maize and the equal infectivity of *Glomus fasciculatum* E3 to onion and maize in another soil.

That the crop species itself can exert a selective effect on which VAM species in a mixed indigenous population become predominant is well illustrated in a study by Schenck and Kinloch.[18] They found marked differences in populations of VAM fungi between different crops grown for 7 years in monoculture at a newly cleared former woodland site. Three species of *Gigaspora* were most prevalent in association with soybean roots, in contrast to cotton and peanut where *Acaulospora* spp. dominated, and Bahia grass where two species of *Glomus* were most numerous. The largest number of VAM species was found with sorghum. More spores occurred with soybean than with the other crops, and there were fewest spores in the original woodland.

The presence of a so-called nonhost seems to be better for a mycorrhizal fungus than no plant at all, as VAM hyphae have been observed colonizing the root surfaces of nonhosts,[17,19] thereby procuring an ecological niche in which to survive in the absence of a host root. Not only does this indicate that these nonhost roots are not releasing toxic factors into the surrounding soil, it also raises the possibility of some saprophytic growth in rhizosphere soil by VAM fungi.

C. Soil Factors and Fungal Activity

For a long time there has been speculation that organic matter added to soil encourages mycorrhizal development. Reed and Frémont,[20] for example, found that VAM infection in citrus in California had dense arbuscular development in soil amended with farmyard manure but not with mineral fertilizer. In alpine plants, Peyronel[21] found considerable VAM infection when there was much humus in the soil; he believed this mitigated the negative effect of low-light intensity on infection. Also, Warner and Mosse[22] recently obtained evidence of saprophytic growth in soil organic matter. At the present time there is speculation that applying mulches to soil may favor growth by improving the mycorrhiza, but the evidence so far is not conclusive. In this regard it may well be useful to examine potting composts for their compatibility with VAM fungi, which could be important for mycotrophic container-grown plants and seedling transplants. Field crops which might benefit in this way are nursery-raised plantation crops such as oil palm, and transplanted crops such as certain vegetables, besides numerous horticultural species. Furthermore, cultivation systems in which the breakdown of organic matter and roots may be slower than in ploughed soil (e.g., with minimum tillage and direct drilling methods) should favor VAM development in field crops. Thus, in the Netherlands, Ruissen[23] observed the most VAM infection in field plots of winter wheat with minimal tillage and the least in plots with traditional tillage; results were similar for 2 successive years. Probably minimal tillage also preserves more of the fine hyphal network that develops from VAM roots in soil. This mycelium can be quite extensive in grassland soils,[24] sometimes attaining several meters of hypha per milliliter of rhizosphere soil,[25] which is probably of considerable importance in stabilizing soil aggregates.

How VAM fungi survive in arable soils will be partly determined by the cropping system. Survival inside roots is an obvious possibility with perennial crops, whereas independent propagules would be needed with annual crops. Thus, according to Butler,[1] VAM infections may be more abundant in orchard and plantation crops than in annual field crops, although recent studies indicate the presence of more spores of VAM fungi in soils growing annual crops. This may be explained by selection pressures acting on a mixed population of VAM fungi, favoring those able to survive fallow periods between crops as resting spores.

D. Effects of Fertilizers

Spore production should be related to the total root-based fungal biomass, i.e., total infected root length, rather than percent of infection. This is a likely explanation for the higher spore numbers reported at intermediate than at high or low levels of phosphate fertilizer in field plots growing barley or potatoes,[26] where presumably there was the most infected root length at the intermediate level, because there is usually less percentage of mycorrhizal infection at high levels of P and less total root growth at low levels of P.

When total root growth is not greatly affected by treatment, percentage of infection and spore numbers can be closely related. This was shown in the negative effects of nitrogen fertilizer on field populations of VAM fungi in wheat,[27] barley,[28] and peanuts,[29] for example. This generally inhibitory effect of nitrogen on VAM contrasts with its tendency to increase infection by root pathogenic fungi, e.g., *Fusarium solani*.[30] However, nitrogen is not always inhibitory to VA mycorrhiza under field conditions, as Sparling and Tinker[31] observed in hill grasslands in northern England. They found no effect of N fertilizer on VAM infection in the native grasses, in contrast to a negative effect of P fertilizer. This negative effect of P fertilizer in hill pastures has also been observed in introduced clover plants.[32]

Bevege[33] found that the effects of N and P on VA mycorrhiza in the field can become rather complex when interactions between the two elements are considered. In field plots of *Araucaria cunninghamii* (hoop pine) in northeast Australia, he found a trend of increased VAM infection with increasing additions of N (448 to 4032 kg N per hectare as urea) at intermediate levels of P (37 and 74 kg P per hectare as superphosphate), but at 148 kg P per hectare infection decreased with increasing N. There were the most spores at high and low levels of N.

Negative effects of compound fertilizer on VA mycorrhiza have been reported in field crop systems. For example, Strzemska[34] observed that NPK fertilizer decreased the intensity of VAM infection within the roots of four cereals in agricultural soils in Poland. The fertilizer had no effect on VAM infection in horse beans, however. This agrees with observations at a farm in the U.K. where N did not affect mycorrhiza in field beans.[35] Nevertheless, fertilizers may have a positive effect on VAM if the initial soil fertility level is very low.[16] Although it is difficult to draw precise conclusions from these field observations, it is clear that there is a need for a more balanced approach to fertilizer application in field crop systems where VA mycorrhiza is having a significant effect.

Apart from inducing quantitative changes in VAM populations in the field, additions of fertilizer can also affect the species composition. Nitrogen fertilizer, for example, depressed numbers of *Glomus* sp. "white reticulate" more than *G. caledonium* "laminate" spores in a sandy loam and a heavy clay loam at two different English farms.[35] Other qualitative changes in the VAM population resulting from soil amendments include a greater reduction in the populations of indigenous endophytes than of introduced *G. mosseae* and *G. fasciculatum* when lime and phosphate were applied to upland grasslands in Wales.[36] In Western Australia the distribution of *Acaulospora laevis* "honey-colored sessile" and *G. mosseae* "yellow vacuolate" spore types was related to soil pH.[37] Wang et al.[38] reported most fine endophyte at low pH in the field, and experiments in pots showed an adaptation of species or strains to soil pH; endophytes were most active in the soil from which they were isolated.

Reticulate spore types seemed to be most common in sandy soils and vacuolate types in clay loams in Pakistan.[39] Other factors still to be elucidated are involved in species distribution, e.g., adjacent fields at Rothamsted, England with the same crop and soil type and similar fertility had a very different endophyte composition.[27] Thus, although certain factors that dictate the size and composition of VA mycorrhizas in field crop systems have been identified, we are still a long way from a satisfactory understanding of the complex ecology of the VAM fungi.

The examples cited above confirm the generally accepted principle that high doses of fertilizer inhibit mycorrhiza. However, this inverse relationship, which exists where a range of fertility levels have been created within a single given site, is less consistent when different locations are compared. For example, observations on five crops at several locations in southern Spain[40] showed no negative correlation between soil fertility (N and P as estimated by chemical analysis) and mycorrhizal development, except for adjacent sites within an individual location. This indicates an inherent natural variability in propagule distribution. Also, some irregular distribution patterns in field soils may be related to suppressive factors in the soil, as occurs with certain root pathogenic fungi. Plant mycotrophy is an important factor, as some field crops can be strongly mycorrhizal even in very fertile soils, e.g., maize in the midwestern U.S. Hence some crops, even at agriculturally realistic levels of soil fertility, develop extensive mycorrhizas in the field.

E. Other Factors

Ecological disturbances induced by pesticides include effects on nontarget organisms such as mycorrhizal fungi. It is not surprising that many fungicides do not distinguish between harmful and beneficial fungi in the soil, and VAM is inhibited by benomyl and other fungicides applied to soil.[28,41,42] Also, some insecticides (e.g., aldrin) inhibit VAM.[43] Nevertheless, some nematicides are reported to increase VAM in some field-grown crops, e.g., cotton.[44] The nematicide-insecticide aldicarb increased VAM in barley in the field, except that high nitrogen levels overrode this effect.[28] The effects of herbicides on mycorrhiza need more extensive investigation; paraquat at the standard rate of 0.5 kg/ha was reported to stimulate *G. fasciculatum* in a tree nursery.[45]

Mycorrhizal populations in field crop systems vary during the growing season. Usually, more spores are found towards the middle or end of the season than at the beginning.[27,46,47] This is attributed to increased spore production as root growth slows down or ceases. VAM root infection percentages commonly reach a maximum towards the end of the growing season,[27] following a three-phase pattern of growth.[10,48] With field-grown *Phaseolus* beans and soybeans, Sutton[10] noted an initial lag phase of 20 to 25 days, attributed to rapid root growth of the seedlings and the time required for spore germination, germ-tube growth, and penetration of the host plant root. In the second phase, lasting 30 to 35 days, extensive mycorrhizal development coincided with most shoot growth and copious spread of external mycelium, leading to multiple infections. During the third phase, from host fruiting to senescence, the proportion of mycorrhizal to nonmycorrhizal roots remained constant. Under some conditions the infection plateau is reached much sooner.[47]

There has been little work on identification of strain differences within a single morphological species, even though strains may become adapted to particular soil factors. Isolates of the same morphological species from different climatic regions will very likely have different temperature optima. In addition, the evolution of strains of *G. mosseae* tolerant of heavy metal contamination in field soil has been described by Gildon and Tinker.[49] A similar effect has been reported for phosphate in Western Australian pasture soils, where the long-term application of superphosphate fertilizer (15 years of 150 kg P per hectare) led to a VAM population that was little affected by subsequent additions of P.[50] It appeared that P-tolerant strains had built up during the 15-year period of superphosphate application because, in experiments with this soil in a glasshouse, endophytes from plots subsequently given the highest rates of P did not differ from those in plots given no P in their ability to infect and increase P uptake and growth of subterranean clover.

Finally, it is difficult to produce a complete ecological picture of VAM fungi in arable soils because of their diversity of propagules. While it is relatively straightforward to extract and count resting spores or to stain and examine root samples from growing plants, it is not so easy to quantify other propagules such as small fragments of infected root and clumps of mycelium. The last two must represent a considerable proportion of the VAM inoculum potential where soil infectivity is high and not related to spore numbers.[51] Baiting soils with suitable host plants can reveal the extent of mycorrhizal infectivity,[51] and the most probable number (MPN) technique can give a good indication of the number of infective propagules in some of the soils where it has been tried.[52] With some refinements, the MPN technique could give a useful, objective measure of inoculum potential.

III. PHYSIOLOGICAL CONSIDERATIONS

A. Phosphorus and Plant Mycotrophy

No discussion on mycorrhizal symbiosis is complete without at least some reference to the key position of phosphorus. Innumerable experiments in the 1970s have shown that a wide range of plants can benefit from infection by mycorrhizal fungi, primarily through an enhanced ability to take up phosphate. Although with some crop species these benefits are only visible under conditions of high nutrient stress, other crops (e.g., white clover) can be improved by mycorrhiza even when given standard dressings of phosphate fertilizer. Certain ones (e.g., cassava and *Stylosanthes*) seem to be obligately mycorrhizal under field conditions.

Inoculation experiments done in pots under controlled conditions have provided useful information on which crop species are most mycotrophic and at what levels of soil P. These experiments have also produced a better understanding of the mechanisms and physiological interactions involved. Hence, our attempts to exploit mycorrhizas in field crop systems in the 1980s and onwards should be less empirical than in the previous decade as new basic principles are incorporated into a practical inoculation technology.

One major principle is the balance between P fertilizer and mycorrhizal activity — too great an input of P can neutralize the beneficial role of the fungal partner. This happens for three reasons: (1) As deduced from experiments using ^{32}P-labeled soil,[53-55] the main role of the fungus is physical — its hyphae extend well beyond the zone of P depletion that forms close to the root when uptake exceeds diffusion through soil, thereby tapping the undepleted soil P. Excess P fertilizer renders this effect of VAM unnecessary by maintaining adequate soluble P at the root surface. (2) The negative effect of phosphate on VAM infection means that too much fertilizer input may raise P levels so high in the plant tissues so as to decrease infection to ineffective amounts.[56] (3) If fertilizer P input is sufficient to meet the needs of the plant but not so great as to decrease infection, then the fungus may cause a carbon drain on the host without any reciprocal effect on P transfer.[57] To avoid overwhelming the mycorrhizal function in this way, the use of less soluble P fertilizers should be considered in some circumstances. Hence, there is interest currently in applying rock phosphate in acid, P-deficient soils, especially in the tropics where so-called "P-fixing" soils are widespread, superphosphate fertilizer is often in short supply and more expensive, and the high temperatures enhance microbial activity.[2,58,59]

If the fungal symbiont is causing a significant carbon drain from the plant symbiont or otherwise upsetting the metabolism of the plant, and the plant is meeting its P requirement from soil or fertilizer sources without depending on its mycorrhiza, the symbiotic balance may tilt from mutualism to parasitism. This was exemplified by Crush,[60] who observed an unfavorable effect of some endophytes on certain pasture legumes at high levels of added superphosphate. A practical implication of this is that any field inoculation program with VAM fungi should utilize endophytes carefully selected for optimum performance at agronomically realistic P levels. Nevertheless, Jensen and Jakobsen[61] discovered that in various agricultural soils in Denmark containing a range of P levels and growing wheat and barley, the symbiotic balance is not necessarily upset. They found no differences in shoot P content under practical field conditions, a low level of soil P apparently being offset by higher mycorrhizal infection and vice versa. Their results suggest that the symbiosis is self-regulatory and that indigenous VAM fungi may be beneficial to the phosphate nutrition of cereals in low P soils rather than deleterious at high P levels. A similar conclusion can be reached from Strzemska's[34] observations that large amounts of fertilizer which markedly decreased mycorrhizal infection did not increase yields of cereals or beans.

B. Effects of Nitrogen and Carbon

The negative effects of nitrogen on VA mycorrhiza in field crops are more variable than those of phosphate. However, a consistent inverse correlation between percent N in plant tissues and percent mycorrhizal infection was noted by Wang and Hayman[62] in onion and clover. The applied N was less inhibitory of mycorrhizal infection in legumes than in nonlegumes, apparently because it only affected N concentrations in the nonlegumes. With lettuces, Hepper[63] stressed the importance of the balance between N and P within the plant, an N/P ratio of above 15:1 leading to more infection. Wang[64] found with onions that the negative effect of NH_4-N was greater than that of NO_3-N, and the lesser growth of plants fed NH_4-N compared to those given NO_3-N in a moderately low-P soil was explained by the marked inhibition of infection by NH_4-N, causing plants to be P deficient in the mycorrhizal but not in plus-P treatments. These results suggest that with other considerations being equal, mycotrophic field crops should be fed nitrate rather than ammonium fertilizers.

The data of Hayman and Wang[65] showing a correlation ($r = 0.81$) between root carbohydrate levels and early mycorrhizal infection in onion seedlings suggest that the "carbohydrate theory" of Bjorkman[66] for ectomycorrhizas also applies to VA mycorrhizas. Root carbohydrate levels were lowered both by reducing the light intensity and by applying ammonium, the latter probably forming amino derivatives of root sugars which were translocated to the shoot, leaving less free carbohydrate for the fungus to utilize. This carbohydrate theory can also partly explain observations that many C-4 plants (e.g., maize, sorghum, and sugar cane) which photosynthesize more efficiently than C-3 plants, are often strongly mycotrophic.[67] It is probably a factor in the special importance of VAM in tropical field crop systems with C-4 plants growing under high-light intensities.

C. Infection and Growth

The amount of VA mycorrhizal infection present is not necessarily directly related to the size of the plant growth response. Fifty percent infection is likely to produce just as much growth improvement as 80% infection. On the other hand, less than 20% infection probably has little effect, but more precise relationships cannot be made. Furthermore, it has been shown that differences between endophytes in their effects on plant growth cannot be predicted simply by measuring their infectivity. For example, Clarke and Mosse[68] found that *Glomus caledonium* increased the yield of field-grown barley more than did *G. fasciculatum* "E3", but it infected less of the root system. Probably differences in external mycelium are important,[69,70] but it is difficult to recover this mycelium from soil and measure it, and even more difficult to estimate its distance from the root and, hence, the volume of soil it is exploiting. In addition, the density of infection or amount of arbuscular development may be more significant than just the percentage of root length infected.

Estimates of VAM infection in the roots are often made when plants are harvested, but it is the amount of early infection that is physiologically important, i.e., when plant demand for P is greatest. In some annual crops in temperate agriculture (e.g., cereals) the establishment of useful levels of infection by indigenous fungi may be delayed until after the main uptake period for P is passed,[13,27] although Jakobsen and Nielsen[47] reported that a plateau of about 50% infection had been reached in spring-sown cereals only 15 days after seedling emergence. Soil temperature exerts a major influence on this, which suggests that cold-tolerant endophytes should be selected where inoculants are to be used in temperate regions, screened for their infectivity at, say, 10°C as well as for their effectiveness at enhancing P uptake at subsequently warmer temperatures.

Another instance of VAM infection levels not reflecting potential growth benefits to the crop is where certain fungicides have been applied. For example, Boatman et al.[71] showed that where benomyl had been applied to soil as a drench, it inhibited the functioning rather than the amount of infection so that plant growth was poor under conditions of P stress. Thus, some plant protection measures against disease organisms may be counterproductive in a field crop which would otherwise derive some benefit from its mycorrhiza. As different VAM endophytes vary in their sensitivity to particular fungicides,[72] fungicide-tolerant ones might be utilized where pesticides are an essential component of the system.

D. Other Factors

There is some evidence that mycorrhiza increases a plant's tolerance of drought. Irrespective of whether this is due to effects on stomatal regulation,[73] nutritional status,[74] movement of phosphate through dry soil,[75] or to actual hyphal transport,[76] it certainly points to another benefit of mycorrhiza in water-deficient field crop systems. The improved ability of seedlings to survive transplanting injury when mycorrhizal is attributed to improved water absorption[77] and may be a useful feature in nurseries for plantation crops.

The overcoming of first-year dormancy of cuttings by VAM inoculation may be a hormonal effect, e.g., in yellow poplar.[78] Hormonal effects may also explain why some plants, e.g., sweetgum *(Liquidambar styraciflua)*[79] and hoop pine *(Araucaria cunninghamii)*,[33] grow better with VA mycorrhiza than with any amount of fertilizer. This suggests a further role of mycorrhiza in tree crops.

Although it is still not clear why some plants are dependent on mycorrhiza while others are not, the former are generally those with coarse root systems and few root hairs.[80] This provides a guide for selecting crop plants to which most effort should be directed in practical mycorrhizal research. Furthermore, symbiotic interactions that reveal some host-endophyte specificity and even more soil-endophyte specificity show there is scope for matching specific mycorrhizas to specific agricultural systems.

IV. INDIGENOUS VA MYCORRHIZAS AND CROP GROWTH

A. General

It is important to identify field situations where the indigenous VAM fungi might be making a substantial contribution to crop productivity. The quantity of effective native strains present will influence decisions on which cultivation procedures will least disrupt their existing ecophysiological habitats and whether the benefits from inoculants superimposed on the native flora will be sufficiently large to make inoculation worthwhile.

The major problem in trying to assess the role of the native VAM population in field crop systems is that it is usually impossible to have mycorrhiza-free controls for comparison. Soil fumigation or other measures to eliminate or suppress the native endophytes also produce changes in the microflora and chemical components of the system. Testing sterilized field soil in pots with and without the reintroduction of native endophytes can provide some indication of their efficiency, but experiments in pots do not necessarily reflect what happens in the field.[5,81] Nevertheless, there is considerable circumstantial evidence that indigenous endophytes can increase yields in the field of some temperate crops by up to around 50% and some tropical crops by several hundred percent.

B. Crops Benefiting from Native Endophytes

Cassava is frequently grown on land in the tropics which is too P deficient to support any other crop. Paradoxically, however, experiments in nutrient solutions show that cassava can grow well only if its high demand for phosphate is satisfied.[82] It has been shown that when a P-deficient soil in which it grows well was fumigated, it grew poorly, whereas under the same conditions soil fumigation was beneficial to Chinese cabbage, a nonmycorrhizal species.[83] These observations suggest that cassava is an extremely mycotrophic crop which grows well in the field, only because it becomes heavily infected by the native VAM fungi.[84]

Stylosanthes spp. are known to thrive in low-phosphate tropical pastures, yet in experiments in pots they can grow poorly in soils containing up to 40 $\mu g/g$ $NaHCO_3$-soluble P (a level which is not deficient for most crops) unless inoculated with VA mycorrhizal fungi.[67] In the field they are heavily mycorrhizal. From tests with 11 tropical soils in pots, Mosse[85] deduced that mycorrhizal inoculum density in the soil, rather than soil P status, was the major determinant of growth responses by *Stylosanthes* to VAM inoculation which were greatest in soils containing few indigenous endophytes.

Rich and Bird[86] reported that VA mycorrhizas are widely associated with cotton in Georgia. They observed VAM infection in cotton in the field 5 days after seedling emergence, and extensive infection (about 70% of the root system) by late August. The logarithmic phase of infection spread occurred between 5 and 25 days after seedling emergence, and there was a significant positive correlation between early-season mycorrhizal infection and vegetative growth and development of the cotton crop. With inoculation experiments in pots, *Endogone (Gigaspora) calospora* stimulated earlier flowering and boll maturation. A further benefit of mycorrhiza was that plants with most VAM infection had the smallest root populations of nematodes.

Sometimes there is an unexpected decrease in crop growth following a cultivation treatment meant to improve growth, and disruption of the indigenous mycorrhiza is inferred. For example, in recent field trials screening new genetic lines of peas in England, treatment of a field with basamid (dazomet) and telone to kill virus-transmitting nematodes resulted in poor growth of semileafless peas compared to pea plants in adjacent untreated soil. This effect was later attributed to elimination of the indigenous mycorrhizal fungi (mainly *Glomus mosseae),* because a natural VAM infection of about 70% had disappeared and subsequent inoculation trials with *G. mosseae* at that site showed marked growth effects; after 9 1/2 weeks dry weights of mycorrhizal plants ranged from 2.2 to 2.6 g/plant, respectively, whereas control plants were *circa* 1 g each and plants given P at 75 and 150 kg/ha averaged 1.1 and 1.8 g, respectively. Plants given nitrogen or *Rhizobium* in the absence of mycorrhiza or phosphate weighed only *circa* 1 g. It was concluded that at that site, where $NaHCO_3$-soluble P was *circa* 20 μg/g soil, semileafless peas depended on the native mycorrhiza for satisfactory growth.[87] In a subsequent larger trial using leafless peas, a higher planting density and five levels of triple superphosphate (0 to 145 kg P per hectare), treatment effects were much less: there was no clear-cut response to P, although plants inoculated with *G. mosseae* were around 10% heavier at nearly all P levels.[88] This reduced response to inoculation was partly explained by observations of some reinvasion by indigenous endophytes, by results of experiments in pots of responses to mycorrhiza by semileafless and conventional but not leafless peas,[139] and by high planting densities annulling VAM effects.[89]

Another example of decreased crop growth where cultivation treatment depressed the native mycorrhizal population occurred with a crop rotation of maize (a strongly mycorrhizal plant) after oil-seed rape (a nonmycorrhizal plant).[90] Maize after rape (RM) grew poorly with marked P-deficiency symptoms, whereas maize after maize (MM) grew well. Root samples collected in early summer showed 12% mycorrhizal infection in maize after rape, in contrast to 71% in maize after maize. Also, leaf P content of MM plants was double that of RM plants. Attempts were made to prove a mycorrhizal effect in different rotations at that site the following year, but maize showed no clear-cut response to inoculation. This was attributed to the widespread development of indigenous mycorrhizal infection (>50% of root length) throughout all plots, inoculated or not, which could have masked any benefits from inoculation. Thus, in cases like this, a beneficial effect of indigenous VAM fungi is highly probable, albeit not strictly proven.

Indigenous mycorrhizal fungi in a field can sometimes be so effective and spread so rapidly that inoculation treatments give no result. Warner et al.[91] sowed barley on a reclaimed gravel pit covered with topsoil that had been stored for 12 years and was believed to be virtually mycorrhiza-free, and the extensive mycorrhizal infection present throughout the crop (40 to 70% of root length) in the summer following autumn sowing was assumed to come from a small residual population or by invasion from surrounding fields. Similar results were found the same year in maize and lucerne at that site.[92]

White clover has been shown to establish and grow better in some upland pastures when inoculated with mycorrhizal fungi.[32,36,93-95] However, these effects have not occurred at some sites even where levels of soil P were similar to those at the responsive sites. One such pasture was examined in detail,[96] and the lack of growth response to either phosphate fertilizer or VAM inoculum despite low levels of soil P was attributed to a particularly efficient indigenous mycorrhizal population. Noninoculated control plants were well infected by mycorrhizal fungi in the field, and experiments with clover grown in pots in a glasshouse with peaty soil from this nonresponsive site and from three responsive sites showed considerable differences in the efficiency of the native endophytes. When the peat from all four sites was sterilized (γ-irradiated), clover growth was poor. It was better in unsterile peat from the three responsive sites, and much better in unsterile peat from the nonresponsive site. Cross-inoculation with endophytes from the nonresponsive site into unsterile peat from one of the responsive sites produced a growth enhancement of about 500%. These results indicate that clover growth in the field cannot always be predicted from analyses of soil P because of the indigenous mycorrhizas. A similar conclusion was reached for leeks grown in a soil range of different native P levels.[97] Powell[98] screened the native endophytes from a range of upland sites in New Zealand and found that most improved growth, although they were significantly better than introduced inoculants in only 4 cases out of 37.

The importance of native mycorrhizal endophytes is well-known from studies in fumigated field systems, especially in horticulture (see Chapter 10). It has also been shown with vegetable crops (e.g., pepper and onion) grown in fumigated fields in Israel.[99] In the latter case indigenous mycorrhizal populations are minimal because of the frequent fumigations needed to kill pathogens which would otherwise build up to unacceptable levels when two or three crops are grown successively in the same season.

Buwalda et al.[100] grew barley and wheat with and without VAM inoculation or P fertilizer in fumigated and nonfumigated field plots. They attempted to assess the value of both indigenous and introduced VAM fungi and concluded that inoculum density of indigenous endophytes was inversely related to growth responses of the crop to introduced inoculants. Reinvasion of fumigated plots by indigenous VAM fungi probably accounted for the decrease of residual effects of inoculation over 2 to 3 years in soils subject to continuous cropping with cereals. They also concluded that natural mycorrhizal infection is important for uptake of P by cereals in the field, but to a lesser extent than for those crops that are particularly dependent on mycorrhiza in arable soils.

C. Conclusions

In summary, the mycorrhizal condition is the natural state for most plants and therefore should not be ignored in field crop systems. There is much evidence to indicate that many crops derive some benefit from indigenous mycorrhizal fungi. The benefit is likely to be small for some crops, but large for others which have a high degree of mycotrophy and are grown in soils of low P status but effective VAM populations. In the latter case, it is clear that possible deleterious side effects on the native VAM fungi of certain cultivation methods, including the use of large amounts of fertilizer and pesticide, should be assessed. When the native population is too small or inefficient to counter P deficiency in a mycotrophic field crop, then inoculation with selected VAM fungi should be considered.

V. CROP RESPONSES TO FIELD INOCULATION

A. General

Inoculation of field crops with mycorrhizal fungi is feasible under certain ecophysiological and economic conditions. Therefore, cropping systems that meet these conditions need to be determined, and most research and development work should be focused there, rather than on efforts to apply mycorrhiza too universally.

Plenchette[101] presented a formula for relative mycorrhizal dependency. Thus, wheat in soil containing 100 μg/g soluble P has zero dependence on mycorrhiza, i.e., $DMRC^{100} = 0\%$ (DMRC = dry weight of mycorrhizal plant minus dry weight of nonmycorrhizal control as a percentage). For carrots in the same soil he gave $DMRC^{100} = 99\%$. These figures are calculated from experiments in pots. To be more applicable to field conditions, an additional figure could be obtained for natural infectivity vs. infectivity of an introduced endophyte. Tests should be done under temperature and light conditions similar to those at sowing time and the time to reach, say, 50% infection in unsterile soil determined. This assumes the native population is symbiotically effective provided its inoculum density is high enough (that could be tested separately in longer experiments). The infectivity figure could be given as time (days) taken for natural infection to reach 50% divided by time taken for the inoculant to achieve that level, i.e., $T = T_n/T_i$. Provided T_i is less than time to reach flowering, then a figure of say $T \geqslant 2$ (this would need to be calibrated in a preliminary test) could be taken as recommending inoculation if the DMRC figure is also high.

B. Cereals

Cereals have been the subject of a number of inoculation trials in small field plots, mainly in arable soils containing very little soluble P. Khan[102] inoculated wheat seedlings with *Glomus mosseae* prior to transplanting to a field deficient in phosphate and native VAM. The inoculant fungus more than doubled shoot dry weight, a bigger effect than obtained by adding 50 kg P per hectare. Very similar results were also reported for maize[102] and barley.[103]

Owusu-Bennoah and Mosse[104] showed a 30% increase in shoot growth of barley following inoculation in a field containing only around 10 μg/g $NaHCO_3$-soluble P. In the same field, Clarke and Mosse[68] obtained a twofold increase in barley ear growth (fresh weight) by inoculation; adding 82.6 kg P per hectare (raising the soluble P to about 40 μg/g soil) increased ear growth 5.5 times without inoculation, four times in the presence of *G. mosseae* and *G. fasciculatum* "E3" inoculum, and 6.5 times with *G. caledonium*. In another P-deficient field on the same farm, Buwalda et al.[100] reported that inoculation increased grain yields from 11 to 37%, dropping to 3 to 16% when 60 kg P per hectare (giving 20 μg/g $NaHCO_3$-soluble P) was added; without inoculation, 60 kg P per hectare increased yield by 14 to 102%. In similar plots fumigated to kill the native mycorrhizal fungi, responses to inoculation were higher, viz. 38 to 46% and 17 to 25% at 0 and 60 kg P per hectare, respectively. They concluded that inoculation of cereals on soils containing small VAM populations can increase grain yield over a fairly wide range of phosphate applications.

Jakobsen[105] emphasized the relevance of field studies on barley, because apart from being widely cultivated in northern Europe, it is usually strongly infected by VAM fungi. He suggested that indirect information on mycorrhizal effects can be obtained by comparing growth and P uptake of plants in fumigated and nonfumigated plots, and by comparing actual plant growth with that expected from the soluble P content of the soil. Direct evidence can be obtained by inoculation in fields containing low native VAM populations resulting from earlier fumigation or fallowing. Thus, in fumigated field plots, Jakobsen[105] found that final dry matter production was increased by an average of 20% by inoculation and 40% by soil P. The final P uptake, by contrast, was increased 100% by inoculation but only 50% by soil P. P uptake during the period from earing to maturity was negligible in uninoculated plots. It was concluded that VAM may play a fundamental role in P uptake during the reproductive phase of barley under certain field conditions.

Black and Tinker[12] increased potato yields by 20% in an infertile field which contained few VAM spores after having been fallow for 2 years. For inoculum they used highly infective soil from an adjacent area growing barley. Adding superphosphate at 82 kg P per hectare increased yields of the uninoculated but not inoculated plants. However, potatoes may not be a suitable crop for commercial inoculation because Stribley[106] suggested that the standard practice of growing the potato crop from whole tubers acts against mycorrhizal infection; in experiments in pots, potato plants became heavily infected in unsterile soil only when grown from true seed or small fragments of tuber bearing a bud.

C. Legumes

Compared to cereals, lucerne (alfalfa) and other forage legumes are particularly responsive to mycorrhiza. For example, a fourfold increase in the growth of lucerne from inoculation with *G. caledonium* was observed in a low-phosphate field where barley was much less responsive to mycorrhiza.[104] Also, a large response by lucerne to inoculation with *G. mosseae* was observed in southern Spain in a field cultivated according to standard practice.[107] In this instance, growth of *Glomus*-inoculated plants was further increased by inoculation with *Rhizobium*, whereas growth of controls without *Glomus* was not, suggesting a positive interaction between symbionts.

In an experiment with red clover in a field containing only 10 μg $NaHCO_3$-soluble P per gram soil, plants showed an early response to superphosphate, but by the end of the second year yields were high in all plots, equivalent to around 15 t/ha dry matter.[108] This result was attributed to one of the introduced endophytes, *G. caledonium,* which had spread and sporulated profusely throughout all the plots (including those inoculated with two other endophytes) and had previously enhanced growth of lucerne at this site. A significant aspect of this experiment is that the ability of VAM fungi to spread and infect in the field may be as important a criterion as efficiency in stimulating plant growth when selecting inoculants. It also suggests that high density inoculum placement in perennial crops may be unnecessary if a suitable endophyte can spread rapidly from scattered infection centers.

White clover is used extensively in temperate regions as a forage legume to supply fixed nitrogen to pasture grasses and protein and minerals to grazing animals. Recent experiments in the agriculturally marginal upland grasslands of Britain and New Zealand have shown that better establishment and growth of white clover in these areas can be achieved by inoculation with selected VAM endophytes. In a series of tests by Powell[93,94] in New Zealand, inoculation with *G. fasciculatum* "E3" produced 50% more growth on a cold south-facing slope, but 42% less growth on a warm north-facing slope; *G. tenuis* and *Gigaspora margarita* increased clover growth by 80% in other experiments, although certain other endophytes were less effective than some of the indigenous ones.

In upland pastures in Wales, Hayman and Mosse[36] found that inoculation of white clover seedlings with a combination of *Glomus mosseae* and *G. fasciculatum* "E3" in field plots given the standard dressing of 90 kg P per hectare as basic slag doubled plant growth and greatly enhanced tissue P content and nodulation. In some trials, however, small uninoculated plants had high P contents, which suggests mycorrhiza was contributing more than just extra P. Growth responses at other sites varied from large to slightly negative, probably governed in part by the effectiveness of the indigenous VAM population.[96] When pellets containing seeds and inoculum were used in a later experiment in these uplands, clover plants grew much better with a combined inoculum *(G. mosseae, G. fasciculatum* "E3", and *G. caledonium)* and superphosphate (90 kg P per hectare) than with superphosphate alone, viz. 954 and 212 kg dm/ha, respectively.[32] The inoculated plots contained about 30% clover in the grass-clover sward within the first growing season. These effects persisted fairly strongly into the third growing season. Nevertheless, no growth response to VAM inoculation occurred in the establishment year at two upland sites in Scotland. By the following year inoculation had not increased growth at a deep peat site, but had increased the amount of shoot dry matter by up to twofold in the presence of 40 kg P per hectare at a brown earth site.[95] The authors concluded that VAM inoculation can have a practical use once responses are made more predictable.

Lotus pedunculatus, which is less mycotrophic than white clover, established slightly better when sown with mycorrhizal than with nonmycorrhizal soil pellets on an eroded hill site.[109]

Cowpea has proven to be mycorrhiza responsive in tropical field crop systems. In the Amazon region of Brazil, La Torraca[110] found that mycorrhizal inoculation increased yield, especially when superphosphate was also added. Similarly, in Nigeria, Islam et al.,[111] who transplanted cowpea seedlings with and without *G. fasciculatum* to the field, observed most growth in crops given both *Glomus* and rock phosphate. This followed earlier experiments in which Islam concluded that rock phosphate might be a better long-term source of phosphate than triple superphosphate, because the indigenous VAM fungi were actively involved in P uptake, and superphosphate, but not rock phosphate, suppressed them. This suppression would create greater crop dependency on superphosphate fertilizer in subsequent years, making it less economical than rock phosphate, quite apart from manufacturing and transport expenses.

Soybeans, which have shown responses to mycorrhiza in a fumigated soil in pots,[112] may be less responsive than cowpeas in the field. For example, soybeans did not respond to field inoculation with *G. fasciculatum* in a nonsterilized field in India.[113] By contrast, Ross and Harper[114] obtained yield increases of 29% and zero in two fumigated fields in North Carolina. Also, in the U.S., Schenck and Hinson[115] found that VAM increased growth of a nodulating soybean line by 53%, but had no significant effect on a nonnodulating isoline where both were grown in fumigated field plots. An interesting aspect of this work is that it suggests that VAM may have indirectly enhanced N_2-fixation in the nodulating line because phosphate was not limiting growth in this soil.

The dual effect of mycorrhiza on both P and N nutrition of legumes could have great practical potential, as it has been shown in controlled experiments with a number of common species, e.g., *Stylosanthes guyanensis,*[116] lucerne,[117] and subterranean clover.[118] Also, the increased P content of plants arising from mycorrhizal infection is significant, because many cultured strains of *Rhizobium* used as inoculants are adapted to culture media containing 100 times more P than is typical of the soil solution[119] and can require at least 0.1% shoot P content to nodulate.[116] On the other hand, some indigenous rhizobia can nodulate at lower concentrations of P in soil to which they are adapted.[85] Another important feature of legumes is that mycorrhizal infection can lead to increased shoot and seed N contents.[112,115] This is of fundamental significance to human nutrition and underlines the need for further studies of mycorrhizal effects on protein levels in legumes grown in field systems.

D. Cassava

As mentioned earlier (Section III), cassava did not respond to phosphate fertilizer in unsterile field plots where it was heavily mycorrhizal, but it did respond in plots that had been fumigated. This indicates a need for inoculation where field-grown mycotrophic plants do not become strongly infected naturally.

E. Woody Plants

A number of woody species (e.g., citrus, peach, apple, sweetgum, and yellow poplar) are very mycorrhiza responsive under some field conditions. Some species have shown considerable benefits from VAM, especially in horticultural systems. Some plantation crops grown in containers before outplanting are improved by mycorrhiza, e.g., coffee.[120] This system could well be expanded in the tropics for other crops similarly grown in plantations.

F. Vegetables

Other field crops where mycorrhiza may be of value are vegetables. Often these have been shown to be very responsive to VAM in experiments in pots, e.g., onions, leeks, celery, asparagus, carrots, cucurbits.[101,121,122] Some are currently being tested in the field. When attempting to utilize mycorrhiza in these crops under field conditions, it now seems that soil water conditions as well as soil P should be controlled. Thus, with field-grown onions inoculated with *G. etunicatus,* high mycorrhizal colonization only occurred when soluble soil P contents were 15 to 20 $\mu g/m\ell$, rising to 30 $\mu g/m\ell$ with a decrease in water availability.[123] Finally, with some crops, shoot growth may not always be in line with yields in terms of response to inoculation. Tomato plants, for example, grew slightly worse when mycorrhizal, but the yield of fruit within the growing period was better than for nonmycorrhizal plants.[124]

VI. FIELD INOCULATION AND INOCULUM TECHNOLOGY

The difficulties involved in producing and applying VAM inoculum in field crop systems preclude any easy usage of mycorrhiza, even where growth benefits are likely. Unlike low-density planting systems, a field of arable crops may require vast quantities of inoculum to achieve adequate infection, one calculation giving 2 to 3 t/ha.[104] Therefore, present studies are directed towards ways of producing inoculum in bulk and better placement of inoculum to reduce the amount needed.

A. Inoculation Methods

Of the various techniques devised for introducing VAM inoculum into a field-grown crop, the use of preinoculated transplants is one of the simplest if it is appropriate for the cropping system in question. A vigorous VAM infection can be established in a seedling well before it is transplanted and exposed to indigenous endophytes of unknown effectiveness. On a practical scale this could be useful for many agronomically important tropical trees, e.g., coffee, rubber, cacao, papaya, oil palm, and also temperate trees such as *Liquidambar styraciflua* (which seems to be obligately mycorrhizal in tree nurseries)[79] and *Liriodendron tulipifera* (vegetatively propagated cuttings that mostly stayed dormant unless made mycorrhizal).[78] Also, a substantial proportion of some vegetable species (e.g., onions and leeks) are transplanted as seedlings from compartmentalized trays; VAM inoculum could be added to the growth medium in the compartment.[122]

Soil inoculum from pot cultures of stock plants contains spores, hyphae, and infected root pieces. This can be incorporated into seed furrows in the field and has been successful in both fumigated and unsterile field plots. Layering inoculum under the seed has proven superior to seed inoculation or banding of the inoculum.[125] Unfortunately, this type of inoculum seems too crude and bulky for large-scale use (cf. calculation above of 2 to 3 t/ha).

Soil inoculum can be concentrated to about one seventh by wet sieving, suspended with germinated seeds in a viscous medium such as 4% methyl cellulose, and applied as a slurry to seed furrows. This method, known as fluid drilling, has been successful with field-grown red clover in soil containing a mixed population of indigenous endophytes.[126] Advantages of this method are that the mycorrhizal inoculum does not significantly increase the total volume of material being drilled, and other inoculants (e.g., *Rhizobium*) can also be incorporated in the fluid.

Adhesives such as methyl cellulose have been used to coat seeds with bacterial inoculants. However, this method is difficult for the much larger propagules of VAM inoculants. Hence, although it has achieved some success with large-seeded crops (e.g., citrus) it is impractical for small seeds, e.g., clover. Therefore, the possibility of incorporating small seeds into a matrix such as lignite or soil-peat-sand mixtures with a suitable binding agent has been tested.[32] The use of such multiseeded pellets is a practical proposition and its large-scale potential is currently being tested.

Increasing the density of mycorrhizal propagules *in situ* can be attempted by growing a heavily mycorrhizal crop in the field in a rotation system. In addition, the top soil can be removed and used elsewhere as crude inoculum,[12] but this requires care because it introduces a new microflora which may contain plant pathogens.

B. Inoculum Production

The large quantities of inoculum needed for field inoculation are usually raised in pot cultures of mycorrhizal plants. Inoculum can be supplied in the form of infective soil, infected roots, and soil sievings, and perhaps one day as axenic cultures. Infected roots grown in NFT cultures (nutrient film technique) are cleaner and less bulky than soil inoculum, but at present there are problems with storage, viability, and longevity, even though they can infect satisfactorily in the field.[127] A special case is *Glomus epigaeum* which produces sporocarps on the soil surface that can be scraped off and used as concentrated inoculum.[128] Methods for storing inoculum (e.g., as refrigerated sievings or lyophilized roots) need further investigation.

With open pot cultures there is a risk of contamination by harmful fungi and nematodes. This risk might be reduced by the development of specific pesticides to apply to the cultures. Also, to prevent carry-over of host-specific plant pathogens, inoculum should be raised on a host species not closely related to the crop to be inoculated in the field, as described by Menge et al.,[129] who used sudan grass as the stock plant and citrus as the field crop. This is one way in which we can take advantage of the generally low host-endophyte specificity in VAM symbioses.

Deliberate contamination of VAM inoculum with microorganisms that are beneficial to plants could be a useful innovation. As the stock cultures are raised in sterilized soil, selected bacteria could be introduced and established before invasion and competition from aerial contaminants become intense. Bacteria which dissolve insoluble phosphates, produce plant growth-promoting compounds, or are antagonistic to specific pathogens might be introduced in this way. Strains of *Rhizobium* can build up in the rhizospheres of mycorrhizal maize or other nonlegumes, thereby forming an integral part of VAM inoculum being raised for inoculation of legumes.

It was shown that microorganisms attached to VAM propagules can have a positive effect on plants, washings ("leachings") from the inoculum increasing plant growth by about 10% over controls which received nothing.[130] Even so, many bacteria may well be so closely bound to the VAM fungi that they are not readily removed by washing. They could readily exist as microcolonies in niches on or in the walls of the large VAM resting spores. This casts some doubt on the accuracy of controls given washings, because microorganisms still remaining with the inoculum may slightly enhance the mycorrhizal effect. Indeed, this may be a factor where curves for growth of mycorrhizal and "control" plants are parallel, but with the mycorrhizal plants some 10% bigger irrespective of whether growth is plotted against increasing P[79,88,101] or increasing soil sterilization,[131] for example. Whether this is true or not, it shows there is scope for establishing selected bacteria in the hyphospheres or sporospheres of VAM fungi to improve inoculum effectiveness.

VII. CONCLUSIONS

As the world reserves of utilizable phosphate are finite, and energy costs continue to escalate, it is clear that more efficient use of applied fertilizer is required for sustainable agricultural systems. From the evidence presented above, it can be seen that one approach to achieving this is to exploit the ability of VA mycorrhizal fungi to maintain acceptable crop yields with lower inputs of high-grade superphosphates or with less costly low-grade fertilizers like rock phosphate. There are parallels with the use of *Rhizobium* inoculants for nitrogen input, but a key difference is that with mycorrhiza there must be some nonmicrobial phosphorus input to prevent exhaustion of the native soil phosphate. Since chemical fertilizers are believed to account for up to one half of the current U.S. agricultural output,[132] economy in the use of P fertilizer without decreasing crop yields is probably where VA mycorrhizas can make their biggest contribution to field crop systems. In addition, secondary effects such as enhanced uptake of trace elements and hormonal effects on crop physiology should be harnessed, also the ability of VAM fungi to increase resistance to drought and to improve aggregation of soil particles.

Although few plants in a field are likely to be totally free of mycorrhiza, natural mycorrhizal development can be sparse at the critical early stages of plant growth when an adequate supply of P is essential. Inoculation can reverse this situation by facilitating early infection, which could be vital for an annual crop. Therefore in soils of low infectivity, low P status, and planted with a mycotrophic crop, inoculation with selected endophytes should be considered. Where indigenous infectivity is high but symbiotically inefficient, native endophytes could be suppressed chemically to reduce competition with the introduced inoculants. There is an obvious need for continuing empirical investigations under realistic agronomic conditions to provide information relevant to identifying potential inoculation situations more readily, and for more fundamental studies to develop underlying concepts and principles.

Protection against disease organisms is another valuable feature of mycorrhiza under certain conditions. In a recent review, Dehne[133] listed 53% of cases where VAM decreased disease, 19% where it increased it, and 28% where there was no effect. Although these results are not consistent enough to be adapted commercially on a broad scale, they could be important in systems where VAM is well established in the root before exposure to unsterile field soils, e.g., in preinoculated transplants and nursery plants growing in sterilized compost. For most field crops, however, these benefits are likely to be subordinate to VAM-enhanced P uptake.

To establish selected VAM fungi in the field, it is important to consider not only their means of introduction, but also their survival and spread in competition with the native ones. Hence, competitiveness should be tested under controlled conditions as part of a screening program to select efficient endophytes. For cropping systems in temperate regions, competitiveness and ability to infect rapidly should be examined at soil temperatures that relate to those in the field at sowing time, instead of the usual procedure of screening endophytes at glasshouse temperatures of over 20°C. It is essential to know whether inoculation can bring forward the time when enough infection is present to affect early crop growth. Inoculum potential of the inoculant will be related to the number of propagules per unit volume and to their infective vigor. The latter will be influenced by the form of inoculum (e.g., infected roots vs. spores[134]) and its nutritional status (perhaps indicated by lipid content) as affected by the physiology and even, perhaps, the species of stock plant.

As high temperatures generally favor VAM establishment in roots, VA mycorrhizas should be more active in tropical than temperate field crop systems. This, together with the widespread cultivation of mycorrhiza-dependent crops, P-deficient soils, and limited fertilizer resources in the tropics, suggests that VA mycorrhizas may be of special significance in tropical agriculture.

In view of the considerable specificity between endophyte and soil, and some specificity between endophyte and host plant in terms of growth effects, the use of mixed inocula may give more consistent results.[135] The survival and compatibility of strains or species in such inocula require further study.

The importance of developing more simple techniques than those now in common use to identify desirable inoculant strains is stressed by Abbott and Robson.[136] They emphasize the need for continual basic studies on the life cycles and biology of a wide range of VAM fungi to define more clearly fungal characteristics associated with extensive colonization of roots and persistence in soil. Better techniques are also needed for monitoring the fate of inoculant fungi in the field. The ecology of both indigenous and introduced VAM fungi in field crop systems has been rather neglected. Their taxonomy is still open to improvement and their genetics are virtually unknown. Hence, reducing these gaps in our knowledge should help make our attempts to harness the VAM symbiosis less empirical and more precise, such that large-scale recommendations can be made for specific, clearly identified situations. As Menge[137] stated, the potential for using VAM fungi on a broad scale in agriculture largely depends on the development of crop growth-promoting strains of VAM that are superior to indigenous soil populations of VAM fungi. The introduction of novel genes into extant strains is one way in which this may be achieved in the future.[138] Although not a general panacea for improving crop growth in infertile field soils, the contribution of VA mycorrhiza cannot be ignored. Therefore, manipulation of this symbiotic association to attain its full ecophysiological and economic potential in field crop systems is an important goal of present and future studies.

REFERENCES

1. Butler, E. J., The occurrences and systematic position of the vesicular-arbuscular type of mycorrhizal fungi, *Trans. Br. Mycol. Soc.*, 22, 274, 1939.
2. Mosse, B., Vesicular-arbuscular mycorrhiza research for tropical agriculture, *Research Bulletin 194, Hawaii Institute of Tropical Agriculture and Human Resources*, 1981, 82 pp.
3. Mosse, B., Stribley, D. P., and LeTacon, F., Ecology of mycorrhizae and mycorrhizal fungi, *Adv. Microb. Ecol.*, 5, 137, 1981.
4. Hayman, D. S., Influence of soils and fertility on activity and survival of vesicular-arbuscular mycorrhizal fungi, *Phytopathology*, 72, 1119, 1982.
5. Tinker, P. B., Mycorrhizas: the present position, in *Whither Soil Research*, Trans. 12th Int. Congr. of Soil Science, New Delhi, India, ISSS/AISS/IBG, New Delhi, 1982.
6. Harley, J. L. and Smith, S. E., *Mycorrhizal Symbiosis*, Academic Press, London, 1983, 483 pp.
7. Trappe, J. M., Synoptic keys to the genera and species of zygomycetous mycorrhizal fungi, *Phytopathology*, 72, 1102, 1982.
8. Azcón, R. and Ocampo, J. A., Factors affecting the vesicular-arbuscular infection and mycorrhizal dependency of thirteen wheat cultivars, *New Phytol.*, 87, 677, 1981.
9. Lambert, D. H., Cole, H., and Baker, D. E., Variation in the response of alfalfa clones and cultivars to mycorrhizae and phosphorus, *Crop Sci.*, 20, 615, 1980.
10. Sutton, J. C., Development of vesicular-arbuscular mycorrhizae in crop plants, *Can. J. Bot.*, 51, 2487, 1973.
11. Iqbal, S. H. and Qureshi, K. S., The influence of mixed sowing (cereals and crucifers) and crop rotation on the development of mycorrhiza and subsequent growth of crops under field conditions, *Biologia (Pakistan)*, 22, 287, 1976.
12. Black, R. L. B. and Tinker, P. B., Interaction between effects of vesicular-arbuscular mycorrhiza and fertilizer phosphorus on yields of potatoes in the field, *Nature (London)*, 267, 510, 1977.
13. Black, R. and Tinker, P. B., The development of endomycorrhizal root systems. II.Effect of agronomic factors and soil conditions on the development of vesicular-arbuscular mycorrhizal infection in barley and on the endophyte spore density, *New Phytol.*, 83, 401, 1979.
14. Sanders, F. E. and Hayman, D. S., The agricultural importance of vesicular-arbuscular mycorrhiza, in *Advances in Agriculture*, Hayes, W. A., Ed., University of Aston, Birmingham, U.K., 1978, 123.
15. McGraw, A. C. and Hendrix, J. W., Host and soil fumigation effects on spore population densities of species of endogonaceous mycorrhizal fungi, *Mycologia*, 76, 122, 1984.
16. Kruckelmann, H. W., Effects of fertilizers, soils, soil tillage, and plant species on the frequency of *Endogone* chlamydospores and mycorrhizal infection in arable soils, in *Endomycorrhizas*, Sanders, F. E., Mosse, B., and Tinker, P. B., Eds., Academic Press, London, 1975, 511.
17. Ocampo, J. A., Martin, J., and Hayman, D. S., Influence of plant interactions on vesicular-arbuscular mycorrhizal infections. I. Host and non-host plants grown together, *New Phytol.*, 84, 27, 1980.
18. Schenck, N. C. and Kinloch, R. A., Incidence of mycorrhizal fungi on six field crops in monoculture on a newly cleared woodland site, *Mycologia*, 72, 445, 1980.
19. Bevege, D. I. and Bowen, G. D., *Endogone* strain and host plant differences in development of vesicular-arbuscular mycorrhizas, in *Endomycorrhizas*, Sanders, F. E., Mosse, B., and Tinker, P. B., Eds., Academic Press, London, 1975, 77.
20. Reed, H. and Frémont, T., Factors that influence the formation and development of mycorrhizal associations in citrus roots, *Phytopathology*, 25, 645, 1935.
21. Peyronel, B., Prime osservazioni sui rapporti tra luce e simbiosi micorrizica, *Annu. Lab. Chanousia Giardino Bot. Alpino Ordine Mauriziano Piccolo San Bernardo*, 4, 1, 1940.
22. Warner, A. and Mosse, B., Independent spread of vesicular-arbuscular mycorrhizal fungi in soil, *Trans. Br. Mycol. Soc.*, 74, 407, 1980.

23. Ruissen, M. A., The Development and Significance of Vesicular-Arbuscular Mycorrhizas as Influenced by Agricultural Practices, dissertation, University of Wageningen, The Netherlands, 1982.
24. Read, D. J., Koucheki, H. K., and Hodgson, J., Vesicular-arbuscular mycorrhiza in natural vegetation systems. I. The occurrence of infection, *New Phytol.,* 77, 641, 1976.
25. Tisdall, J. M. and Oades, J. M., Stabilization of soil aggregates by the root systems of ryegrass, *Aust. J. Soil Res.,* 17, 429, 1979.
26. Hayman, D. S., Johnson, A. M., and Ruddlesdin, I., The influence of phosphate and crop species on *Endogone* spores and vesicular-arbuscular mycorrhiza under field conditions, *Plant Soil,* 43, 489, 1975.
27. Hayman, D. S., *Endogone* spore numbers in soil and vesicular-arbuscular mycorrhiza in wheat as influenced by season and soil treatment, *Trans. Br. Mycol. Soc.,* 54, 53, 1970.
28. Ocampo, J. A. and Hayman, D. S., Effects of pesticides on mycorrhiza in field-grown barley, maize and potatoes, *Trans. Br. Mycol. Soc.,* 74, 413, 1980.
29. Porter, D. M. and Beute, M. K., *Endogone* species in roots of Virginia type peanuts, *Phytopathology,* 62(Abstr.), 783, 1972.
30. Toussoun, T. A., Nash, S. M., and Snyder, W. C., The effect of nitrogen sources and glucose on the pathogenesis of *Fusarium solani* f. *phaseoli, Phytopathology,* 50, 137, 1960.
31. Sparling, G. P. and Tinker, P. B., Mycorrhizal infection in pennine grassland. I. Levels of infection in the field, *J. Appl. Ecol.,* 15, 943, 1978.
32. Hayman, D. S., Improved establishment of white clover in hill grasslands by inoculation with mycorrhizal fungi, in *Forage Legumes,* Thomson, D. J., Ed., *British Grasslands Society,* Hurley, U.K., 1984.
33. Bevege, D. I., Vesicular-Arbuscular Mycorrhizas of *Araucaria:* Aspects of Their Ecology and Physiology and Role in Nitrogen Fixation, Ph.D. thesis, University of New England, Armidale, N.S.W., Australia, 1971.
34. Strzemska, J., Mycorrhiza in farm crops grown in monoculture, in *Endomycorrhizas,* Sanders, F. E., Mosse, B., and Tinker, P. B., Eds., Academic Press, London, 1975, 527.
35. Hayman, D. S., The occurrence of mycorrhiza in crops as affected by soil fertility, in *Endomycorrhizas,* Sanders, F. E., Mosse, B., and Tinker, P. B., Eds., Academic Press, London, 1975, 495.
36. Hayman, D. S. and Mosse, B., Improved growth of white clover in hill grasslands by mycorrhizal inoculation, *Ann. Appl. Biol.,* 93, 141, 1979.
37. Abbott, L. K. and Robson, A. D., The distribution and abundance of vesicular arbuscular endophytes in some Western Australian soils, *Aust. J. Bot.,* 25, 515, 1977.
38. Wang, G. M., Stribley, D. P., and Tinker, P. B., Soil pH and mycorrhizal fungi, *Rothamsted Rep. 1982,* Part 1, 269, 1983.
39. Khan, A. G., Occurrence of *Endogone* spores in West Pakistan soils, *Trans. Br. Mycol. Soc.,* 56, 217, 1971.
40. Hayman, D. S., Barea, J. M., and Azcón, R., Vesicular-arbuscular mycorrhiza in Southern Spain: its distribution in crops growing in soil of different fertility, *Phytopathol. Mediterranea,* 15, 1, 1976.
41. Bailey, J. E. and Safir, G. R., Effect of benomyl on soybean endomycorrhizae, *Phytopathology,* 68, 1810, 1978.
42. Menge, J. A., Effect of soil fumigants and fungicides on vesicular-arbuscular fungi, *Phytopathology,* 72, 1125, 1982.
43. Kruckelmann, H. W., Die vesikulär-arbuskuläre Mykorrhiza und ihre Beeinflussung in landwirtschaftlichen Kulturen, Dissertation, University of Braunschweig, West Germany, 1973.
44. Bird, G. W., Rich, J. R., and Glover, S. U., Increased endomycorrhizae of cotton roots in soil treated with nematicides, *Phytopathology,* 64, 48, 1974.
45. Pope, P. E. and Holt, H. A., Paraquat influences development and efficacy of the mycorrhizal fungus *Glomus fasciculatus, Can. J. Bot.,* 59, 518, 1981.
46. Mason, D. T., A survey of numbers of *Endogone* spores in soil cropped with barley, raspberry and strawberry, *Hortic. Res.,* 4, 98, 1964.
47. Jakobsen, I. and Nielsen, N. E., Vesicular-arbuscular mycorrhiza in field-grown crops. I. Mycorrhizal infection in cereals and peas at various times and soil depths, *New Phytol.,* 93, 401, 1983.
48. Saif, S. R., The influence of stage of host development on vesicular-arbuscular mycorrhizae and Endogonaceous spore population in field-grown vegetable crops. I. Summer-grown crops, *New Phytol.,* 79, 341, 1977.
49. Gildon, A. and Tinker, P. B., A heavy metal-tolerant strain of a mycorrhizal fungus, *Trans. Br. Mycol. Soc.,* 77, 648, 1981.
50. Porter, W. M., Abbott, L. K., and Robson, A. D., Effect of rate of application of superphosphate on populations of vesicular arbuscular endophytes, *Aust. J. Exp. Agric. Anim. Husb.,* 18, 573, 1978.
51. Hayman, D. S. and Stovold, G. E., Spore populations and infectivity of vesicular-arbuscular mycorrhizal fungi in New South Wales, *Aust. J. Bot.,* 27, 227, 1979.

52. Porter, W. M., The 'most probable number' method for enumerating infective propagules of vesicular arbuscular mycorrhizal fungi in soil, *Aust. J. Soil Res.*, 17, 515, 1979.
53. Hayman, D. S. and Mosse, B., Plant growth responses to vesicular-arbuscular mycorrhiza. III. Increased uptake of labile P from soil, *New Phytol.*, 71, 41, 1972.
54. Sanders, F. E. and Tinker, P. B., Phosphate flow into mycorrhizal roots, *Pestic. Sci.*, 4, 385, 1973.
55. Rhodes, L. H. and Gerdemann, J. W., Phosphate uptake zones of mycorrhizal and non-mycorrhizal onions, *New Phytol.*, 75, 555, 1975.
56. Sanders, F. E., The effect of foliar-applied phosphate on the mycorrhizal infections of onion roots, in *Endomycorrhizas*, Sanders, F. E., Mosse, B., and Tinker, P. B., Eds., Academic Press, London, 1975, 261.
57. Stribley, D. P., Tinker, P. B., and Rayner, J. H., Relation of internal phosphorus concentration and plant weight in plants infected by vesicular-arbuscular mycorrhizas, *New Phytol.*, 86, 261, 1980.
58. Mikola, P., Ed., *Tropical Mycorrhiza Research*, University Press, Oxford, 1980.
59. Gianinazzi-Pearson, V. and Diem, H. G., Endomycorrhizae in the tropics, in *Microbiology of Tropical Soils and Plant Productivity*, Dommergues, Y. R. and Diem, H. G., Eds., Nijhoff-Junk, The Hague, 1982, 209.
60. Crush, J. R., Endomycorrhizas and legume growth in some soils of the Mackenzie Basin, Canterbury, New Zealand, *N.Z. J. Agric. Res.*, 19, 473, 1976.
61. Jensen, A. and Jakobsen, I., The occurrence of vesicular-arbuscular mycorrhiza in barley and wheat grown in some Danish soils with different fertilizer treatments, *Plant Soil*, 55, 403, 1980.
62. Wang, S. R. and Hayman, D. S., Effect of nitrogen on mycorrhizal infection, *Rothamsted Rep. 1981*, Part 1, 211, 1982.
63. Hepper, C. M., The effect of nitrate and phosphate on the vesicular-arbuscular mycorrhizal infection of lettuce, *New Phytol.*, 93, 389, 1983.
64. Wang, S. R., Effect of Nitrogen and Other Factors on Plant Growth Responses to Vesicular-Arbuscular Mycorrhiza, Ph.D. thesis, University of London, London, 1984.
65. Hayman, D. S. and Wang, S. R., Spread of mycorrhizal infection in developing root systems, *Rothamsted Rep. 1983*, Part 1, 166, 1984.
66. Björkman, E., Forest tree mycorrhiza — the conditions for its formation and the significance for tree growth and afforestation, *Plant Soil*, 32, 589, 1970.
67. Hayman, D. S., The physiology of vesicular-arbuscular endomycorrhizal symbiosis, *Can. J. Bot.*, 61, 944, 1983.
68. Clarke, C. and Mosse, B., Plant growth responses to vesicular-arbuscular mycorrhiza. XII. Field inoculation responses of barley at two soil P levels, *New Phytol.*, 87, 695, 1981.
69. Bethlenfalvay, G. J., Brown, M. S., and Pacovsky, R. S., Relationships between host and endophyte development in mycorrhizal soybeans, *New Phytol.*, 90, 537, 1982.
70. Graham, J. H., Linderman, R. G., and Menge, J. A., Development of external hyphae by different isolates of mycorrhizal *Glomus* spp. in relation to root colonization and growth of Troyer citrange, *New Phytol.*, 91, 183, 1982.
71. Boatman, N., Paget, D., Hayman, D. S., and Mosse, B., Effects of systemic fungicides on vesicular-arbuscular mycorrhizal infection and plant phosphate uptake, *Trans. Br. Mycol. Soc.*, 70, 443, 1978.
72. Spokes, J. R., Macdonald, R. M., and Hayman, D. S., Effects of plant protection chemicals on vesicular-arbuscular mycorrhizas, *Pestic. Sci.*, 12, 346, 1981.
73. Levy, Y. and Krikun, J., Effect of vesicular-arbuscular mycorrhiza on *Citrus jambhiri* water relations, *New Phytol.*, 85, 25, 1980.
74. Safir, G. R., Boyer, J. S., and Gerdemann, J. W., Nutrient status and mycorrhizal enhancement of water transport in soybean, *Plant Physiol.*, 49, 700, 1972.
75. Nelsen, C. E. and Safir, G. R., Increased drought tolerance of mycorrhizal onion plants caused by improved phosphorus nutrition, *Planta*, 154, 407, 1982.
76. Allen, M. F., Influence of vesicular-arbuscular mycorrhizae on water movement through *Bouteloua gracilis* (H.B.K.) Lag ex Steud, *New Phytol.*, 91, 191, 1982.
77. Menge, J. A., Davis, R. M., Johnson, E. L. V., and Zentmyer, G. A., Mycorrhizal fungi increase growth and reduce transplant injury in avocado, *Calif. Agric.*, 32, 6, 1978.
78. Kormanik, P. P., Bryan, W. C., and Schultz, R. C., Endomycorrhizal inoculation during transplanting improves growth of vegetatively propagated yellow poplar, *Plant Propagator*, 23, 4, 1979.
79. Schultz, R. C., Kormanik, P. P., Bryan, W. C., and Brister, G. H., Vesicular-arbuscular mycorrhiza influence growth but not mineral concentrations in seedlings of eight sweetgum families, *Can. J. For. Res.*, 9, 218, 1979.
80. Baylis, G. T. S., Root hairs and phycomycetous mycorrhizas in phosphorus-deficient soil, *Plant Soil*, 33, 713, 1970.
81. Hayman, D. S., Practical aspects of vesicular-arbuscular mycorrhiza, in *Advances in Agricultural Microbiology*, Subba Rao, N. S., Ed., Oxford and IBH Publishing, New Delhi, 1982, 325.

82. Howeler, R. H., Asher, C. J., and Edwards, D. G., Establishment of an effective endomycorrhizal association on cassava in flowing solution culture and its effects on phosphorus nutrition, *New Phytol.*, 90, 229, 1982.
83. Vander Zaag, P., Fox, R. L., De La Pena, R. S., and Yost, R. S., P nutrition of cassava, including mycorrhizal effects on P, K, S, Zn and Ca uptake, *Field Crops Res.*, 2, 253, 1979.
84. Kang, B. T., Islam, R., Sanders, F. E., and Ayanaba, A., Effect of phosphate fertilization and inoculation with VA-mycorrhizal fungi on performance of cassava (*Manihot esculenta* Crantz) grown on an alfisol, *Field Crops Res.*, 3, 83, 1980.
85. Mosse, B., Plant growth responses to vesicular-arbuscular mycorrhiza. X. Responses of *Stylosanthes* and maize to inoculation in unsterile soils, *New Phytol.*, 78, 277, 1977.
86. Rich, J. R. and Bird, G. W., Association of early-season vesicular-arbuscular mycorrhizae with increased growth and development of cotton, *Phytopathology*, 64, 1421, 1974.
87. Mosse, B., Hayman, D. S., and Snoad, B., Vesicular-arbuscular mycorrhiza, field inoculation studies: semi-leafless peas, John Innes, *Rothamsted Rep. 1979*, Part 1, 186, 1980.
88. Hayman, D. S., Grace, C. A., and Snoad, B., Vesicular-arbuscular mycorrhiza: peas at the John Innes Institute, *Rothamsted Rep. 1981*, Part 1, 209, 1982.
89. Bååth, E. and Hayman, D. S., Effect of soil volume and plant density on mycorrhizal infection and growth response, *Plant Soil*, 77, 373, 1984.
90. Hayman, D. S., Grace, C. A., Spokes, J. R., and O'Shea, J., Vesicular-arbuscular mycorrhiza: maize at Grassland Research Institute, *Rothamsted Rep. 1981*, Part 1, 210, 1982.
91. Warner, A., Gee, P., and Fyson, A., The spread of an indigenous VA mycorrhizal population in a field soil, *Rothamsted Rep. 1981*, Part 1, 210, 1982.
92. Hayman, D. S., Spokes, J. R., and Grace, C. A., Vesicular-arbuscular mycorrhiza: lucerne and maize at Panshanger, *Rothamsted Rep. 1981*, Part 1, 210, 1982.
93. Powell, C. Ll., Mycorrhizas in hill country soils. III. Effect of inoculation on clover growth in unsterile soils, *N.Z. J. Agric. Res.*, 20, 343, 1977.
94. Powell, C. Ll., Inoculation of white clover and ryegrass seed with mycorrhizal fungi, *New Phytol.*, 83, 81, 1979.
95. Rangeley, A., Daft, M. J., and Newbould, P., The inoculation of white clover with mycorrhizal fungi in unsterile hill soils, *New Phytol.*, 92, 89, 1982.
96. Hayman, D. S. and Hampson, K. A., VA mycorrhiza. Field inoculation trial (white clover in Welsh upland soil), *Rothamsted Rep. 1978*, Part 1, 238, 1979.
97. Stribley, D. P., Tinker, P. B., and Snellgrove, R. C., Effect of vesicular-arbuscular mycorrhizal fungi on the relations of plant growth, internal phosphorus concentration and soil phosphate analyses, *J. Soil Sci.*, 31, 655, 1980.
98. Powell, C. Ll., Mycorrhizal fungi stimulate clover growth in New Zealand hill country soils, *Nature (London)*, 264, 436, 1976.
99. Dodd, J., Krikun, J., and Haas, J., Relative effectiveness of indigenous populations of vesicular-arbuscular mycorrhizal fungi from four sites in the Negev, *Isr. J. Bot.*, 32, 10, 1983.
100. Buwalda, J. G., Stribley, D. P., and Tinker, P. B., Effects of mycorrhizas on cereal growth in the field, *Rothamsted Rep. 1982*, Part 1, 271, 1983.
101. Plenchette, C., Les endomycorhizes à vésicules et arbuscules (VA): un potentiel à exploiter en agriculture, *Phytoprotection (Quebec)*, 63, 86, 1982.
102. Khan, A. G., Growth effects of VA mycorrhiza on crops in the field, in *Endomycorrhizas*, Sanders, F. E., Mosse, B., and Tinker, P. B., Eds., Academic Press, London, 1975, 419.
103. Saif, S. R. and Khan, A. G., The effect of vesicular-arbuscular mycorrhizal associations on growth of cereals. III. Effects on barley growth, *Plant Soil*, 47, 17, 1977.
104. Owusu-Bennoah, E. and Mosse, B., Plant growth responses to vesicular-arbuscular mycorrhiza. XI. Field inoculation responses in barley, lucerne and onion, *New Phytol.*, 83, 671, 1979.
105. Jakobsen, I., Vesicular-arbuscular mycorrhiza in field-grown crops. II. Effect of inoculation on growth and nutrient uptake in barley at two phosphorus levels in fumigated soil, *New Phytol.*, 94, 595, 1983.
106. Stribley, D. P., Effects of mycorrhizas on plant growth and internal P concentration in the field, *Rothamsted Rep. 1980*, Part 1, 251, 1981.
107. Azcón-G. de Aguilar, C., Azcón, R., and Barea, J. M., Endomycorrhizal fungi and *Rhizobium* as biological fertilizers for *Medicago sativa* in normal cultivation, *Nature (London)*, 279, 325, 1979.
108. Hayman, D. S., Page, R. J., and Clarke, C. A., Vesicular-arbuscular mycorrhiza: field inoculation studies with Red clover, Sawyers I, *Rothamsted Rep. 1980*, Part 1, 201, 1981.
109. Hall, I. R., Growth of *Lotus pedunculatus* Cav. in an eroded soil containing soil pellets infested with endomycorrhizal fungi, *N.Z. J. Agric. Res.*, 23, 103, 1980.
110. La Torraca, S., Effects of Inoculation with VA Mycorrhiza on Growth and Nodulation of *Vigna unguiculata* (L.) Walp in Three 'Terra Firme' Soils, dissertation, INPA, Manaus, Brazil, 1979.

111. Islam, R., Ayanaba, A., and Sanders, F. E., Response of cowpea *(Vigna unguiculata)* to inoculation with VA mycorrhizal fungi and to rock phosphate fertilization in some unsterilized Nigerian soils, *Plant Soil,* 54, 107, 1980.
112. Ross, J. P., Effect of phosphate fertilization on yield of mycorrhizal and nonmycorrhizal soybeans, *Phytopathology,* 61, 1400, 1971.
113. Bagyaraj, D. J., Manjunath, A., and Patil, R. B., Interaction between a vesicular-arbuscular mycorrhiza and *Rhizobium* and their effects on soybean in the field, *New Phytol.,* 82, 141, 1979.
114. Ross, J. P. and Harper, J. A., Effect of *Endogone* mycorrhiza on soybean yields, *Phytopathology,* 60, 1552, 1970.
115. Schenck, N. C. and Hinson, K., Response of nodulating and nonnodulating soybeans to a species of *Endogone* mycorrhiza, *Agron. J.,* 65, 849, 1973.
116. Mosse, B., Powell, C. Ll., and Hayman, D. S., Plant growth responses to vesicular-arbuscular mycorrhiza. IX. Interactions between VA mycorrhiza, rock phosphate and symbiotic nitrogen fixation, *New Phytol.,* 76, 331, 1976.
117. Smith, S. E. and Daft, M. J., Interactions between growth, phosphate content and nitrogen fixation in mycorrhizal and non-mycorrhizal *Medicago sativa, Aust. J. Plant Physiol.,* 4, 403, 1977.
118. Smith, S. E., Nicholas, D. J. D., and Smith, F. A., Effect of early mycorrhizal infection on nodulation and nitrogen fixation in *Trifolium subterraneum* L., *Aust. J. Plant Physiol.,* 6, 305, 1979.
119. Munns, D. N. and Mosse, B., Mineral nutrition of legume crops, in *Advances in Legume Science,* Summerfield, R. J. and Bunting, A. H., Eds., Her Majesty's Stationery Office, London, 1980, 115.
120. Lopes, E. S., Oliveira, E., Neptune, A. M. L., and Moraes, F. R. P., Efeito da inoculação do cafeeiro com diferentes especies de fungos micorrizicos vesicular-arbusculares, *Rev. Bras. Cien. Solo,* 7, 137, 1983.
121. Hayman, D. S., Collins, R. I., and Hampson, K. A., Influence of VA mycorrhiza on plant growth: arable crops in UK soils, *Rothamsted Rep. 1977,* Part 1, 240, 1978.
122. Stribley, D. P. and Snellgrove, R. C., Mycorrhizas and growth of transplanted onions, *Rothamsted Rep. 1983,* Part 1, 166, 1984.
123. Bolgiano, N. C., Safir, G. R., and Warncke, D. D., Mycorrhizal infection and growth of onion in the field in relation to phosphorus and water availability, *J. Am. Soc. Hortic. Sci.,* 108, 819, 1983.
124. McGraw, A.-C. and Schenck, N. C., Growth stimulation of citrus, ornamental, and vegetable crops by select mycorrhizal fungi, *Proc. Fla. State Hortic. Soc.,* 93, 201, 1980.
125. Jackson, N. E., Franklin, R. E., and Miller, R. H., Effects of vesicular-arbuscular mycorrhizae on growth and phosphorus content of three agronomic crops, *Soil Sci. Soc. Am. Proc.,* 36, 64, 1972.
126. Hayman, D. S., Morris, E. J., and Page, R. J., Methods for inoculating field crops with mycorrhizal fungi, *Ann. Appl. Biol.,* 99, 247, 1981.
127. Elmes, R. P., Hepper, C. M., Hayman, D. S., and O'Shea, J., The use of vesicular-arbuscular mycorrhizal roots grown by the nutrient film technique as inoculum for field sites, *Ann. Appl. Biol.,* 104, 437, 1983.
128. Daniels, B. A. and Menge, J. A., Evaluation of the commercial potential of the vesicular-arbuscular mycorrhizal fungus *Glomus epigaeus, New Phytol.,* 87, 345, 1981.
129. Menge, J. A., Lembright, H., and Johnson, E. L. V., Utilization of mycorrhizal fungi in citrus nurseries, *Proc. Int. Soc. Citric.,* 1, 1977, 129.
130. Mosse, B., Mycorrhiza and plant growth, in *Structure and Functioning of Plant Populations,* Koninklijke Nederlandse Akademie van Wetenschappen, Holland, 1978, 269.
131. Jakobsen, I., Mycorrhizal infectivity of soils eliminated by low doses of ionizing radiation, *Soil Biol. Biochem.,* 16, 281, 1984.
132. Lockeretz, W., *Agriculture and Energy,* Academic Press, New York, 1977.
133. Dehne, H. W., Interaction between vesicular-arbuscular mycorrhizal fungi and plant pathogens, *Phytopathology,* 72, 1115, 1982.
134. Hall, I. R., Response of *Coprosma robusta* to different forms of endomycorrhizal inoculum, *Trans. Br. Mycol. Soc.,* 67, 409, 1976.
135. Daft, M. J. and Hogarth, B. G., Competitive interactions amongst four species of *Glomus* on maize and onion, *Trans. Br. Mycol. Soc.,* 80, 339, 1983.
136. Abbott, L. K. and Robson, A. D., The role of vesicular arbuscular mycorrhizal fungi in agriculture and the selection of fungi for inoculation, *Aust. J. Agric. Res.,* 33, 389, 1982.
137. Menge, J. A., Utilization of vesicular-arbuscular mycorrhizal fungi in agriculture, *Can. J. Bot.,* 61, 1015, 1983.
138. Hirsch, P., Improved crop plant productivity through genetic manipulation of mycorrhizal fungi?, *Chem. Ind.,* 23, 833, 1984.
139. Estaun, V., Calvet, C., and Hayman, D. S., unpublished, 1984.

Chapter 10

VA MYCORRHIZAE IN HORTICULTURAL SYSTEMS

Stanley Nemec

TABLE OF CONTENTS

I. INTRODUCTION

Horticultural crop production is more varied than ever before, with new introductions and clones available to the public, advances being made in the development of soilless container media, and the use of slow-release and other specialty fertilizers. Horticultural crops, more than any other groups of crops, are produced in a diverse number of systems that range from hydroponics to field production. Plants are grown in containers of all types and sizes, in greenhouses, in raised and ground beds, and in field plantings. Horticultural crops are grown for flower, foliage, and food production, as well as for aesthetic values.

Most horticulturally produced plants are high cash-value crops that require expensive facilities for their production, specialized equipment, extensive use of labor, and various cultural practices that often involve soil fumigation. Production methods use many time- and cost-saving procedures. Currently, the potential to overcome the disadvantages of soil fumigation by inoculating soils, potting mixes, and plants with vesicular-arbuscular mycorrhizal (VAM) fungi is being evaluated with a wide range of horticultural plants. The growth-promoting effects of these fungi are recognized by many growers and researchers as a means to reduce production costs. Growth of a diverse group of horticultural plants has been stimulated by inoculation. Inoculation has resulted in growth increases with such herbaceous plants as the Easter lily,[1] geranium,[2] and sunflower;[3] and semiherbaceous poinsettia,[4] various woody ornamentals,[5] and fruit crops.[6-8]

Many environmental and cultural factors are known to affect the response of a plant to inoculation with VAM fungi, but perhaps the most important factor is the plant's root structure. It is often possible to predict the potential for a plant to respond to inoculation based on its root geometry according to the hypothesis proposed by Baylis.[9] Baylis grouped plant roots into three classes: magnolioid for those plants that are coarsely branched so that the ultimate roots are rarely less than 0.5 mm in diameter, graminoid for plants with ultimate root branches of less than 0.1 mm in diameter, and intermediate types that fall between the two extremes. He considered the magnolioid root the most dependent on mycorrhizae for phosphorus uptake in soils with low levels of phosphorus. Woody species consisting of trees and shrubs would be allied closely to the magnolioid type of root, and many annuals, herbaceous perennials, and lawn grasses could be considered more naturally associated with the graminoid root. However, some annuals and herbaceous perennials probably are more intermediate in their VAM fungus dependency. These latter two groups of plants have been studied less intensively than other horticultural crops in mycorrhizal studies. Carefully planned experiments will more clearly define their requirements for VAM fungi. "Baylis' hypothesis" and his comments about root characteristics and their relationship to VAM infection have evolved as one of the most reliable measures of host dependency.

Fruit crops more than any other group of horticultural plants have received the greatest attention in work with VAM fungi. Many advances in the study of mycorrhizae have been made with fruit crops, partly because of their dependency on these fungi and economic importance, but also because they were one of the first groups of horticultural plants in which VAM fungus infection was first explored. Fruit crops were also used in pioneering experiments by leaders in the field of this research. Many such studies dating back to the 1920s and 1930s defined infection characteristics of endophytes in the root systems of fruit crops.[10-13] Some of this work evolved from studies on important root diseases. However, the presence of the endophyte in diseased tissue was sometimes interpreted as the cause of the disease. Later research by Mosse clearly demonstrated that typical vesicular-arbuscular infections could be produced in aseptically grown strawberry and apple by inoculating them with spores excised from soil containing VAM fungus fructifications.[14] Another achievement occurred when Kleinschmidt and Gerdemann proved that the lack of these fungi was responsible for stunting of citrus in fumigated citrus nursery soil.[8] Until then, this condition had been attributed to "soil toxicity". Because a large body of information has been generated on mycorrhizae and fruit crops, this group of horticultural types will receive the most emphasis in the following discussion.

To fully address certain topics in various sections of this chapter, a limited amount of literature on crops other than horticultural types will be presented. In addition, key and significant papers dealing specifically with the fungi will be included to also clarify discussions in various sections.

II. ASSOCIATIONS OF VAM WITH HORTICULTURAL CROPS

Horticultural crop production occurs world-wide, but is concentrated in countries with temperate to subtropical climates and is near many regions with a high population density. Few horticultural plants and their fruit are harvested mechanically; therefore, a reliable supply of labor is needed for plant and fruit production. With modern transportation, plants and fruit can be shipped quickly to nearby markets or air transported rapidly to other states and countries where the product is in demand.

With increasing frequency, more horticultural crop-producing areas are reporting the presence of VAM in soils or in root systems of crop plants. These reports include such associations as *Glomus mosseae* with almonds in Spain,[15] *G. fasciculatum* with grapes in Italy,[16] *G. fasciculatum* with avocados in South Africa,[17] and various VAM fungi with citrus in China.[18] Early literature on these associations is limited in its interpretation because the fungi were referred to as either species of *Endogone* or *Rhizophagus*.[19-22] Since 1974, when the taxonomy of VAM fungi was revised,[23] many members of the Endogonaceae have been found on horticultural crops. Members of the genus *Glomus* are the most common. In California, *G. fasciculatum, G. constrictum,* and *G. microcarpum* were encountered the most frequently in citrus groves, while in Florida, *Gigaspora margarita, Glomus etunicatum, G. fasciculatum,* and *G. macrocarpum* were most frequently present in citrus groves.[24] *Glomus microcarpum, G. etunicatum,* and *G. fasciculatum* were most commonly associated with citrus in Texas,[25] and in Taiwan *G. fasciculatum, G. monosporum, G. vesiculifer, Gigaspora calospora,* and *Sclerocystis rubiformis* were the most frequent species in maize and citrus soils.[26] *Glomus* species resembling those of *G. mosseae, G. macrocarpum,* or *G. fasciculatum* were predominant in soils of wild raspberries.[46]

The frequent association of an endophyte with a host does not necessarily mean the endophyte is effective in increasing growth of that host. *Gigaspora margarita,* which is common to Florida citrus soils, infects citrus seedlings poorly and probably is a more frequent resident in weeds which are common in citrus groves. *Glomus fasciculatum,* which is common to citrus soils, has been found on seven different fruit crops grown in seven different countries. Horticultural plants that have been shown to respond with growth increases after inoculation are listed in Table 1.

Although estimates of species have been determined from spore numbers collected from soils, most studies of this type have only reported gross spore numbers. Roldan-Fajardo et al.[15] provided specific spore counts for *G. mosseae, G. macrocarpum* var. *geosporum,* and *G. fasciculatum* under almond. Spore numbers peaked for all species in autumn. Populations of spores were nearer the soil surface than in the subsurface layers.[16,24] More spores were present in soil of mulched vineyards than in soil around grape plants that had been weeded.[16] *Endogone* spore numbers ranged between 250 and 800/ℓ soil in July and August under raspberry and generally around 500/ℓ under

Table 1
HORTICULTURAL CROPS THAT RESPONDED WITH A GROWTH INCREASE AFTER INOCULATION WITH VESICULAR-ARBUSCULAR MYCORRHIZAL FUNGI

Field or outdoor container-grown plants			Greenhouse-grown plants					
			Fruit crops			Ornamental crops		
Plant	Fungus	Ref.	Plant	Fungus	Ref.	Plant	Fungus	Ref.
Geranium	*Glomus mosseae*	145	Citrus	*Endogone mosseae*	149	Marigold	*G. monosporum*	32
	G. etunicatum	145		*E. epigaeum*	146	*Acacia*	*E. calospora*	160
Sweetgum	*G. mosseae*	147		*G. mosseae*	28, 29, 146	*Rhododendron*	*Glomus* spp.	161
	G. fasciculatum	152		*G. fasciculatum*	29, 37, 38, 58,	Tuliptree	*E. fasciculatum*	162
Peach	*G. fasciculatum*	41			87, 146, 150	Viburnum	*G. fasciculatum*	5
Citrus	*Endogone mosseae*	8		*G. etunicatum*	29		*G. mosseae*	5
	G. mosseae	42	Raspberry	*G. mosseae*	101	*Pittosporum*	*G. fasciculatum*	5
	G. etunicatum	42		*G. fasciculatum*	101		*G. mosseae*	5
	G. caledonium	148		*G. tenuis*	101	*Podocarpus*	*G. fasciculatum*	5
	G. fasciculatum	148, 151	Avocado	*G. fasciculatum*	17, 40		*G. mosseae*	5
	Gigaspora margarita	148	Apple	*G. epigaeum*	31, 73	Sugar Maple	*Glomus* spp.	34
	Endogone macrocarpum	33		*G. macrocarpum*	31	Southern magnolia	*G. fasciculatum*	53
Yellow	*G. fasciculatum*	153, 154		*G. monosporum*	31	*Chrysanthemum*	*G. fasciculatum*	163
poplar	*G. mosseae*	154		*Gigaspora calospora*	31	Poinsettia	*Gigaspora margarita*	4
	G. caledonium	154		*G. mosseae*	6, 59, 155			
Black	*Glomus* spp.	68		*G. fasciculatum*	6			
cherry			Peach	*Endogone* spp.	7			
Red maple	*Glomus* spp.	68		*G. margarita*	156			
Sugar maple	*Glomus* spp.	68		*G. etunicatum*	156			
Sycamore	*Glomus* spp.	68	Strawberry	*G. epigaeus*	31			
				Endogone sp.	19, 157			
			Almond	*G. mosseae*	15			
				G. fasciculatum	15			
			Papaya	*G. calospora*	158			
				G. macrocarpum	158			
			Grape	*Endogone* sp.	159			

strawberry.[20] They were as high as 86/g soil under strawberry in Ontario,[21] and usually ranged between 50 and 100/g soil under strawberry in Illinois.[27] Most of these studies concluded that spore production increased with the onset of root senescence or with fruit production. The frequent reports of increases in spore numbers in the fall may be linked to root senescence in many horticultural crops.

Frequent associations of one or several VAM fungus species with a host plant suggest that under the prevailing soils and cultural conditions, a balance is probably established between the host and endophyte. Indeed, when a local crop plant is tested for its reaction to an endemic species, a favorable growth response often occurs.[15,19,24,28,29] Unlike the specificity which occurs among other obligate fungi and certain basidiomycetes, a range of endophyte effectiveness occurs on most hosts inoculated with VAM fungal species. In tests where selected hosts were inoculated with two or more endophytes, one or two of the fungi were usually the least effective in enhancing growth or other parameters of symbiosis.[6,30-33] Specificity for a particular host may be modified by such factors as soil mixes,[34] pH,[35] and fertilizer,[36] which have either an indirect or direct effect on the endophyte.

Characteristics within the same host genus influence the behavior of the plant toward the endophyte. Citrus rootstocks are known to vary in their dependency to VAM.[28,29,37,38] Sour orange, Cleopatra mandarin, and rough lemon exhibited a higher level of dependency on VAM fungi than Rangpur lime and sweet orange. Troyer and Carrizo citranges, which result from a cross of *Poncirus trifoliata* and *Citrus sinensis*, were the least dependent on *Glomus* species. Results from these studies showed that cultivars with a low degree of dependency produced more roots, generally had higher root/shoot ratio or root/leaf ratios, or had thinner roots.

Significant growth increases due to VAM fungus infection have been measured in various ways. Growth habits vary among horticultural plants, and this fact must be considered when growth is evaluated. Weight analysis, a commonly used technique, was used by Gilmore[39] and Mataré and Hattingh[17] to demonstrate growth increases of 3.4X and 2.8X for mycorrhizal peach and avocado, respectively. Menge et al.[37] calculated mycorrhizal dependency of various citrus rootstocks from weight data. When plants cannot be sacrificed or large numbers of plants are in tests, especially in field studies, other measures of growth should be used. Growth rate analysis,[40] stem height,[29] and trunk caliper[41] may be more useful under those conditions. Nemec found stem caliper to be a discriminating measure of growth in field-planted citrus.[42] Root length can be used as another parameter when plants are removed from pots at the end of a test.[30] Stolon length measurements in a plant such as strawberry may be a more sensitive measure of growth than weight analysis.

III. SOIL AND NUTRIENT FACTORS AFFECTING PLANT GROWTH AND MINERAL UPTAKE BY VAM FUNGI

A. Soils and Soil Mixes

Most studies of VAM fungus-host interactions have been done in sand or sandy loam soil deficient in nutrients, especially phosphorus. Few horticultural crops are grown in purely sandy soils; most are grown in loamy soils or artificial soils containing mixes of vermiculite, perlite, peat, and other potting media. Some potting mixes contain hardwood bark, sawdust, and other materials.

Artificial potting mixes have been developed to minimize problems with root disease fungi.[43] Survival and growth-promoting capabilities of VAM fungi in artificial soils have a significant impact on the usefulness of these organisms in crop production. This issue is one which is currently under intensive study.

Colonization of geranium and subclover by VAM was reduced and shoot dry weight and phosphorus uptake correspondingly not increased in soilless media such as peat, perlite, vermiculite, and bark.[44] However, partial amendment of natural soils with peat or vermiculite diminished the inhibitory properties of either alone.[44,45] Growth increases of three ornamentals inoculated with *G. fasciculatum* and *G. mosseae* were also achieved in a 1:1 (by volume) peat/sand mixture.[5] A significant plant growth response due to *Glomus*-inoculated citrus occurred in four of nine commercial potting mixes with a high organic matter content;[47] growth increases in inoculated potting mixes were less than those achieved in various inoculated soils. Many peats have a low natural pH, and this may have a tendency to depress mycorrhizae formation.[47] Menge et al.[48] reported that the percentage of citrus roots infected with *G. fasciculatum* was reduced in potting mixes containing more than 50% redwood shavings or peat moss. Ultimately, he and others recommended a medium with only 3% organic matter for maximum growth of citrus inoculated with VAM fungi.[49] More recent information implies that peats can be suitable for growth media of inoculated plants and for spore production.[50] VAM fungi are known to have longer hyphal lengths in the presence of organic particles,[51] which may serve as a source of energy for limited saprophytic activity; and increasing pH by liming peats has afforded good infection of plants grown in them.[52]

Organic matter in the form of peat may not be needed in container media if suitable replacements are available. Plant growth of horticultural crops has been stimulated by VAM fungus inoculation in media, such as sugar maple compost,[34] composted hardwood bark/expanded shale (2:1, by volume),[53] a calcined montmorillonite clay,[54] and a sand/bark/soil mixture (1:1:1, by volume).[55] Typical VAM fungus infection was reported in various crop plants and vegetation growing on volcanic ash-derived soils in Chile.[56] Such volcanic ash products may be suitable in mixes containing other products which also support VAM fungus infection in plants. Survival and growth-promoting effectiveness of VAM fungi occur in a wide range of potting mix materials. Therefore, it is probable that a lightweight, economical mix can be devised which will support the growth of many horticultural crops and the activity of selected VAM fungal species. Results of these experiments suggest that fungus activity is optimum in a mix of two or more materials rather than a medium of one specific substrate. Critical factors affecting the fungi in these mixes are pH, nutrient content, biological activity, and pore space characteristics.

B. Nutrition

Enhanced plant growth by VAM fungi is determined by the dependency of the plant on the fungi, the ability of the fungi to forage away from the root for nutrients, especially P, the nutrition level of the growth medium, the type of medium itself, and other factors. It is now recognized that plant growth increases due to VAM fungus species are often the result of increased P nutrition. Mycorrhizal roots utilize the same sources of soil phosphate as nonmycorrhizal roots and do not appear to mobilize insoluble soil phosphate to any considerable extent.[57] However, when sparingly soluble phosphates, such as bone meal, rock phosphate, apatite, $FePO_4$, $AlPO_4$ and iron, and calcium phytates, were added to soil, results from various studies showed that phosphate was taken up more readily from all these sources when plants were mycorrhizal.[57] Better utilization of sparingly soluble phosphates does not mean that mycorrhizal fungi mobilized soil phosphate not otherwise available to plant roots, but that mycorrhiza utilized the same sources of soil P as uninfected roots and were able to obtain more of it. Soil P levels have a major effect on the host-endophyte relations. In soil amended with 0 to 556 ppm superphosphate, weight of mycorrhizal fungus-infected sour orange seedlings was significantly greater than noninfected seedlings when fertilized with 0, 6, 28, and 56 ppm P, and weight of mycorrhizal fungus-infected Troyer citrange seedlings was significantly greater than noninfected seedlings only when fertilized with 0 and 6 ppm P.[58] Menge et al. showed similar diversity of inoculation responses for six citrus cultivars at three P nutrient levels.[37] The point at which the P level affords the mycorrhizal plant no additional growth advantage over the nonmycorrhizal plant depends on the plant species. Various VAM fungi apparently can reach the same level of colonization in apple and marigold provided the appropriate levels of soluble P are used,[54] thus providing additional evidence that each host plant has its own P requirements. Excess or luxury uptake of P by mycorrhizal fungus-infected plants should be determined in order to calculate minimal fertilizer requirements for optimum growth of plant species. Increasing soil levels of P, often more than required by mycorrhizal plants, have been shown to decrease fungus spores in soil around citrus[58] and decrease root fungus colonization in marigold[54] and apple.[54,59] Yet in soil, *G. mosseae* and *G. caledonium* germinated and grew normally at levels of $NaHCO_3$-soluble P up to 982 mg kg^{-1} soil.[60] Menge et al.[61] suggested that plants become immune to infection when their internal P content becomes too high for fungus development in roots. High internal plant P levels result in lower membrane permeability and reduced exudation of amino acids and reducing sugars which may limit fungus colonization and root penetration.[62] However, recent data indicate that sugar and organic acid concentrations in cortical cells are more important in restricting root penetration than internal P content.[63]

There is no evidence that mycorrhiza play any part in the nitrogen (N) uptake of plants.[57] In general, N content in aerial parts of nonmycorrhizal plants is higher than in mycorrhizal plants because of an accumulation of amino acid-N.[64] Occasionally, improved N is reported in aerial parts of plants.[19,30] Hughes et al.[30] suggested that this may be due to some mechanism which accelerates other parts of the N uptake process. Nitrogen has had favorable and nonfavorable effects on spore production and root colonization by the endophyte. It is generally agreed that excess N and complete fertilizer can reduce mycorrhizal infection in plants,[36,65,66] yet in vitro germination studies with *Gigaspora margarita* showed that a mixture of NPK salts improved spore germination and germ tube growth.[63] Only when N was used alone, germ tube growth was depressed by 55%. Daniels and Trappe also showed that spore germination of *Glomus epigaeum* was not influenced by up to 200 ppm NH_4NO_3 in in vitro tests.[67] Other studies support the results of the in vitro tests. Infection increased in six of eight hardwood tree species fertilized with three rates of a 10-10-10 fertilizer.[68] Evidently, nitrogen effects on the fungus are influenced directly or indirectly by the other elements.Manures have had a favorable effect on fungus development in the root. Better fungus development occurred in citrus and lavender grown in soil fertilized with manure than with $NaNO_3$ or $CaCO_3$ alone or in soil not fertilized.[69-71] Elements other than N in both manures and complete fertilizers favorably affect host growth, but also probably allow better spore germination and root colonization by the fungus. Also, elements other than N may diminish the effects of pathogenic fungi and endophyte competitors, making it easier for the endophyte to become established on and in the root. Ames and Linderman demonstrated in their high-fertilizer treatments on Easter lily that root rot due to *Fusarium oxysporum* was more severe in the high-fertilizer treatments than in low-fertilizer treatments, and that VAM fungus infections were also the lowest in the high-fertilizer treatments.[1] Immediate effects on VAM fungi with soluble fertilizers may be reduced with slow-release fertilizers. Maronek et al.[53] reported that incorporating slow-release fertilizers into a container medium did not inhibit mycorrhizal formation in southern magnolia. Even soluble fertilizer effects on VAM fungi may be buffered when plants are grown in certain mixes. The incidence of VAM fungi in maple compost was favorably affected by added NH_4NO_3 and sugar maple roots were 80% mycorrhizal in this mix.[72] Although it is known that VAM fungi

Table 2
OPTIMUM PH REQUIREMENTS FOR VA-MYCORRHIZAL HOST GROWTH AND PH REQUIREMENTS FOR FUNGUS GROWTH AND GERMINATION

Host plant and pH optima	VA fungus	Fungus pH range or optima		Ref.
		Germination	Germ tube growth	
Tagetes minuta, pH 4.3	*Glomus macrocarpum*	—	—	35
Liquidambar styraciflua, pH 5.1 and 5.9	*G. fasciculatum*	—	—	165
L. styraciflua, pH 6—8.1	*G. mosseae*	—	—	165
Paspalum notatum, pH 4.8—5.0	Honey-colored *Endogone* sp.	—	—	164
—	*Gigaspora margarita*	6.0	6.0	63
—	*G. heterogama*	6.0	—	166
—	*G. coralloidea*	5.0	—	166
—	*G. mosseae*	7.0—8.0	—	166
—	*G. epigaeum*	7.0—7.4	—	67
Glycine max, pH 5.1	*G. gigantea*	—	—	167
Glycine max, pH 6.2	*G. mosseae*	—	—	167

improve uptake of certain minor elements, such as Cu in apple and citrus,[28,73] Zn in peach,[7,41] and Cu in citrus,[8,28] little information has been presented on effects of minor elements in soils on VAM fungi. Some information relative to this subject will be discussed in the section devoted to cultural practices and pesticides.

IV. ENVIRONMENTAL AND CULTURAL FACTOR INFLUENCES ON VA MYCORRHIZAE

A. Ecological, Edaphic, and Physical Factor Influences on VA Mycorrhizae

One of the most difficult subjects to address regarding VAM fungus activity and survival is the effect that changes in soil ecology and natural edaphic and stress factors have on the fungus. Some information is available on this topic, but little of these data have been derived from work on horticultural crops. This issue is particularly important in horticultural crop production because of the wide range of soil and management conditions under which horticultural crops are produced. Adapting mycorrhizal fungi into production schemes will require knowledge on their survival, temperature and pH requirements, tolerance of water, salinity, and oxygen and other gaseous limitations.

Soil pH has been one of the most thoroughly studied aspects of VA mycorrhizae ecophysiology. A number of in vitro studies has evaluated pH on spore germination and germ tube growth, or on growth of the host (Table 2). Soil pH has a decided effect on fungus germination and on its efficacy in promoting plant growth. Most VAM-fungus species germinate in a pH range favorable for growth of most plants. Soil pH optima for growth of infected plants were generally under slightly acid conditions (Table 2). However, Read et al.[74] found that spore numbers and viability in natural soils declined with increasing acidity. Lambert and Cole reported that six isolates of *Glomus tenue* differed in their ability to form mycorrhizae at low pH.[75] They reported that an isolate of *Gigaspora gigantea* failed to infect at low pH. Soil pH can affect P uptake by mycorrhizal fungi, and this, in an indirect way, affects plant vigor and health.[76] Variations in the spread of VAM fungi from a point of inoculation in field plots have been reported,[77] and soil pH may be one of the variables affecting spread. For instance, *Glomus mosseae* will not generally colonize soils below pH 5.6, whereas *Acaulospora laevis* does not usually occur in neutral or alkaline soils.[57]

Increasing salinity content in many agricultural soils, especially heavily irrigated soils in southern California and along the Colorado River watershed, has had an unfavorable impact on crop production. Many fruit crops are grown under irrigation in those regions, and studies should be undertaken to assess the degree to which roots become mycorrhizal in these saline soils. In Israel, increased salinity of irrigation-water reduced VAM fungi in deep subsoils.[78] Although a specific ion effect could not be demonstrated conclusively, germination rates of *Gigaspora margarita* were reduced in the presence of Cl^- and Na^+.[79] The effect of Cl^- on reducing germination appeared to be relatively greater than the Na^+ effect. Hirrel and Gerdemann suggested that plants infected with mycorrhizal fungi may be more salt tolerant than uninfected nonmycorrhizal plants.[80]

Soil water is another factor known to affect VAM fungi. Gerdemann pointed out that most aquatic plants and plants growing in very wet places were most likely to be nonmycorrhizal.[81] Decreasing water tables[82] and increasing intervals between irrigation[78] improved the mycorrhizal condition of several plants. Nonmycorrhizal plants growing under wet conditions often become mycorrhizal if they are transplanted to well-drained soil.[83] Lowland ecotypes of *Nyssa sylvatica* can establish mycorrhizae with *Glomus mosseae* under continuously flooded conditions.[84] However, less mycorrhizae developed in more distal roots and it was suggested that this was due to limited oxygen transport to these roots. Water effects have been studied directly on the fungi. Sylvia and Schenck reported the maximum germination of *G. clarum,*[85] *G. etunicatum,* and *G. macrocarpum* occurred at a matric water potential of −100 bars. In another study, Daniels and Trappe reported that *G. epigaeus* germination was favored in soil at or above field capacity, but decreased with decreasing water potentials below field capacity.[67] Sieverding showed that the development of mycorrhizae in sorghum was better in conditions of water deficiency than in well-watered plants.[86] Establishment of mycorrhizas on roots should be facilitated in soils that are not excessively watered and which are well drained.

Many horticultural crops are produced in containers in greenhouses. Temperature and light can be regulated to favor endophyte activity in greenhouses in climates that have long periods with short days and overcast weather. VAM-fungus infection and growth of *Citrus sinensis* were increased by long-day photoperiods.[87] Reduction in day length and irradiance depressed the growth more in mycorrhizal than in nonmycorrhizal maize plants, and in *Rhizobium*- and *Glomus*-infected alfalfa plants, a day length of 16 hr produced the highest carbon contents and C/N ratios.[88] Even low-light intensities can favor fungus activity. Furlan and Fortin reported that development of the intramatrical components of *G. calospora* was most extensive and rapid under low-light intensities.[89] In general, high-light intensities favor the establishment of VAM fungi on their hosts, low-light intensities may simulate conditions similar to decreasing day length in the fall, and under these conditions, sporulation may be favored.

Temperatures higher than ambient are usually preferred for enhanced development of VAM fungi in plant roots. In growth chamber studies, Furlan and Fortin showed that to obtain maximum infection,[90] number of spores, and growth enhancement, a high temperature (21°C day/26°C night) regime was necessary. Daniels and Trappe found that maximum germination of *G. epigaeum* occurred at 18 to 25°C.[67] Data from work by Schenck et al.[91] implied that strains of *Glomus* may be adapted to temperature common to their environment. They found that two *Glomus* isolates from Florida germinated best at 34°C and one from Washington had an optimum of 20°C. Schenck and Smith later tested six VAM-fungus species at four soil temperatures on infection of soybeans,[92] sporulation, and enhancement of growth. The highest mean value for all fungus and plant variants was 30°C. Optimum spore production for the species ranged from 24 to 30°C. Mosse et al.[93] suggested that because most species of VAM fungi are world-wide in their distribution, it seems likely that temperature adaptation is common.

The ability of VAM fungi to adapt has made it possible to find these fungi in a variety of ecosystems. *Glomus* and *Gigaspora* are reported as common on trees in the semiarid zones of Senegal,[94] and in the northern Negev of Israel,[95] in tussock grassland soils with no fertilizer history in New Zealand,[96] and in heathland soils.[97] The fungi are able to survive in dune soils,[98] coal spoils,[99] and in marshes.[100] The ubiquitous nature of these fungi suggests that a wide range of horticultural crops with differing environmental and soil requirements would be natural hosts of these fungi.

B. Cultural Practices and Pesticides

1. Cultural Practices

One of the least understood and most deficient areas of our knowledge on VAM fungus systems concerns cultural practices and their effects on these organisms. Some of this topic was addressed in the discussion on potting mixes and fertilizers, but the subject of cultural impact on these fungi covers many facets of plant production. Of this subject, areas often overlooked include effects caused by heavy elements, irrigation practices, crop rotations, pesticides, cover cropping, and weed control.

Transplant stress occurs with nearly all plants when they are dug from soil and repotted. Transplant shock has been visibly reduced in mycorrhizal avocado[40] and is less noticeable on axenically propagated mycorrhizal raspberry plants transplanted into soil.[101] Apparently, mycelia of the fungi reestablish a "water-bridge" from the plant to the soil more quickly than can occur in nonmycorrhizal plants after transplanting. Stomatal regulation is controlled by VAM fungi in citrus,[102] and this factor may be modified by the fungi during transplant stress.

Soil populations and root infection by VAM fungi have been affected by cultural practices. Weeding compared to mulching surface soils markedly reduced spore numbers in all layers of a vineyard soil.[16]

Concord grape seedlings grown in vineyard soil amended with old grape roots contained higher levels of VAM fungus infection than those grown in unamended soil.[103] Corn, millet, sudex, and sorghum were all effective cover crops for increasing inoculum in soils planted with sweetgum seed; resultant seedlings did not differ in size on the various cover crop plots, but 89% were of a size recommended for outplanting.[104] Mycorrhizal infection in lavender and lettuce plants grown in sterilized soil inoculated with *Glomus mosseae* was not depressed by the presence of roots of nonhost plants grown previously in the soil as compared with the controls.[105] Toxins, if produced by nonhost plants, were not produced on a scale large enough to reduce inoculum potentials in the soil.

Irrigation plays a role in fungus infection of citrus roots. Infection in the 0- to 60-cm layer of soil was enhanced by decreased soil moisture, but irrigation intervals of 40 days further decreased soil moisture and fungus infection. In deeper soils, these extended irrigation intervals improved infection in roots.[78] In deep soils planted to citrus, soil water potential is buffered against the fluctuations that occur in surface soils.[106] Shorter irrigation schedules of 18 days and a higher preirrigation availability of water repressed infection in citrus roots.[78] This decreased infection may be due to lowered soil O_2 levels or to water potentials, or both.[107] The effects of O_2 and water potential on VAM fungi were addressed in the section on ecological and edaphic factors.

Some practices unrelated to plant growth may have a detrimental effect on the host and its endophyte. Application of deicing salts for ice control in northeastern states often damages roadside sugar maple plantings. Damaged trees had diminished root systems, particularly in the surface layers of soil where mycorrhizas were prevalent.[108]

2. Pesticides and Toxic Metals

VAM fungi are affected by a wide array of soil treatments, pesticides, and soil chemicals. Steam[8] and heat[109] effectively kill these fungi in soils, resulting in stunting and deficiency symptoms in the nonmycorrhizal hosts. Steam treatment, still used on soils and mixes for production of many types of container-grown horticultural crops, will efficiently kill indigenous VAM fungus species along with pathogenic organisms it was designed to control.

Fumigation of soil with biocides such as methyl bromide, chloropicrin, formaldehyde, Mylone, Vapam, and Vorlex effectively kill endophytes in the treatment zone.[110-112] Fortunately, these fungi reinvade most fumigated soils within several years. Unlike the general biocides, most nematicides such as 1,3-dichloropene (1,3-D), 1,2-dibromoethane (EDB), 1,2-dibromo-3-chloropropane (DBCP), and some of the organophosphates (phenamiphos) and organocarbamates (aldicarb) at recommended rates exhibit no effect or only a slight inhibitory effect on these fungi.[109,113,168] Some of these nematicides, such as DBCP and 1,3-D, have stimulated root infection in host plants.[114] This type of response is believed to be due to control of competitive and plant-pathogenic microflora, to possible stimulation of exudates from host plant roots, or to other factors.

Fungicides exhibit a range of activity toward VAM fungi. The systemic fungicides as a group are more toxic to these fungi than most other compounds.[115,116] The most fungitoxic compounds in this group are thiabendazole, benomyl, and triadimefon. Pentachloronitrobenzene, which is not systemic, is also highly toxic. Other fungicides such as captan and metalaxyl apparently do not affect these fungi.[115] Each class of fungicide affects physiological functions in fungi. Captan, a trichloromethylthio compound, reacts with thiols to form thiocarbonyl chloride (thiophosgene), whereas benomyl, a benzimidazole, inhibits mitosis.[117] Evidently, mitosis is either more sensitive in these fungi than sulfhydryl (SH) group conversion, or captan is not physiologically active on these fungi for various reasons. Many fungicides are probably only fungistatic to these fungi, because few effected complete control in studies done in pots.

Most research with pesticides and VAM fungi has been undertaken with soil fumigants, nematicides, and fungicides. Limited research has been done with herbicides, but little information is available on insecticides and their behavior toward these fungi. Paraquat and simazine appear to have a moderate to high toxicity to these fungi, but trifluralin, bromacil, and diuron do not appear to be active against them.[118] The insecticides Metasystox and aldrin differ in their activity on VAM fungi; the former was less toxic than the latter.[119]

Many pesticides contain heavy metals,[120] and the presence of such metals in soils may be responsible for poor germination of these fungi.[121] In highly acid soils, especially those of the tropics, manganese and aluminum may be more soluble and reach concentrations that limit growth of these fungi. Some heavy-metal toxicity has been documented in soils planted to trees. Apples planted in soils containing high levels of arsenic formed very few mycorrhizae.[122] Harris and Jurgensen found copper Cu in mine tailings inhibitory to formation of *Populus* mycorrhizae.[123] However, Cu applied at rates up to 224 kg/ha was more phytotoxic to citrus than it was fungitoxic on its endophyte *G. etunicatum*.[115] Some citrus soils contain levels higher than 224 kg/ha after years of using copper sprays to control foliage diseases. Heavy metal toxicity to VAM fungi may be expected to be more serious where mine tailings have been used as fill, where sewage sludge has been applied as fertilizers, and where pesticides containing these metals have been used for extended periods of time.

V. PHYSIOLOGICAL AND ANATOMICAL ASPECTS OF INFECTION

A large amount of literature has been published regarding the mechanisms of phosphorus uptake by VAM fungi and the role certain minor elements perform in the host-endophyte interaction. Other aspects of host and fungus physiology are equally important, but due to the fact that the endophyte has not been cultured, and that the fungus mycelia is difficult to obtain in quantity and separate from soil, few of these studies have been done with the endophyte alone.

Research has revealed many of the characteristics of carbon assimilation by the endophyte, and studies using the electron microscope have defined the association of host and arbuscule membranes which allow the bidirectional flow of elements between the host and its endophyte. It is well-defined that carbon derived from host-photosynthate is transferred from the host to the fungus.[124] No sugar distinctive for VAM fungi has been found,[125,126] and this distinguishes it from the carbon transfer mechanisms of ectomycorrhizas in which carbohydrate of the host is converted to specific fungal carbohydrates, trehalose and mannitol, and ultimately the storage polysaccharide glycogen.[127] Bevege reported that sucrose was the major component, followed by glucose, in VAM fungi and noninfected hoop pine roots, but that extracts of infected and noninfected hoop pine and clover roots contained similar levels of the same sugars.[125] However, Safir[128] and Nemec and Guy[129] reported an increase in reducing sugars in infected roots of onion and citrus, respectively. If these reducing sugars were composed of principally glucose, their increase may be due, in part, to the hydrolysis of starch which is depleted in cells occupied by the fungus.

Host photosynthate is apparently synthesized rapidly into lipid in fungus structures. Significantly more triglycerides and phospholipids occurred in mycorrhizal roots of six citrus rootstocks compared to the nonmycorrhizal controls.[130,131] Histochemical tests[132] confirm that triglyceride is the major lipid stored in vesicles and hyphae. Fatty acid composition of mycorrhizal citrus roots[133] differs from that of nonmycorrhizal roots. The fatty acids of mycorrhizal roots were present primarily in the triglycerides and were more closely related to acids of fungal origin than plant origin. Campesterol was the major sterol found in *Glomus mosseae* chlamydospores.[133] Synthesis of lipid by the fungus has been suggested as an alternate storage sink for the plant's photosynthates.[134] The magnitude and amounts of carbon diverted to VAM fungi have been difficult to determine. In citrus, at least 3 to 5% of the whole plant ^{14}C-labeled photosynthesis was allocated to the mycorrhizae side of plants grown with a split-root system.[135] Synthesis of this lipid by the fungal endophyte probably also serves as a form of stored energy for the fungi. Future studies should focus on the mechanisms of carbon allocation in these sinks and its ultimate fate in healthy and senescing root systems.

A variety of horticultural crops, principally trees and fruit crops, has been utilized in studies to determine mineral uptake mechanisms in the fungus and host, as well as host-fungus interface interactions at the ultrastructural level. Polyphosphate, the suggested form in which VAM fungi translocate P to the host,[136] was found in hyphae and vesicles of VAM in sweetgum[137] and citrus.[132] Glycogen bodies, which may be a stage in the conversion of host starch to the fungus, were present in hyphae and vesicles of VAM fungi in raspberry,[138] grape,[139] and yellow poplar.[140] Typical intracellular, mature arbuscular branches are surrounded by an interfacial matrix or space[139,140] and an extrahaustorial host membrane.[141] Walls of the intracellular arbuscule have been reported to be osmiophilic,[132,139,140] as well as possessing acidic and PAS reactivity.[132]

Arbuscule walls of *G. fasciculatum* in grape were shown to contain polysaccharide in addition to protein.[142] Intracellular hyphal walls of *G. tenue* in raspberry react positively for polysaccharides and appear to have a two-layered structure.[138] Recent recognition of DES-sensitive ATPase activity along the host (sycamore) plasmalemma in the living arbuscule branches strongly suggests that an active phosphate transport system is located at this membrane.[143] This work with ATPase supports other observations that the interfacial matrix is not an inert space, but is a physiologically complex zone containing host membranes,[144] vesicles, polysaccharides, ATPase, and neutral phosphatase activities.

VI. CONCLUSION

Studies of VAM fungi and horticultural crops eventually lead to the question: how can these organisms be adapted to commercial plant production? This question is naturally a valid one, because many horticultural plants are considered high cash-value crops and are also slow growing. Any method adapted to enhance production of these crops may mean increased profits and lowered costs. This question has been raised by many people, and in some quarters it is now being addressed. Some nurserymen in California and Utah are making inocula available to the trade. Eventually, production and distribution of inocula will be expanded to a large number of crops in various regions where they are produced. This report has discussed a wide array of horticultural plants that have enhanced growth because of VAM fungus infection. Use of VAM fungi in commercial application may be practical with only a limited number of these plants. Some plants, because of the way they are produced, may be effectively grown with well-designed fertilizer programs alone. Crops that are grown on a short-term basis or for flower production may not need VAM fungi. Slow-growing, container-grown perennials and field-produced cultivars may benefit from inoculation. Plants that naturally become infected (strawberry daughter plants) before they are dug and sold may not require inoculation.

Many basic studies about VAM fungi have provided the foundation for manipulating them in agricultural endeavors. Some species have been recognized as more aggressive than others, and through a selection process, superior strains can be targeted for commercial purposes. In the near future, studies are needed on the shelf-life characteristics of VAM fungi. Work needs to be initiated on factors that can extend shelf life of potentially useful species. Specific soil mixes need to be developed for VAM fungi and for their hosts. This work is being undertaken by a number of researchers, but should also be a project developed by companies which market potting mix media.

Basic studies continue to provide us with the knowledge needed to conserve VAM fungi in field plantings. Studies, especially those with pesticides, have indicated which ones may be safer to use in field production schemes. Data from pesticide research have already accounted for the appearance of growth irregularities in field plots and commercial fields where VAM fungi were killed. Considerable additional knowledge is needed on factors that stress and upset the ecological balance these fungi establish with the plant. Knowledge needed to conserve VAM fungi in soils will facilitate progress being made to adapt them to commercial applications.

REFERENCES

1. Ames, R. N. and Linderman, R. G., The growth of Easter lily *(Lilium longiflorum)* as influenced by vesicular-arbuscular mycorrhizal fungi, *Fusarium oxysporum,* and fertility level, *Can. J. Bot.,* 56, 2773, 1978.
2. Rhodes, L. H. and Powell, C. C., Response of florists' geranium in outplant sites to prior inoculation with vesicular-arbuscular mycorrhizal fungi, *Phytopathology,* 69 (Abstr.), 1024, 1979.
3. Jodicc, R., Nappi, P., and Luzzati, A., Influenza delle micorrize vescicola-arbuscolari e della sostanza organica sulla nutrizione borica del girasole, *Allionia,* 24, 43, 1980.
4. Barrows, J. B. and Roncadori, R. W., Endomycorrhizal synthesis by *Gigaspora margarita* in poinsettia, *Mycologia,* 69, 1173, 1977.
5. Crews, C. E., Johnson, C. R., and Joiner, J. N., Benefits of mycorrhizae on growth and development of three woody ornamentals, *HortScience,* 13, 429, 1978.
6. Bensen, N. R. and Covey, R. P., Jr., Response of apple seedlings to zinc fertilization and mycorrhizal inoculation, *HortScience,* 11, 252, 1976.
7. Gilmore, A. E., The influence of endotropic mycorrhizae on the growth of peach seedlings, *J. Am. Soc. Hortic. Sci.,* 96, 35, 1971.
8. Kleinschmidt, G. D. and Gerdemann, M. J., Stunting of citrus seedlings in fumigated nursery soils related to the absence of endomycorrhizae, *Phytopathology,* 62, 1447, 1972.
9. Baylis, G. T. S., The magnolioid mycorrhiza and mycotrophy in root systems derived from it, in *Endomycorrhizas,* Sanders, F. E., Mosse, B., and Tinker, P. B., Eds., Academic Press, London, 1975, 373.
10. O'Brien, D. G. and McNaughton, E. J., The endotrophic mycorrhiza of strawberries and its significance, *West Scotland Agric. Coll. Res. Bull.,* 1, 1, 1928.
11. Bouwens, H., Investigations about the mycorrhiza of fruit trees, especially quince *(Cydonia vulgaris)* and of strawberry plants *(Fragaria vesca), Zentralbl. Bakteriol. Abt.,* 97, 34, 1937.
12. Hildebrand, A. A. and Koch, L. W., A microscopical study of infection of the roots of strawberry and tobacco seedlings by micro-organisms of the soil, *Can. J. Res. Sect. C,* 14, 11, 1936.
13. Muller, H. R. A., Mycorrhiza van citrus, *Landbouw,* 12, 1, 1936.
14. Mosse, B., Fructifications of an *Endogone* species causing endotrophic mycorrhiza in fruit plants, *Ann. Bot.,* 20(78), 349, 1956.
15. Roldan-Fajardo, B. E., Barea, J. M., Ocampo, J. A., and Azcon-Aguilar, C., The effect of season on VA mycorrhiza of the almond tree and of phosphate fertilization and species of endophyte on its mycorrhizal dependency, *Plant Soil,* 68, 361, 1982.
16. Nappi, P., Jodice, R., and Kofler, A., Vesicular-arbuscular mycorrhizas in vineyards, given different soil treatments, in southern Tyrol, *Allionia,* 24, 27, 1980.
17. Mataré, R. and Hattingh, M. J., Effect of mycorrhizal status of avocado seedlings on root rot caused by *Phytophthora cinnamomi, Plant Soil,* 49, 433, 1978.
18. Zhen-yao, T. and Zhao-liang, W., The effects of inoculation of mycorrhizal fungi on phosphorus nutrition and growth of citrus seedlings, *Acta Pedol. Sin.,* 17, 336, 1980.
19. Holevas, C. D., The effect of a vesicular-arbuscular mycorrhiza on the uptake of soil phosphorus by strawberry *(Fragaria* sp. var. Cambridge Favourite), *J. Hortic. Sci.,* 41, 57, 1966.
20. Mason, D. T., A survey of numbers of *Endogone* spores in soil cropped with barley, raspberry and strawberry, *Hortic. Res.,* 4, 98, 1964.
21. Sutton, J. C. and Barron, G. L., Population dynamics of *Endogone* spores in soil, *Can. J. Bot.,* 50, 1909, 1972.
22. Wilhelm, S., Parasitism and pathogenesis of root-disease fungi, in *Plant Pathology, Problems and Progress 1908—1958,* Holton, C. S. et al., Eds., University of Wisconsin Press, Madison, 1959, 356.
23. Gerdemann, J. W. and Trappe, J. M., The Endogonaceae in the Pacific Northwest, *Mycol. Mem.,* 5, 1, 1974.
24. Nemec, S., Menge, J. A., Platt, R. G., and Johnson, E. L. V., Vesicular-arbuscular mycorrhizal fungi associated with citrus in Florida and California and notes on their distribution and ecology, *Mycologia,* 73, 112, 1981.
25. Davis, R. M., Mycorrhizal fungi associated with citrus in south Texas, *J. Rio Grande Valley Hortic. Soc.,* 35, 127, 1982.
26. Tzean, S. and Huang, Y., The occurrence and formation of vesicular-arbuscular mycorrhiza of citrus and maize, *Bot. Bull., Acad. Sin.,* 21, 119, 1980.
27. Nemec, S., Populations of *Endogone* in strawberry fields in relation to root rot infection, *Trans. Br. Mycol. Soc.,* 62, 45, 1974.
28. Krikun, J. and Levy, Y., Effect of vesicular-arbuscular mycorrhiza on citrus growth and mineral composition, *Phytoparasitica,* 8, 195, 1980.

29. Nemec, S., Response of six citrus rootstocks to three species of *Glomus,* a mycorrhizal fungus, in *Proc. Florida State Horticulture Soc.,* Tait, W. L., Ed., E. O. Painter, De Leon Springs, Fla., 1978, 10.
30. Hughes, M., Martin, L. W., and Breen, P. J., Mycorrhizal influence on the nutrition of strawberries, *J. Am. Soc. Hortic. Sci.,* 103, 179, 1978.
31. Plenchette, C., Furlan, V., and Fortin, J. A., Effects of different endomycorrhizal fungi on five host plants grown on calcined montmorillonite clay, *J. Am. Soc. Hortic. Sci.,* 107, 535, 1982.
32. Plenchette, C., Furlan, V., and Fortin, J. A., Responses of endomycorrhizal plants grown in a calcined montmorillonite clay to different levels of soluble phosphorus. II. Effect on nutrient uptake, *Can. J. Bot.,* 61, 1384, 1983.
33. Schenck, N. C. and Tucker, D. P. H., Endomycorrhizal fungi and the development of citrus seedlings in Florida fumigated soils, *J. Am. Soc. Hortic. Sci.,* 99, 284, 1974.
34. Guttay, A. J. R., The growth of three woody plant species and the development of their mycorrhizae in three different plant composts, *J. Am. Soc. Hortic. Sci.,* 107, 324, 1982.
35. Graw, D., The influence of soil pH on the efficiency of vesicular-arbuscular mycorrhiza, *New Phytol.,* 82, 687, 1979.
36. Chambers, C. A., Smith, S. E., and Smith, F. A., Effects of ammonium and nitrate ions on mycorrhizal infection, nodulation and growth of *Trifolium subterraneum, New Phytol.,* 85, 47, 1980.
37. Menge, J. A., Johnson, E. L. V., and Platt, R. G., Mycorrhizal dependency of several citrus cultivars under three nutrient regimes, *New Phytol.,* 81, 553, 1978.
38. Mehraveran, H., Mycorrhizal Dependency of Six Citrus Cultivars, Ph.D. thesis, University of Illinois, Urbana, 1977.
39. Gilmore, A. E., The influence of endotrophic mycorrhizae on the growth of peach seedlings, *J. Am. Soc. Hortic. Sci.,* 96, 35, 1971.
40. Menge, J. A., Davis, R. M., Johnson, E. L. V., and Zentmeyer, G. A., Mycorrhizal fungi increase growth and reduce transplant injury in avocado, *Calif. Agric.,* 32, 6, 1978.
41. Larue, J. H., McClellan, W. D., and Peacock, W. L., Mycorrhizal fungi and peach nursery nutrition, *Calif. Agric.,* 19, 6, 1975.
42. Nemec, S., Inoculation of citrus in the field with vesicular-arbuscular mycorrhizal fungi in Florida, *Trop. Agric.,* 60, 97, 1983.
43. Hoitink, H. A. J., Composted bark: a lightweight growth medium with fungicidal properties, *Plant Dis.,* 64, 142, 1980.
44. Biermann, B. J. and Linderman, R. G., Effect of growth medium on establishment and performance of vesicular-arbuscular mycorrhizae on geranium and subclover, *5th N. Am. Conf. Mycorrhizae (Quebec),* p. 22, 1981.
45. Fardelmann, D. and McNabb, H. S., Jr., Specificity of Endogonaceae and seed size relationships in black walnut seedling production, *5th N. Am. Conf. Mycorrhizae (Quebec),* p. 23, 1981.
46. Gianinazzi-Pearson, V., Trouvelot, A., Morandi, D., and Marocke, R., Ecological variations in endomycorrhizas associated with wild raspberry populations in the Vosges region, *Acta Oecol. Oecol. Plant,* 1, 111, 1980.
47. Nemec, S., Growth of mycorrhizal and nonmycorrhizal citrus in natural and artificial soils, *5th N. Am. Conf. Mycorrhizae (Quebec),* p. 55, 1981.
48. Menge, J., Labanauskas, C. K., Johnson, E. L. V., and Sibert, D., Problems with the utilization of mycorrhizal fungi in the production of containerized citrus in the nursery or greenhouse, *4th N. Am. Conf. Mycorrhizae (Ft. Collins),* Abstr., 1979.
49. Menge, J. A., Jarrell, W. M., Labanauskas, C. K., Ojola, J. C., Huszar, C., Johnson, E. L. V., and Sibert, D., Predicting mycorrhizal dependency of Troyer citrange on *Glomus fasciculatus* in California citrus soils, *Soil Sci. Soc. Am. J.,* 46, 762, 1981.
50. Menge, J. A., Utilization of vesicular-arbuscular mycorrhizal fungi in agriculture, *Can. J. Bot.,* 61, 1015, 1983.
51. St. John, T. V., Coleman, D. C., and Reid, C. P. P., Association of vesicular-arbuscular mycorrhizal hyphae with soil organic particles, *Ecology,* 64, 957, 1983.
52. Hepper, C. M. and Warner, A., Role of organic matter in growth of a vesicular-arbuscular mycorrhizal fungus in soil, *Trans. Br. Mycol. Soc.,* 81, 155, 1983.
53. Maronek, D. M., Hendrix, J. W., and Kiernan, J., Differential growth response to the mycorrhizal fungus *Glomus fasciculatus* of southern magnolia and Bar Harbor juniper grown in containers of composted hardwood bark-shale, *J. Am. Soc. Hortic. Sci,* 105, 206, 1980.
54. Plenchette, C., Furlan, V., and Fortin, J. A., Responses of endymycorrhizal plants grown in a calcined montmorillonite clay to different levels of soluble phosphorus. I. Effect of growth and mycorrhizal development, *Can. J. Bot.,* 61, 1377, 1983.
55. Schultz, R. C., Kormanik, P. P., and Bryan, W. C., Nutrient concentrations in V. A. and nonmycorrhizal hardwood seedlings grown with various symbionts and fertilizer regimes, *4th N. Am. Conf. Mycorrhizae (Ft. Collins),* Abstr., 1979.

56. Barie, F., Ocampo, J. A., and Barea, J. M., V-A mycorrhizal populations in volcanic ash-derived soils from Chile ("Trumaos"), *5th N. Am. Conf. Mycorrhizae (Quebec),* p. 66. 1981.
57. Mosse, B., Vesicular-arbuscular mycorrhiza research for tropical agriculture, *Res. Bull.* 194, Hawaii Institute of Trop. Agr. and Human Resources.
58. Menge, J. A., Labanauskas, C. K., Johnson, E. L. V., and Platt, R. G., Partial subsitution of mycorrhizal fungi for phosphorus fertilization in the greenhouse culture of citrus, *Soil Sci. Soc. Am. J.,* 42, 926, 1978.
59. Hoepfner, E. F., Koch, B. L., and Covey, R. P., Enhancement of growth and phosphorus concentrations in apple seedlings by vesicular-arbuscular mycorrhizae, *J. Am. Soc. Hortic. Sci.,* 108, 207, 1983.
60. Hepper, C. M., Effect of phosphate on germination and growth of vesicular-arbuscular mycorrhizal fungi, *Trans. Br. Mycol. Soc.,* 80, 487, 1983.
61. Menge, J. A., Steirle, D., Bagyaraj, D. J., Johnson, E. L. V., and Leonard, R. T., Phosphorus concentrations in plants responsible for inhibition of mycorrhizal infection, *New Phytol.,* 80, 575, 1978.
62. Ratnayake, M., Leonard, R. T., and Menge, J. A., Root exudation in relation to supply of phosphorus and its possible relevance to mycorrhizal formation, *New Phytol.,* 81, 543, 1978.
63. Siqueira, J. O., Hubbell, D. H., and Schenck, N. C., Spore germination and germ tube growth of a vesicular-arbuscular mycorrhizal fungus *in vitro, Mycologia,* 74, 952, 1982.
64. Nemec, S. and Meredith, F. I., Amino acid content of leaves in mycorrhizal and non-mycorrhizal citrus rootstocks, *Ann. Bot.,* 47, 351, 1981.
65. Hayman, D. S., Influence of soils and fertility on activity and survival of vesicular-arbuscular mycorrhizal fungi, *Phytopathology,* 72, 1119, 1982.
66. Mosse, B., Advances in the study of vesicular-arbuscular mycorrhiza, *Annu. Rev. Phytopathol.,* 11, 171, 1973.
67. Daniels, B. A. and Trappe, J. M., Factors affecting spore germination of the vesicular-arbuscular mycorrhizal fungus, *Glomus epigaeus, Mycologia,* 72, 457, 1980.
68. Schultz, R. C., Kormanik, P. P.,and Bryan, W. C., Effects of fertilization and vesicular-arbuscular mycorrhizal inoculation on growth of hardwood seedlings, *Soil Sci. Soc Am. J.,* 45, 961, 1981.
69. Reed, H. S. and Frémont, T., Factors that influence the formation and development of mycorrhizal associations in citrus roots, *Phytopathology,* 25, 645, 1935.
70. Sabet, Y. S., Reaction of citrus mycorrhizae to manurial treatment, *Proc. Egyptian Acad. Sci. (Cairo),* p. 21, 1945.
71. Azcon, R., Barea, J. M., and Montoya, E., Fertilizacion biologica con micorrizas "VA" Y fosfabacterias. II. Influencia del estercolado y epoca de aplicacion de fosfato sobre la micorrizacion de *Lavandula spica* L. en semillero, *An. Edafol. Agrobiol.,* 37.5, 99, 1978.
72. Guttay, J. R., The interaction of fertilizers and vesicular-arbuscular mycorrhizae in composted plant residues, *J. Am. Soc. Hortic. Sci.,* 108, 222, 1983.
73. Grainger, R. L., Plenchette, C., and Fortin, J. A., Effect of a vesicular arbuscular (VA) endomycorrhizal fungus *(Glomus epigaeum)* on the growth and leaf mineral content of two apple clones propagated *in vitro, Can. J. Plant Sci.,* 63, 551, 1983.
74. Read, D. J., Koucheki, H. K., and Hodgson, J., Vesicular-arbuscular mycorrhiza in natural vegetation systems. I. The occurrence of infection, *New Phytol.,* 77, 641, 1976.
75. Lambert, D. H. and Cole, H., Jr., Effects of mycorrhizae on establishment and performance of forage species in mine spoils, *Agron. J.,* 72, 257, 1980.
76. Mosse, B., Specificity in VA mycorrhizas, in *Endomycorrhizas,* Sanders, F. E., Mosse, B., and Tinker, P. B., Eds., Academic Press, New York, 1975, 469.
77. Mosse, B., Warner, A., and Clarke, C. A., Plant growth responses to vesicular-arbuscular mycorrhiza. XIII. Spread of an introduced VA endophyte in the field and residual growth effects of inoculation in the second year, *New Phytol.,* 90, 521, 1982.
78. Levy, Y., Dodd, J., and Krikun, J., Effect of irrigation, water salinity and rootstock on the vertical distribution of vesicular-arbuscular mycorrhiza in citrus roots, *New Phytol.,* 95, 397, 1983.
79. Hirrel, M. C., The effect of sodium and chloride salts on the germination of *Gigaspora margarita, Mycologia,* 73, 610, 1981.
80. Hirrel, M. C. and Gerdemann, J. W., Improved growth of onion and bell pepper in saline soils by two vesicular-arbuscular mycorrhizal fungi, *Soil Sci. Soc. Am. J.,* 44, 654, 1980.
81. Gerdemann, J. W., Vesicular-arbuscular mycorrhiza and plant growth, *Ann. Rev. Phytopathol.,* 6, 397, 1968.
82. Mejstrik, V., Study on the development of endotrophic mycorrhiza in the association of *Cladietum marisci,* in *Plant Microbes Relationships,* Czech. Academy of Science, 1965, 283.
83. Maeda, M., The meaning of mycorrhiza in regard to systematic botany, *Kumamoto J. Sci. Ser. B,* 3, 57, 1954.

84. Keeley, J. E., Endomycorrhizae influence growth of blackgum seedlings in flooded soils, *Am. J. Bot.*, 67, 6, 1980.
85. Sylvia, D. M. and Schenck, N. C., Germination of chlamydospores of three *Glomus* species as affected by matric potential and fungal contamination, *Mycologia*, 95, 30, 1983.
86. Sieverding, E., Influence of soil water regimes on the efficacy of the VA-mycorrhizae, *Angew. Bot.*, 53, 91, 1979.
87. Johnson, C. R., Menge, J. A., Schwab, S., and Ting, I. P., Interaction of photoperiod and vesicular-arbuscular mycorrhizae on growth and metabolism of sweet orange, *New Phytol.*, 90, 665, 1982.
88. Daft, M. J. and El-Giahmi, A. A., Effect of arbuscular mycorrhiza on plant growth. VIII. Effects of defoliation and light on selected hosts, *New Phytol.*, 80, 365, 1978.
89. Furlan, V. and Fortin, J. A., Effects of light intensity on the formation of vesicular-arbuscular endomycorrhizas on *Allium cepa* by *Gigaspora calospora*, *New Phytol.*, 79, 335, 1977.
90. Furlan, V. and Fortin, J. A., Formation of endomycorrhizae by *Endogone calospora* on *Allium cepa* under three temperature regimes, *Nat. Can.*, 100, 467, 1973.
91. Schenck, N. C., Graham, S. O., and Green, N. E., Temperature and light effect on contamination and spore germination of vesicular-arbuscular mycorrhizal fungi, *Mycologia*, 67, 1189, 1975.
92. Schenck, N. C. and Smith, G. S., Responses of six species of vesicular-arbuscular mycorrhizal fungi and their effects on soybean at four soil temperatures, *New Phytol.*, 92, 193, 1982.
93. Mosse, B., Stribley, D. P., and LeTacon, F., Ecology of mycorrhizae and mycorrhizal fungi, in *Advances in Microbial Ecology*, Vol. 5, Alexander, E. M., Ed., Plenum Press, New York, 1981, 137.
94. Diem, H. G., Gueye, I., Gianinazzi-Pearson, V., Fortin, J. A., and Dommergues, Y. R., Ecology of VA mycorrhizae in the tropics: the semiarid zone of Senegal, *Acta Oecol. Oecol. Plant*, 2, 53, 1981.
95. Dodd, J., Krikun, J., and Haas, J., Relative effectiveness of indigenous populations of vesicular-arbuscular mycorrhizal fungi from four sites in the Negev, *Isr. J. Bot.*, 32, 10, 1983.
96. Crush, J. R., Occurrence of endomycorrhizas in soils of the MacKenzie Basin, Canterbury, New Zealand, *N. Z. J. Agric. Res.*, 18, 361, 1975.
97. Sward, R. J., Hallam, N. D., and Holland, A. A., *Endogone* spores in a heathland area of south-eastern Australia, *Aust. J. Bot.*, 26, 29, 1978.
98. Koske, R. E., Sutton, J. C., and Sheppard, B. R., Ecology of *Endogone* in Lake Huron sand dunes, *Can. J. Bot.*, 53, 87, 1975.
99. Daft, M. J., Hackskaylo, E., and Nicolson, T. M., Arbuscular-mycorrhizas in plants colonizing coal spoils in Scotland and Pennsylvania, in *Endomycorrhizas*, Sanders, F. E., Mosse, B., and Tinker, P. B., Eds., Academic Press, London, 1975, 561.
100. Read, D. J., Koucheki, H. K., and Hodgson, J., Vesicular-arbuscular mycorrhiza in natural vegetation systems. I. The occurrence of infection, *New Phytol.*, 77, 641, 1976.
101. Morandi, D., Gianinazzi, S., and Gianinazzi-Pearson, V., Importance of using endomycorrhizae for the establishment and growth of axenically propagated raspberry plants after transplanting, *Ann. Amélior Plantes*, 29, 623, 1979.
102. Levy, Y. and Krikun, J., Effect of vesicular-arbuscular mycorrhiza on citrus *Citrus jambhiri* water relations, *New Phytol.*, 85, 25, 1980.
103. Deal, R., The Microflora, Nematodes, Mycorrhiza and Root Ecology of Grape in Replant Situations, Ph.D. thesis, Cornell University, Ithaca, N. Y., 1969.
104. Kormanik, P. P., Bryan, W. C., and Schultz, R. C., Increasing endomycorrhizal fungus inoculum in forest nursery soil with cover crops, *South. J. Appl. For.*, 4, 151, 1980.
105. Ocampo, J. A., Effect of crop rotations involving host and nonhost plants on vesicular-arbuscular mycorrhizal infection of host plants, *Plant Soil*, 56, 283, 1980.
106. Nemec, S., Oxygen, temperature, and water potential in shallow and deep soils of a citrus grove with blight, in *Proc. Florida Soil and Crop Science Soc.*, Vol. 42, Grierson, W., Ed., E. O. Painter, De Leon Springs, Fla., 1983, 85.
107. Saif, S. R., The influence of soil aeration on the efficiency of vesicular-arbuscular mycorrhizae. I. Effect of soil oxygen on the growth and mineral uptake of *Eupatorium odoratum* L. inoculated with *Glomus macrocarpus*, *New Phytol.*, 88, 649, 1981.
108. Guttay, A. J. R., Impact of de-icing salts upon the endomycorrhizae of roadside sugar maple, *Soil Sci. Soc. Am. J.*, 40, 952, 1976.
109. Menge, J. A., Johnson, E. L. V., and Minassian, V., Effect of heat treatment and three pesticides upon the growth and reproduction of the mycorrhizal fungus *Glomus fasciculatus*, *New Phytol.*, 82, 473, 1979.
110. O'Bannon, J. H. and Nemec, S., Influence of soil pesticides on vesicular-arbuscular mycorrhizae in a citrus soil, *Nematropica*, 8, 56, 1978.
111. Nesheim, O. N. and Linn, M. B., Deleterious effects of certain fungi-toxicants on the formation of corn by *Endogone fasciculata* and on corn root development, *Phytopathology*, 59, 297, 1969.

112. Menge, J. A., Munnecke, D. E., Johnson, E. L. V., and Carnes, D. W., Dosage response of the vesicular-arbuscular mycorrhizal fungi *Glomus fasciculatus* and *G. constrictus* to methyl bromide, *Phytopathology*, 68, 1368, 1978.
113. Nemec, S. and O'Bannon, J. H., Response of *Citrus aurantium* to *Glomus etunicatus* and *G. mosseae* after soil treatment with selected fumigants, *Plant Soil*, 53, 351, 1979.
114. Bird, G. W., Rich, J. R., and Glover, S. U., Increased endomycorrhizae of cotton roots in soil treated with nematicides, *Phytopathology*, 64, 48, 1974.
115. Nemec, S., Effect of 11 fungicides on endomycorrhizal development in sour orange, *Can. J. Bot.*, 58, 522, 1980.
116. Menge, J. A., Effect of soil fumigants and fungicides on vesicular-arbuscular fungi, *Phytopathology*, 72, 1125, 1982.
117. Hall, R., Fungitoxicants and fungal taxonomy, *Bot. Rev.*, 45, 1, 1979.
118. Nemec, S. and Tucker, D., Effects of herbicides on endomycorrhizal fungi in Florida citrus (*Citrus* spp.) soils, *Weed Sci.*, 31, 427, 1983.
119. Kruckelmann, H. W., Die Vesikulärarbusculare Mykorrhiza and ihre Beeinflussung in Landwirtschaft lichen Kulturen, *Diss. Naturwiss. Fak. Tech.*, Universitat Carolo-Wilhelmina, Braunschweig, 1973.
120. McGrath, H., Chemicals for plant disease control, *Agric. Chem.*, 19, 18, 1964.
121. Hepper, C. M. and Smith, G. A., Observations on the growth of *Endogone* spores, *Trans. Br. Mycol. Soc.*, 66, 189, 1976.
122. Trappe, J. M., Stahly, E. A., Benson, N. R., and Duff, D. M., Mycorrhizal deficiency of apple trees in high arsenic soils, *HortScience*, 87, 52, 1973.
123. Harris, M. M. and Jurgensen, M. F., Development of *Salix* and *Populus* mycorrhizae in metalic mine tailings, *Plant Soil*, 47, 509, 1977.
124. Hayman, D. S., The physiology of vesicular-arbuscular endomycorrhizal symbiosis, *Can. J. Bot.*, 61, 944, 1983.
125. Bevege, D. I., Bowen, G. D., and Skinner, M. F., Comparative carbohydrate physiology of ecto- and endomycorrhizas, in *Endomycorrhizas*, Sanders, F. E., Mosse, B., and Tinker, P. B., Eds., Academic Press, London, 1975, 149.
126. Hayman, D. S., Plant growth responses to vesicular-arbuscular mycorrhiza. VI. Effect of light and temperatures, *New Phytol.*, 73, 71, 1974.
127. Marks, G. C. and Kozlowski, T. T., *Ectomycorrhizae*, Marks, G. C. and Kozlowski, T. T., Eds., Academic Press, London, 1973, 444.
128. Safir, G. E., The Influence of Vesicular-Arbuscular Mycorrhiza on the Resistance of Onion to *Pyrenochaeta terrestris*, M. S. thesis, University of Illinois, Urbana, 1968.
129. Nemec, S. and Guy, C., Carbohydrate status of mycorrhizal and nonmycorrhizal citrus rootstocks, *J. Am. Soc. Hortic. Sci.*, 107, 177, 1982.
130. Cooper, K. M. and Losel, D. M., Lipid physiology of vesicular-arbuscular mycorrhiza. I. Composition of lipids in roots of onion clover and ryegrass infected with *Glomus mosseae*, *New Phytol.*, 80, 143, 1978.
131. Nagy, S., Nordby, H. E., and Nemec, S., Composition of lipids in roots of six citrus cultivars infected with the vesicular-arbuscular mycorrhizal fungus, *Glomus mosseae*, *New Phytol.*, 85, 377, 1980.
132. Nemec, S., Histochemical characteristics of *Glomus etunicatus* infection of *Citrus limon* fibrous roots, *Can. J. Bot.*, 59, 609, 1981.
133. Nordby, H. E., Nemec, S., and Nagy, S., Fatty acids and sterols associated with citrus root mycorrhizae, *J. Agric. Food Chem.*, 19, 396, 1981.
134. Cox, G., Sanders, F. E., Tinker, P. B., and Wild, J. A., Ultrastructural evidence relating to host-endophyte transfer in a vesicular-arbuscular mycorrhiza, in *Endomycorrhizas*, Sanders, R. E., Mosse, B., and Tinker, P. B., Eds., Academic Press, London, 1975, 297.
135. Koch, K. E. and Johnson, C. R., Photosynthate partitioning in split-root citrus seedlings with mycorrhizal and nonmycorrhizal root systems, *Plant Physiol.*, in press.
136. Callow, J. A., Capaccio, L. C. M., Parish, G., and Tinker, P. B., Detection and estimation of polyphosphate in vesicular-arbuscular mycorrhizas, *New Phytol.*, 80, 125, 1978.
137. Ling-Lee, M., Chilvers, G. A., and Ashford, A. E., Polyphosphate granules in three different kinds of tree mycorrhiza, *New Phytol.*, 75, 551, 1975.
138. Gianinazzi-Pearson, V., Morandi, D., Dexheimer, J., and Gianinazzi, S., Ultrastructural and ultracytochemical features of a *Glomus tenuis* mycorrhiza, *New Phytol.*, 88, 633, 1981.
139. Bonfante-Fasolo, P., Some ultrastructural features of the vesicular-arbuscular mycorrhiza in the grapevine, *Vitis*, 17, 386, 1978.
140. Kinden, D. A. and Brown, M. F., Electron microscopy of vesicular-arbuscular mycorrhizae of yellow poplar. III. Host-endophyte interactions during arbuscular development, *Can. J. Microbiol.*, 21, 1930, 1975.

141. Strullu, D. G., Histologie et cytologie des endomycorhizes, *Physiol. Veg.*, 16, 657, 1978.
142. Bonfante-Fasolo, P. and Grippiolo, R., Ultrastructural and cytochemical changes in the wall of a vesicular-arbuscular mycorrhizal fungus during symbiosis, *Can. J. Bot.*, 60, 2303, 1982.
143. Marx, C., Dexheimer, J., Gianinazzi-Pearson, V., and Gianinazzi, S., Enzymatic studies on the metabolism of vesicular-arbuscular mycorrhizas. IV. Ultracytoenzymological evidence (ATPase) for active transfer processes in the host-arbuscule interface, *New Phytol.*, 90, 37, 1982.
144. Gianinazzi, S., Dexheimer, J., Gianinazzi-Pearson, V., and Marx, C., Role of the host-arbuscule interface in the VA mycorrhizal symbiosis: ultracytological studies of processes involved in phosphate and carbohydrate exchange, *Plant Soil.*, 71, 211, 1983.
145. Chatfield, J. A., Rhoades, L. H., and Powell, C. C., Jr., Response of florists'geranium in outplant sites to prior inoculation with vesicular-arbuscular mycorrhizal fungi, *Phytopathology*, 69 (Abstr.), 1024, 1979.
146. Daniels, B. A. and Menge, J. A., Evaluation of the commercial potential of the vesicular-arbuscular mycorrhizal fungus, *Glomus epigaeus, New Phytol.*, 87, 345, 1981.
147. Bryan, W. C. and Kormanik, P. P., Mycorrhizae benefit survival and growth of sweetgum seedlings in the nursery, *South. J. Appl. For.*, 1, 21, 1977.
148. Lee, A. T. C., Von Broembsen, L. A., and Hattingh, M. J., Endomycorrhizal fungi enhance growth of rough lemon seedlings in fumigated nursery soil, *Citrus Subtrop. Fruit J.*, 536, 5, 1978.
149. Marx, D. H., Bryan, W. C., and Campbell, W. A., Effect of endomycorrhizae formed by *Endogone mosseae* on growth of citrus, *Mycologia*, 63, 1222, 1971.
150. Timmer, L. W. and Leyden, R. F., Stunting of citrus seedlings in fumigated soils in Texas and its correction by phosphorus fertilization and inoculation with mycorrhizal fungi, *J. Am. Soc. Hortic. Sci.*, 103, 533, 1978.
151. Ferguson, J. J., The use of mycorrhizal fungi in citrus nurseries, *Citrus Ind.*, 63, 8, 1982.
152. Kormanik, P. P., Bryan, W. C., and Schultz, R. C., Effects of three vesicular-arbuscular mycorrhizal fungi on sweetgum seedlings from nine mother trees, *For. Sci.*, 27, 327, 1981.
153. Kormanik, P. P., Bryan, W. C., and Schultz, R. C., Endomycorrhizal inoculation during transplanting improves growth of vegetatively propagated yellow-poplar, *Plant Propagator*, 23, 4, 1977.
154. Starkey, D. A. and Brown, M. F., Effect of vesicular-arbuscular (VA) mycorrhizal fungi on growth of yellow poplar seedlings in forest nursery soil, *Trans. Ill. State Acad. Sci.*, 70, 190, 1977.
155. Koch, B. L., Covey, R. P., and Larsen, H. J., Response of apple seedlings in fumigated soil to phosphorus and vesicular-arbuscular mycorrhiza, *HortScience*, 17, 232, 1982.
156. Strobel, N. E., Hussey, R. S., and Roncadori, R. W., Interactions of vesicular-arbuscular mycorrhizal fungi, *Melodogyne incognita*, and soil fertility on peach, *Phytopathology*, 72, 690, 1982.
157. Paget, D. K., The effect of *Cylindrocarpon* on plant growth responses to vesicular-arbuscular mycorrhiza, in *Endomycorrhizas*, Sanders, F. E., Mosse, B., and Tinker, P. B., Eds., Academic Press, London, 1975, 593.
158. Ramirez, B. N., Mitchell, D. J., and Schenck, N. C., Establishment and growth effects of three vesicular-arbuscular mycorrhizal fungi on papaya, *Mycologia*, 67, 1039, 1975.
159. Possingham, J. V. and Obbink, J. G., Endotrophic mycorrhiza and the nutrition of grape vines, *Vitis*, 10, 120, 1971.
160. Johnson, C. R. and Michelini, S., Effect of mycorrhizae on container grown Acacia, in *Proc. Florida State Horticulture Soc.*, Vol. 87, Grierson, W., Ed., E. O. Painter, De Leon Springs, Fla., 1974, 520.
161. Johnson, C. R., Joiner, J. N., and Crews, C. E., Effects of N, K, and Mg on growth and leaf nutrient composition of three container grown woody ornamentals inoculated with mycorrhizae, *J. Am. Soc. Hortic. Sci.*, 105, 286, 1980.
162. Gerdemann, J. W., Vesicular-arbuscular mycorrhizae formed on maize and tulip tree by *Endogone fasciculata, Mycologia*, 57, 562, 1965.
163. Johnson, C. R. and Hummel, R. L., Influence of solution pH, ammonium/nitrate ratio and mycorrhizal fungus on growth of chrysanthemum, *HortScience*, 17 (Abstr.), 55, 1982.
164. Mosse, B., The influence of soil type and *Endogone* strain on the growth of mycorrhizal plants in phosphte deficient soils, *Rev. Ecol. Biol. Sol.*, 9, 529, 1972.
165. Davis, E. A., Young, J. L., and Linderman, R. G., Soil lime level (pH) and VA mycorrhiza effects on growth responses of sweetgum seedlings, *Soil Sci. Soc. Am. J.*, 47, 251, 1983.
166. Green, N. E., Graham, S. O., and Schenck, N. C., The influence of pH on the germination of vesicular-arbuscular mycorrhizal spores, *Mycologia*, 68, 929, 1976.
167. Skipper, H. D. and Smith, G. W., Influence of soil pH on the soybean — endomycorrhiza symbiosis, *Plant Soil*, 53, 559, 1979.
168. Nemec, S., unpublished data.

INDEX

C

D

E

F

G

H

M

N

O

P

R

S

T

U

V

W

X

Y

Z